The Perversity of Things

Electronic Mediations

Series Editors: N. Katherine Hayles, Peter Krapp, Rita Raley, and Samuel Weber
Founding Editor: Mark Poster

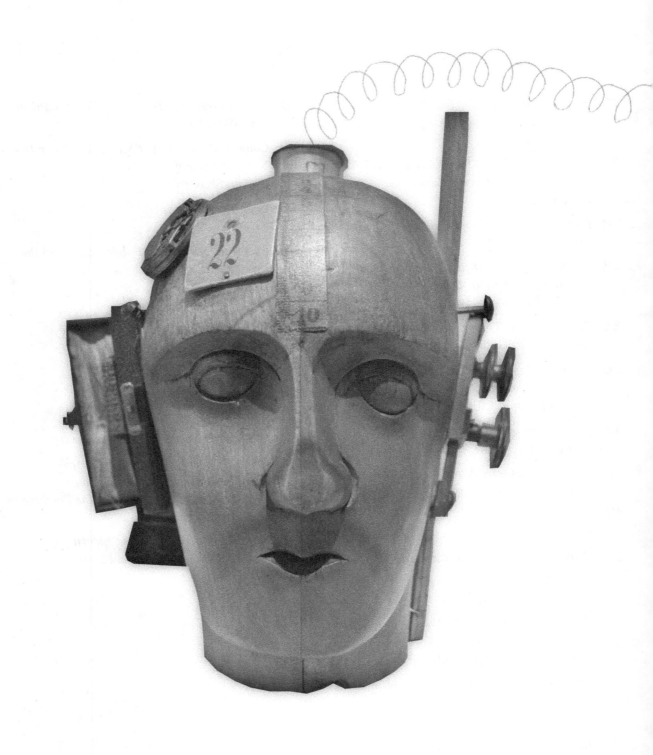

The Perversity of Things

HUGO GERNSBACK
on Media, Tinkering, and Scientifiction

Edited by GRANT WYTHOFF

Electronic Mediations 52
University of Minnesota Press
MINNEAPOLIS • LONDON

This publication was made possible in part by the Barr Ferree Foundation Fund for Publications, Department of Art and Archaeology, Princeton University.

A portion of the Introduction was published in an earlier version as "Aerophone, Telephot, Hypnobioscope: Hugo Gernsback's Media Theory," *Wi: Journal of Mobile Media* 8, no. 2 (2014), Future Media edition.

Unless otherwise noted, all illustrations in the volume were provided courtesy of Syracuse University Libraries, Special Collections Research Center.

Published by the University of Minnesota Press
111 Third Avenue South, Suite 290
Minneapolis, MN 55401-2520
http://www.upress.umn.edu

Library of Congress Cataloging-in-Publication Data
Names: Gernsback, Hugo, 1884–1967, author. | Wythoff, Grant, editor.
Title: The perversity of things : Hugo Gernsback on media, tinkering, and scientifiction / Hugo Gernsback ; edited by Grant Wythoff.
Minneapolis : University of Minnesota Press, 2016. | Series: Electronic mediations ; 52 | Includes bibliographical references and index.
Identifiers: LCCN 2016022208 | ISBN 978-1-5179-0084-7 (hc) | ISBN 978-1-5179-0085-4 (pb)
Subjects: | BISAC: Literary Criticism / Science Fiction & Fantasy. | Literary Collections / Essays. | Science / History.
Classification: LCC PS3513.E8668 A6 2016 | DDC 813/.54—dc23
LC record available at https://lccn.loc.gov/2016022208

for SARA

Thematic Contents

Chronological Contents

How to Use This Book

This collection contains writings that are representative of Hugo Gernsback's thinking across an incredibly prolific and varied career, as well as pieces that are uniquely interesting documents in the history of science fiction and media technology. The book has been designed to invite several different points of entry for different kinds of readers. In order to preserve a sense of how the Gernsback magazines seamlessly toggled between fantastic futures and hard technical detail, the book is arranged chronologically. Much like the way in which each issue of a Gernsback title contained an impressive variety of topics from one page to the next, this book read chronologically encourages a similarly kaleidoscopic reading experience. As a navigational aid, the reader will find an additional table of contents that is arranged thematically. These categories in fact bleed into one another, with articles rarely touching solely on regulation or solely on television, for instance. Nevertheless, this thematic table of contents allows readers interested in particular aspects of Gernsback's career to drop in at those points. Finally, the book's electronic Manifold edition, which will be available early in 2017 at manifold.umn.edu, makes available the complete magazine issues in which each of these pieces first appeared and will serve as a resource for those interested in digging further.

A few notes on the text: pieces that are included in this collection are referenced throughout the book with their titles in bold and can be found alphabetically using the index. Any footnotes included in the original magazines are prefaced by "Gernsback: . . . ," followed by the note. Finally, Gernsback experimented with many idiosyncratic spellings throughout his career as a magazine editor, including words like *publisht, diafram, altho,* and *slipt.* I have chosen to retain these spellings. The writings are reproduced as closely as possible to their original versions, including odd capitalizations, punctuation, omitted words, and the like.

Acknowledgments

Research for this project began in the Department of English at Princeton University, and I am grateful for the warm community of scholars and friends who supported both me and this book throughout my time there. In particular: Eduardo Cadava, Anne Cheng, Bill Gleason, Thomas Y. Levin, Meredith Martin, and Benjamin Widiss. A Summer Prize from the Program in American Studies allowed me to conduct some of the foundational research for this book at the Hugo Gernsback Papers in Syracuse, where I luxuriated in a collection containing every issue ever of Gernsback's magazines.

My colleagues at the Columbia Society of Fellows saw this project through several iterations and provided invaluable feedback on the manuscript at many points along the way; thanks especially to Vanessa Agard-Jones, Teresa Bejan, Maggie Cao, William Derringer, Brian Goldstone, David Gutkin, Hidetaka Hirota, Murad Idris, Ian McCready-Flora, Dan-el Padilla Peralta, Carmel Raz, and Rebecca Woods. Christopher Brown and Eileen Gillooly are brilliant, devoted, and incredibly generous leaders of this community, and the quality of the work that comes out of the program is a testament to their leadership.

I express my gratitude to staff and curators like Bruce Roloson at the Antique Wireless Association Museum in Bloomfield, New York, for actually letting me touch artifacts that for years I was able only to read about. During this trip, I met Jim and Felicia Kreuzer, collectors and aficionados who told me the entire history of radio through a show-and-tell that I will never forget, and who are wonderful hosts as well. Thanks to Patrick Belk of the Pulp Magazines Project and David Gleason of American Radio History for their help in securing digital copies of Gernsback issues.

Many others provided input, encouragement, links, and rejoinders along the way, including Mike Ashley, Andrew Baer, Nolan Baer, John Cheng, Richard Dienst, Eric Drown, Bruce Franklin, Kristin Gallerneaux, James Gleick, Sean X. Goudie, Ben Gross, Michael Holley, Paul Israel, Eugenia Lean, Paul Lesch, Mara Mills, Phillip Polefrone, Sean Quimby, Eric Schockmel, Bernhard Siegert, Steve Silberman, Jim Steichen, Priscilla Wald, Lisa Yaszek, and Siegfried Zielinski.

Susan Lehre was immensely helpful in applying for funding from the Barr-Ferree Publication Fund at Princeton University, support that made possible the publication of such a richly illustrated volume. Nicolette Dobrowolski and Nicole Dittrich at the Syracuse

University Special Collections Research Center tirelessly followed the trail of breadcrumbs I left for them in hunting down almost all of the images you see here. Special thanks to Doug Armato, Erin Warholm-Wohlenhaus, Mike Stoffel, Caitlin Newman, and Jeff Moen at the University of Minnesota Press for their patience and hard work throughout this entire process. Their belief in this wacky project gave me the confidence to write exactly the kind of book I wanted. The anonymous readers who provided comments on early drafts of the manuscript helped me to see this book as a whole for the first time. I am grateful for their insights.

Dave Prout provided timely and much-needed feedback on the index. And book designer Judy Gilats truly captured the spirit of Gernsback's magazines as well as the idea behind this book; Hugo would be proud.

Special thanks to the close friends who reviewed the manuscript front to back at a moment when I thought that I would never be able to let it go: Sand Avidar-Walzer, Ben Breen, Michael Johnduff, Sara J. Grossman, Jeffrey Kirkwood, and Hannes Mandel.

Above all, this book is a product of my family's support. From the love of my parents, to my brother Dan Wythoff, who shares not only a passion for science fiction but also his books, to my sister Erin Wythoff, who loaned her design sensibilities to the making of *The Perversity of Things*, none of this would have been possible without them. And finally, to Sara: this is one of those rare moments that I can't ask for the benefit of your brilliance and care to guide my writing, so of course I'm at a loss for words. I hope it's enough to say how insanely lucky I am to share a world with you.

Introduction

t's the summer of 1906 in downtown Manhattan, and Louis Coggeshall hovers over a bucket of dimes in the back room of the Electro Importing Company's retail store, filing them down to a coarse powder. Sitting amid shelves of electrolytic detectors, circuit switches, ammeters, and Geissler tubes ready for sale to the city's growing community of amateur experimenters, he lets the metal filings fall into a small cardboard box at his feet. Making a coherer, one of the earliest forms of radio receiver, required a certain amount of alchemical improvisation to find the proportion of metals that would produce the strongest signal. At the time, dimes were minted on 90 percent silver, so Coggeshall next began to mix the filings with the perfect ratio of iron powder and finally to pour that mixture into a small glass vial.[1] When a radio frequency wave comes into contact with this coherer, the metal filings cling together, allowing a signal to flow between electrodes connected to either end of the vial. Depending on how long the telegraph key on the transmitting end of that radio wave is depressed, the signal produces a *dot* or a *dash* in Morse code. Mounting this final element onto a wooden base wired with other handmade components, Coggeshall completes the construction of another one of the Electro Importing Company's flagship products: the Telimco wireless telegraph set, one of the first fully assembled radios ever sold to the American public.[2]

Meanwhile in the offices upstairs, the founder of Electro Importing, Hugo Gernsback, writes increasingly breathless advertising copy for the Telimco, promising it to be a means of upward mobility. In one issue of the *Electro Importing Company Catalog*—a mail-in marketplace that provided access to "Everything for the Experimenter"— Gernsback claimed that with the Telimco, "We give you the opportunity to tick yourself up to the head of a future wireless telegraph company as did Marconi, De Forest and others." Priced at $7.50, the set was an attractive proposition from a company that claimed to be "the largest makers of experimental Wireless Material in the world," requiring little more than a working knowledge of Morse code in order to get started. First advertised in the November 25, 1905, issue of *Scientific American,* the Telimco appeared thereafter every two weeks, quickly becoming one of Electro Importing's best-selling items. In order to reach a wider public, Gernsback—a twenty-one-year-old Jewish immigrant from Luxembourg, who at that point had been in the United States for only two years—visited retailers around the city like Macy's, Gimbels, Marshall Field's, and FAO Schwartz,

1. Guglielmo Marconi's original "recipe" for the coherer called for one part silver to nineteen parts nickel (as opposed to iron). Thomas H. Lee, "A Nonlinear History of Radio," in *The Design of CMOS Radio-Frequency Integrated Circuits* (Cambridge: Cambridge University Press, 2004), 4. Coggeshall, who began his career as a telegraph operator for the Erie Railroad, first met Gernsback at a boardinghouse on 14th Street where they were both staying. Sam Moskowitz, *Explorers of the Infinite: The Shapers of Science Fiction* (New York: The World Publishing Company, 1963), 231. Coggeshall also did the cover art for the first issues of the *Electro Importing Company Catalog.* Hugo Gernsback, "The Old E.I. Co. Days," *Radio-Craft* 9, no. 9 (March 1938): 573–74.

2. "Telimco" is a portmanteau of "The Electro Importing Company." "Telimco Wireless Telegraph," *Radio Museum* (2002), http://www.radiomuseum.org/r/electro_im_telimco_wireless_telegraph.html. Hugo Gernsback, "50 Years of Home Radio," *Radio-Electronics* 27, no. 3 (March 1956): 42–43.

FIGURE I.1. A replica of the Telimco receiver made by Gernsback for the Ford Museum in 1957. The coherer is the glass tube mounted between the two binding posts next to the bell. Other components include the spark gap and the large cylindrical battery. From the Collections of The Henry Ford. Gift of Hugo Gernsback.

giving demonstrations of the device to incredulous salespeople who had no conceptual framework by which to understand "wireless." And for good reason: not only was wireless still in its infancy and largely unknown to the public, it was a technology whose successful transmission of information through the air was not yet fully understood even by the scientific community. From the physical substrate of metal filings to an entirely new concept of communication that would soon be in every home, Gernsback and his colleagues were in the process of transmuting one medium into a medium of another kind.

The Perversity of Things is the story of a literary genre's emergence. But it is likely not one that the reader is familiar with. Hugo Gernsback is remembered today as the founding editor of the first science fiction magazine, *Amazing Stories,* a large-format title printed on thick pulp paper that debuted on newsstands in early March 1926. *Amazing Stories* gave a name to fiction treating the speculative and the otherworldly through the lens of systematic realism: scientifiction.[3] And it established a forum for fans of the genre to debate and influence the future of its development. In recognition of this legacy, Gernsback's name adorns the awards given out each year to the best works in the genre, the Hugo Awards. Many scholars use the launch of *Amazing Stories* in 1926 to date the invention of modern science fiction.

This book is devoted to the idea that the project of science fiction as Gernsback understood it in fact had its origins in an earlier context: as a series of interlinking devices, debates, and visions shared by a community of tinkerers that formed around Gernsback's electrical supply shop and technology magazines. Largely thanks to the iconography

3. An early abbreviation for scientifiction among fans was *stf,* pronounced *stef.* Good stories were described with the adjective *st(e)fnal* and devoted fans were *st(e)fnists.* See the fanzine edited by Dick Eney and Jack Speer, *Fancyclopedia 2* (Alexandria, Va.: Operation Crifanac, 1959). Available online at http://fancyclopedia .wikidot.com/stf.

and standardized plots codified by *Amazing Stories,* the term "science fiction" today conjures up images of bug-eyed monsters, ray guns, and starships. But in the opening decades of the twentieth century, before the accretion of a hundred years' worth of narratives, images, and clichés, that which was not yet called "science fiction" consisted of a number of concrete practices all geared toward reckoning with technological revolutions in the fabric of everyday life. Before it was a particular kind of story or plot, science fiction was a way of thinking about and interacting with emerging media.

What began with the *Electro Importing Catalog* and its miscellany of strange devices like the Telimco soon expanded into a number of companion periodicals for the amateur tinkerer like *Modern Electrics* (first published in 1908), *The Electrical Experimenter* (1913), and *Radio News* (1919). In these richly illustrated magazines, one could find blueprints for a home-brewed television receiver (well before the technology was feasible) alongside a literary treatise on how scientifiction stories should be structured. Giving equal space to the soberly technical and

FIGURE I.2. Electro Importing Company advertisement in *Modern Electrics* (1908).

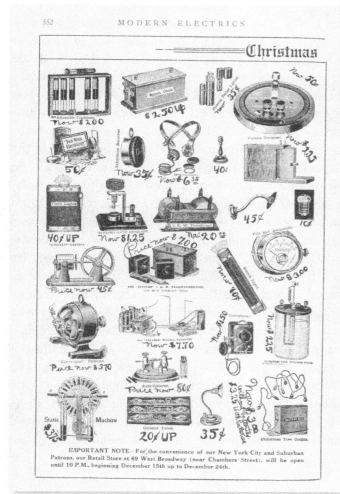

the wildly utopian meant that Gernsback's translation of an influential German handbook titled *The Practical Electrician,* for instance, could run alongside a speculative article on what it would take to provide a global system of free electricity powered by ocean currents. Each issue showcased designs submitted by readers, their own personal "wrinkles, recipes, and formulas," that would be taken up and debated by others through letters printed in subsequent issues, much like the famous letter columns in later science fiction magazines.[4] Long before Gernsback founded *Amazing Stories,* these magazines and their readers used speculative thought to find a language suited to the analysis of emerging media like radio, television, or the more exotic osophone and telegraphone. While Gernsback has for better or worse been enshrined as science fiction's founding figure (fans refer to him as "Uncle Hugo"), less well known are the ways in which he is "the father of American electronic culture," in the words of Franz Pichler, curator of a recent exhibit on Gernsback.[5] To come to terms with these two nascent discourses, scientifiction and media culture, we have to see them as emerging from a single continuum.

One of the challenges in recovering an understanding of science fiction *avant la lettre* is the fact that many of the variables its stories revolved around—science, media, and technology—were concepts still very much in flux in the early twentieth century. Science as it was understood in the public sphere was a highly variable entity and had no settled explanations for the accomplishments of new technologies like wireless telegraphy. These technical media were only just beginning to be understood as *media* in the modern sense of mass communication, and even the term "technology" itself wasn't used in the American English vernacular until the 1920s, as we will see later.[6] The essays that follow, therefore, represent a media theory in the making, one in which *media* are addressed primarily not as mass-cultural forms like cinema or television programming but as the affordances of and possibilities inherent in the smallest individual components: the selenium-coated plate, the tungsten lamp, the chromic plunge battery. In his monthly editorials, feature articles, and short fiction, Gernsback pioneered a kind of writing that combined hard technical description with an openness to the fantastic. Using interwoven descriptive and narrative frameworks to describe a particular device, experience, or vision of the future, Gernsback followed the smallest of technological developments through to their furthest conclusions: the increased availability of a light-sensitive alloy implied that the coming of visual telephones was near, and the number of amateurs sending in their own designs for primitive television receivers only served to confirm the imminence of this new mode of communication.

4. For a history of the science fiction fandom that has constituted what Samuel R. Delany calls a "vast tributary system of informal criticism," see Sam Moskowitz, *Immortal Storm: A History of Science Fiction Fandom* (Westport, Conn.: Hyperion, 1974). Delany, "Reflections on Historical Models," in *Starboard Wine: More Notes on the Language of Science Fiction* (Middletown, Conn.: Wesleyan University Press, 2012), 229.

5. Co-curated by Pichler and the media theorist Peter Weibel, this exhibit at the Zentrum für Kunst und Medientechnologie (Center for Art and Media) in Karlsruhe, Germany, was one of two recent museum showcases of Gernsback's life and work. The other was held at Luxembourg's Centre National de Littérature in 2011. Franz Pichler, *Hugo Gernsback und Seine Technischen Magazine* (Linz: Rudolf Trauner Verlag, 2013).

6. John Guillory, "Genesis of the Media Concept," *Critical Inquiry* 36, no. 2 (January 2010): 321–62; Leo Marx, "Technology: The Emergence of a Hazardous Concept," *Technology and Culture* 51, no. 3 (2010): 561–77.

Regardless of how advanced the devices detailed in *Modern Electrics* and *The Electrical Experimenter* seemed—solar cells, automobile-mounted radiotelephones, electric keyboards powered by vacuum tubes—Gernsback and his staff reported on them as if they required only a combination of already existing electrical principles and components to be built. These new media appeared as little more than the sum of individual building blocks that one could pick and choose from the pages of the *Electro Importing Catalog*. Even long-term projections like thought wave recorders and videophones figured as handicraft futures that would come to pass with just a little more tinkering. As Samuel R. Delany writes,

> The new American SF took on the practically incantatory task of naming nonexistent objects, then investing them with reality by a host of methods, technological and pseudotechnological explanations, embedding them in dramatic situations, or just inculcating them by pure repetition.[7]

These gadgets appeared so frequently and in such diverse contexts—as props in short stories, as homemade designs in letters to the editor, as profiles of similar developments across Europe—that one gets the sense paging through the magazines that they are all part of a coherent fictional world, built up across many years and many issues. Given the pace of technological change in the early twentieth century, it seemed as if any element of this fictional world could bleed into everyday life at any moment.

It's perhaps this penchant in his work to dwell in the extremes of both technical detail and fantastic speculation that accounts for Gernsback's absence in the pages of science fiction studies and from the history of technology. For the former, he is too obsessed with material details, and as an editor and writer merely produces stylistically bland lists of new devices, what are dismissively referred to as "gadget stories." For historians of technology, Gernsback's inventions and technical writings are never able to back up their promises, being far too concerned with future contingencies to merit serious attention. But it is precisely the novel coexistence of these diametrically opposed modes throughout Gernsback's work—soberly technical and wildly speculative—that led to the explosive popularity of his ideas. Across the thirty-year period covered by this book, the reader will find technical precision and utopian speculation in varying proportions, and articles that range from one end of this spectrum to the other with the ease of a tuning knob.

By the 1910s and '20s, the American magazine-reading public was familiar with the idiom of popular science reportage. While

7. Samuel R. Delany, "Critical Methods/Speculative Fiction," in *The Jewel-Hinged Jaw: Notes on the Language of Science Fiction* (Middletown, Conn.: Wesleyan University Press, 2009), 17–28.

nineteenth-century scientific periodicals in the United States took the form of highly specialized *Proceedings* or *Transactions* or *Reports* of academic research organizations, the *fin de siècle* saw a proliferation of titles aimed at a much broader audience. *Scientific American* (first published in 1845), for instance, the best known of these publications and one still in print, attracted a readership that included credentialed researchers, industrial manufacturers, and an interested general public, as well as a "nebulous community of inventors (ranging from the local tinkerer to manufacturer and professional inventor/technologist)."[8] Science in these new magazines was made accessible to a growing number of readers seeking to educate themselves or simply desiring to remain informed about recent developments. But while Gernsback's magazines may seem familiar to us today through their progeny such as *Wired, Popular Science,* and *Popular Mechanics* (the latter two of which are direct successors to Gernsback titles through mergers and acquisitions), they perhaps have more in common with a longer tradition of popular science that blurred the lines between illusion and truth, skepticism and belief. From medieval displays of the natural world's wondrous curiosities to nineteenth-century phantasmagoria and other audiovisual spectacles, "the positive sciences and the fantastic arts [have been] linked in a dialectic of doubt and certainty," as many historians of science have recently shown.[9] Unseen forces were at work in the miracle of wireless telegraphy, a literally unbelievable technology that meant disembodied thoughts from around the world could be skimmed from the air in the comfort of your home. If this miracle were possible, what else might be?

For Gernsback and his staff, rapid developments in the electrical arts made the speculative sciences of antigravitation and "thought waves" seem within reach. From a belief in the presence of the luminiferous ether, to an argument that gravity is an electrical phenomenon, to the suggestion that humans may be able to tap into the so-called sixth senses of animals, or that the earth's core is made of radium and drives recurring cycles of life's evolution, Gernsback's purportedly "scientific" titles reveled in the extraordinary.[10] And while competing magazines like *The Wireless Age, QST,* and *Popular Science Monthly* were gradually opening up the specialist orientation of the sciences to a wider public through sober reportage, Gernsback addressed a growing readership who found the products of science forming an increasingly bewildering part of their everyday lives. "As time goes on it becomes more apparent that our senses are becoming more and more involved directly due to scientific progress," he wrote. Listing a miscellany of sensory illusions that had become commonplace, from "blazing names being written out in the night sky" by moving electric

8. Matthew D. Whalen and Mary F. Tobin, "Periodicals and the Popularization of Science in America, 1860–1910," *Journal of American Culture* 3, no. 1 (March 1, 1980): 195–203. For a history of the view that a scientifically educated public was an imperative for American democracy in the early twentieth century, a period during which the concept of "science" itself was being determined largely by secondary school curricula, see Andrew Jewett, *Science, Democracy, and the American University: From the Civil War to the Cold War* (repr., Cambridge: Cambridge University Press, 2014), especially chapters 4 and 7.

9. John Tresch, "The Prophet and the Pendulum: Sensational Science and Audiovisual Phantasmagoria around 1848," *Grey Room* 43 (April 2011): 18; Lorraine Daston and Katharine Park, *Wonders and the Order of Nature, 1150–1750* (New York; Cambridge, Mass.: Zone Books; Distributed by the MIT Press, 1998); Stefan Andriopoulos, *Ghostly Apparitions: German Idealism, the Gothic Novel, and Optical Media* (New York: Zone Books, 2013); Tom Gunning, "An Aesthetic Astonishment: Early Film and the (In)Credulous Spectator," in *Viewing Positions: Ways of Seeing Film*, ed. Linda Williams, 114–33 (New Brunswick, N.J.: Rutgers University Press, 1995).

10. Hugo Gernsback, "Ether," *Electrical Experimenter* 4, no. 2 (June 1916): 75; Hugo Gernsback, "Gravitation and Electricity," *Electrical Experimenter* 4, no. 2 (February 1918): 659; Hugo Gernsback, "The Unknown Senses," *Electrical Experimenter* 7, no. 10 (February 1920): 970; Hugo Gernsback, "Radium and Evolution," *Electrical Experimenter* 4, no. 7 (November 1916): 467.

signs, to the "radio illusion" of reproduced sound, to the simulated effects of motion in new carnival rides, Gernsback warns that "we should never trust our senses too much in these latter days of scientific progress."[11] Given the rate of this "progress," the reader should not be surprised if the next great scientific advance, seemingly impossible today, should become commonplace tomorrow. As we will see over the course of Gernsback's writings, this almost spiritual faith in technoscientific progress remains remarkably unscathed in the wake of the First World War and the Great Depression. It was a faith that an American technologist could claim only at a remove from the horrors of the trenches and by doubling down with the idea that social well-being could be technocratically managed by machines and their engineers (**Human Progress; Wonders of the Machine Age**). "Progress in science is as infinite as time, it is inconceivable how either would stop" (**Imagination versus Facts**).

For Gernsback, projections of the future or progress and its wonders were never simply bewildering. They were the occasion for a material education in the way things worked. As one reader of and contributor to *Amazing Stories*, G. Peyton Wertenbaker, wrote in a letter to the editor with characteristically pulpy prose, "Beauty is a groping of the emotions towards realization of things which may be unknown only to the intellect" (**Fiction versus Facts**). The archaeologist and historian of radio Michael Schiffer refers to Gernsback and his writers as "techno-mancers," arguing that their depictions of wide-ranging futures created a "cultural imperative" for inventors, engineers, and scientists to make these dreams a reality.[12] And in a sense, this is true, with many of the speculative ideas described by Gernsback retrospectively seeming like "predictions" of modern technologies (**Television and the Telephot; What to Invent; Predicting Future Inventions**). But what makes sense as a neat, two-step model for literature as a source of inspiration and invention was actually a much messier process in practice, like a melody played so fast that the individual notes become indistinguishable.[13] The Gernsback titles were unique in their willingness to tackle fantastic topics that other publications wouldn't, and did so through the lens of hands-on technical know-how. Behind the gadgets of **Ralph 124C 41+**, Gernsback's famous serial novel of 1911, was a marketplace of evocative objects, a forum for amateurs experimenting with these strange new things, and an emerging consensus vision of their possible applications. Over the course of long-running serial novels and the exchange of reader correspondence, the material basis of scientifiction's marvelous futures was gradually refined.

This book won't attempt to offer a resolution to long-standing debates over whether science fiction had its origins in the pulps or

11. Hugo Gernsback, "Modern Illusions," *Science and Invention* 15, no. 3 (July 1927): 201.

12. Michael B. Schiffer, *The Portable Radio in American Life* (Tucson: University of Arizona Press, 1991), 136.

13. This two-part model of invention forms a common thread in science fiction criticism, with the genre providing a think tank of sorts in magazines, novels, and films, continuously replenishing a reservoir of ideas for technologists to draw upon in their own inventive process. See, for instance, Thomas M. Disch, *The Dreams Our Stuff Is Made Of: How Science Fiction Conquered the World*, 1st ed. (New York: Free Press, 1998); Nathan Shedroff and Christopher Noessel, *Make It So: Interaction Design Lessons from Science Fiction*, 1st ed. (New York: Rosenfeld Media, 2012).

is best considered in relation to a deeper tradition with precursors in Mary Shelley, Edgar Allan Poe, Jules Verne, even John Wilkins's *The Discovery of a World Inside the Moone* (1638) or Thomas More's *Utopia* (1516). Reexamining the works of Hugo Gernsback is not meant to suggest that the cultural, sexual, and ecological complexities of novels like Samuel R. Delany's *Dhalgren* (1975), Kathy Acker's *Empire of the Senseless* (1988), or Paolo Bacigalupi's *The Windup Girl* (2009) can be directly traced to the techno-utopian texts first published in *Amazing Stories*. As Carl Freedman writes, "current Anglo-American science fiction draws on far more than the pulp tradition that constitutes *one* of its filiations."[14] *The Perversity of Things* instead takes seriously Delany's idea that literature and science fiction are two entirely separate phenomena that obey different rules. In order to fully understand "a tradition of writers who considered themselves craftsmen first and artists secondarily, if at all," we have to move beyond literary critical frameworks that are poorly suited to thinking about craft, materiality, and collective practice.[15] Gernsback's handicraft futures seemed so seductive because they were futures that you, the reader, could build yourself. With the hands-on experience promised by these magazines, the reader could pick up any object and sense its affordances rather than being continuously frustrated by what Gernsback called "the perversity of things." For Gernsback, objects exert an influence on thought. The question was how to understand that influence as an inherent promise or beauty, rather than a stubbornness or recalcitrance. Because the complex things of the modern world seemed to impinge further on daily life with each passing year, it was Gernsback's lifelong project to educate a public in the tenets of "science" through a kind of instinctual know-how or "knack"—a term he returns to in many of his writings. The idea was to enable the public to contribute to the making of things rather than allow them to be overwhelmed by the perversity of things.

We will begin in the next section of this introduction, "'up-to-date technic': Hugo Gernsback's Pulp Media Theory," with an overview of Gernsback's life, the development of his periodicals and their international circulation, as well as the idiosyncratic figurations of "science" that were deployed throughout. Unfortunately, the prevailing approach in science fiction studies has been to dismiss the Gernsback magazines as embarrassingly simplistic, tasteless, and even detrimental to the eventual emergence of a mature literature. This is an ironic and all-too-casual judgment of a Jewish immigrant who throughout his life was in search of the respect as a technologist and editor that always seemed to elude him.[16] A certain tone seems to have been set early on by the spectacularly racist H. P. Lovecraft's moniker for

14. Carl Howard Freedman, *Critical Theory and Science Fiction* (Hanover, N.H.: Wesleyan University Press, 2000), 15.

15. Samuel R. Delany, *Silent Interviews: On Language, Race, Sex, Science Fiction, and Some Comics* (Hanover, N.H.: Wesleyan University Press, 1994), 200. See also Delany, "Critical Methods/ Speculative Fiction."

16. For a history of the relationship between Jewish identity and science fiction, see Danny Fingeroth, *Disguised as Clark Kent: Jews, Comics, and the Creation of the Superhero* (New York: Continuum, 2009).

Gernsback: "Hugo the Rat." The overwhelming attention that many science fiction critics give, to this day, to the low rates he paid his writers (a common source of tension between *all* pulp writers and their editors) leads to a misplaced derision of Gernsback's literary quality that often carries with it an explicit disgust at his perceived character.[17] One can be forgiven for wondering why such singular attention has gone toward bankruptcy proceedings, profits, and wages in works of literary scholarship.

This situation is beginning to change, however, with new research on early science fiction. Joining the voices of Mike Ashley and Gary Westfahl, tireless champions of Gernsback's work for decades, scholars like John Cheng, Eric Drown, Justine Larbalestier, and John Rieder are painting a richer portrait of these magazines and their sociocultural milieu. Thanks to this scholarship, we are learning that the overlapping publics attracted to Gernsback's magazines opened a space for voices that were otherwise explicitly occluded from public discourse.[18] The following section of this introduction, "'a perfect Babel of voices': Communities of Inquiry and Wireless Publics," contributes to the ongoing rediscovery of these communities and their role in the development of science fiction as well as new media technologies. As Steve Silberman has shown, "Both amateur radio and science fiction fandom offered ways of gaining social recognition outside traditional channels."[19] These magazines promised a populist education in new scientific and technical principles that would offer social inroads for wage laborers hoping to secure middle-class careers, women who took advantage of the anonymity of Morse code and of

17. For Brian Stableford, Gernsback's "cowardly" launch of *Amazing Stories* was "purely a commercial adventure, which Gernsback undertook because he thought his existing subscription lists might help establish a secure commercial base for such a magazine." Brian Stableford, "Creators of Science Fiction 10: Hugo Gernsback," *Interzone* 126 (December 1997): 47–50. Malcolm J. Edwards argues that Gernsback "bestowed upon his creation provincial dogmatism and an illiteracy that bedeviled US SF for years." Malcolm J. Edwards, "Gernsback, Hugo," in *The Encyclopedia of Science Fiction*, ed. John Clute and Peter Nicholls (New York: St. Martin's Griffin, 1995), 491. Richard Bleiler sees him as "a disastrous (if not a pernicious) figure, a man whose stultifying vision and lack of literary taste led to the establishment of a literature that for too many years was considered a laughingstock, that emphasized other elements than literary quality, and, perhaps worse [*sic*] of all, that paid the majority of its writers badly." Richard Bleiler, "Hugo Gernsback and His Writers," in *Sense of Wonder*, ed. Leigh Ronald Grossman (Holicong, Penn.: Wildside Press, 2011), 65.

18. Mike Ashley, *The Time Machines: The Story of the Science-Fiction Pulp Magazines from the Beginning to 1950*, The History of the Science-Fiction Magazine 1 (Liverpool, U.K.: Liverpool University Press, 2000). Mike Ashley, *The Gernsback Days: A Study of the Evolution of Modern Science Fiction from 1911 to 1936* (Holicong, Penn.: Wildside Press, 2004). Gary Westfahl, *The Mechanics of Wonder: The Creation of the Idea of Science Fiction*, Liverpool Science Fiction Texts and Studies 15 (Liverpool, U.K.: Liverpool University Press, 1998). Gary Westfahl, *Hugo Gernsback and the Century of Science Fiction* (Jefferson, N.C.: McFarland & Company, 2007). Work by this new generation of scholars includes John Cheng, *Astounding Wonder: Imagining Science and Science Fiction in Interwar America* (Philadelphia: University of Pennsylvania Press, 2012). Eric Drown, "Usable Futures, Disposable Paper: Pulp Science Fiction and Modernization in America, 1908–1937" (PhD Diss., University of Minnesota, 2001). Eric Drown, "'A Finer and Fairer Future': Commodifying Wage Earners in American Pulp Science Fiction," *Endeavour* 30, no. 3 (August 2006): 92–97. Eric Drown, "Business Girls and Beset Men in Pulp Science Fiction and Science Fiction Fandom," *Femspec* 7, no. 1 (September 2006): 5–35. Justine Larbalestier, *The Battle of the Sexes in Science Fiction* (Middletown, Conn.: Wesleyan University Press, 2002). John Rieder, *Colonialism and the Emergence of Science Fiction* (Middletown, Conn.: Wesleyan University Press, 2008). See also chapter 3, "Getting Out of the Gernsback Continuum," in Andrew Ross, *Strange Weather: Culture, Science, and Technology in the Age of Limits* (London: Verso, 1991).

19. Steve Silberman, *NeuroTribes: The Legacy of Autism and the Future of Neurodiversity*, 1st ed. (New York: Avery, 2015), 240.

the power to reach an audience of countless listeners from home, as well as those "who found it nearly impossible to communicate through speech" due to a range of neurological conditions, such as the one we might associate today with autism.[20]

In the next two sections, "'phone and code: Dynamophone, Radioson, and Other Emerging Media" and "'certain future instrumentalities': The Mineral Proficiencies of Tinkering," Gernsback's inventions and his theorization of tinkering take center stage. At a moment in which the energies of American invention were shifting from independent amateurs to corporate research labs, Gernsback's devices increasingly incorporated metaphor, narrative, and a sense of humor. As technical solutions to some of the most pressing problems in the development of radio, television, and other emerging media, they relied on his theory of amateur tinkering as an activity distinct from, and even superior to, "invention" by credentialed researchers and engineers. These arguments anticipate a recent movement in academic media studies known as critical making, which advocates a material engagement with technology as a means of understanding it. Making things allows us "to bridge the gap between creative, physical and conceptual exploration."[21] Media studies scholars will find many other forms of argumentation and analysis throughout Gernsback's writing to be familiar. This is perhaps no surprise, given the long-standing fascination in media studies with science fictional metaphors like virtuality, prosthetic extensions, and cyberspace. But the pages that follow suggest something more at work than mere echoes in terminology: the Gernsback magazines constitute a forgotten and decidedly pulpy point of origin for today's academic discipline. Science fiction in its early days wasn't just a literary form, it was a means of understanding media.

The introduction's final section, "'we exploit the future': Scientifiction's Debut," profiles Gernsback's contributions to the science fiction genre. The literary-historical gambit is to recover the radical sense of openness that greeted not only the basement tinkerer working through the feasibility of transmitting images over a wire, but also the author of scientifiction stories who possessed a highly sophisticated awareness of the fact that "two hundred years ago, stories of this kind were not possible" (**A New Sort of Magazine**). Often, these individuals were one and the same, weaving together functional and fictional devices in a manner that served for them as a form of scientific discovery in itself.

The Perversity of Things thus seeks to provide a reappraisal of the communities of amateur experimenters that participated in the early twentieth-century emergence of new media, of the "hard" technical

20. Ibid., 244.
21. Matt Ratto, "Critical Making: Conceptual and Material Studies in Technology and Social Life," *The Information Society* 27, no. 4 (July 2011): 252–60. See also Michael Dieter, "The Virtues of Critical Technical Practice," *differences* 25, no. 1 (January 2014): 216–30.

roots of American science fiction, and of the highly imaginative orientation toward media technologies that was prevalent during this period. This isn't the story of an Edison or a Jobs, inventors whose creations have changed for better or worse the way we inhabit the world. Rather, Gernsback's career left us with a way of participating in that change, imagining its possible futures, and debating which future it is that we should live in. This is the story of the development of a community, of a series of practices, and of a way to approach the technologized world.

"up-to-date technic": Hugo Gernsback's Pulp Media Theory

Born Hugo Gernsbacher in 1884, the third son of German émigrés to Luxembourg, Gernsback was raised in Hollerich, a tiny suburb of the nation's capital. His parents, Moritz and Bertha Gernsbacher (née Dürlacher), raised him in comfortable circumstances thanks to Moritz's successful wine wholesaling business. Growing up before Hollerich was connected to Luxembourg City's new electrical grid, the young Gernsback's passion for technology began with the battery after a handyman employed by his father, Jean-Pierre Görgen, taught him at six years old how to wire a series of bells to a Leclanché cell. Gernsback recalls being instantly enchanted by the bell "ringing amid a shower of wonderful green sparks" and would soon acquire a reputation for wiring homes and businesses in the area with telephones and these ringers.[22] According to a story he often told later in life, Gernsback received special permission from Pope Leo XIII at thirteen years old to enter the Carmelite convent of Luxembourg City to install a series of electric call bells for the nuns there.[23]

Although this story is partially verifiable through a certificate of thanks from the convent found by the Luxembourgeois Centre National de Littérature, it is worth pointing out that the portrait of Gernsback's life we are left with in the historical record consists largely of a series of self-propagated stories that border on braggadocio: that a police officer intruded on the Electro Importing Company offices to interrogate them for fraud with the Telimco set in 1906, for "no wireless combination could be sold at this low price";[24] that Gernsback coined the term "television" in his December 1909 editorial **Television and the Telephot** (he didn't); that his recommendations were incorporated "word for word" into the Radio Act of 1912 (while the recommendation was, the wording wasn't—see **The Alexander Wireless Bill** and **Wireless and the Amateur**). Once the influential

22. Paul O'Neil, "The Amazing Hugo Gernsback, Prophet of Science, Barnum of the Space Age," *LIFE Magazine* 55, no. 4 (July 1963): 63–68. Gernsback, "The Old E.I. Co. Days."

23. Mark Richard Siegel, *Hugo Gernsback, Father of Modern Science Fiction: With Essays on Frank Herbert and Bram Stoker*, 1st ed., The Milford Series 45 (San Bernardino, Calif: Borgo Press, 1988), 16.

24. Reportedly, the officer shot back after the demonstration: "I still think youse guys is a bunch of fakers. This ad here says that you are selling a *wireless* machine. Well, if you do, what are all them wires for?" Gernsback, "The Old E.I. Co. Days."

science fiction historian and friend of Gernsback, Sam Moskowitz, recorded these stories in his many profiles of Gernsback, they became established as gospel truth, with Moskowitz playing the apostle to Gernsback's prophecies. Evidence contrary to or even in excess of the received doxa is hard to come by, with many press and literary critical accounts barely bothering to rephrase Moskowitz's prose.[25]

We do know that despite his precociousness, Gernsback was by most measures a terrible student, falling at the bottom of his class at an industrial school near home from the ages of twelve to fifteen. When he left to attend the Technikum in Bingen, Germany (now the Fachhochschule Bingen), from seventeen to eighteen, he regularly skipped classes and received poor grades in all subjects save electricity and physics. It was during these teenage years that Gernsback acquired a penchant for gambling away the money he earned on various electrical jobs in poker games, though his tendency to be cleaned out by older players seemed to keep him from falling too deep into this habit.[26] Outside of school, Gernsback gravitated to American culture from an early age. He was a fan of John Philip Sousa's military marches and even composed his own patriotic Luxembourgish piece in the style of Sousa titled *Rŏd, Wêis, Blo* that continued to be performed by the Military Band at the Place d'Armes in Luxembourg into the 1930s.[27] He was a devoted reader of cowboy stories and such a fan of Mark Twain that at seventeen he wrote a now-lost novel of his own, *Der Pechvogel*, under the name Huck Gernsbacher. But it was stories of the otherworldly that truly fired his imagination. The astronomer Percival Lowell's book *Mars,* with its fusion of the hard sciences and speculation about alien life, served as nine-year-old Gernsback's introduction to a literature he would later come to think of as the distinct genre of scientifiction. He dove headfirst into the work of Jules Verne and H. G. Wells, claiming to have nearly memorized many of their novels while still very young.[28]

Despite his predilection for intellectual journeys to unmapped frontiers, whether the American West or Martian canals, Gernsback remained tied to home even after leaving, if not geographically then through a meticulous self-fashioning. Gernsback cultivated a uniquely Luxembourgish identity throughout his life. Though raised by German parents, he grew up in Luxembourg at a moment in which the tiny country's national identity was becoming more developed than ever before. The anti-German sentiments of Gernsback's editorials and short fiction published during World War I (especially **The Magnetic Storm**) largely echo the growing importance already sensed by "a good part of the Luxembourgish people at the end of the Nineteenth century to demarcate between [Germany and Luxembourg] when faced with the more or less well marked pan-Germanic designs

25. Beginning with Sam Moskowitz, *Hugo Gernsback: Father of Science Fiction* (New York: Criterion Linotyping & Printing Co, 1959).

26. Luc Henzig, Paul Lesch, and Ralph Letsch, *Hugo Gernsback: An Amazing Story* (Mersch: Centre National de Littératur, 2010), 15.

27. Ibid., 15–16.

28. Moskowitz, *Explorers of the Infinite,* 229.

of Germany."[29] But the many identities attributed to Gernsback throughout his life—he is described variously in the press as Prussian, German, Belgian, French, a "multilingual dandy"—also seemed to allow him a kind of ambiguity that he relished.[30]

> In the era of tie-dye and sandals, Gernsback continued to dress like a visiting dignitary. For evenings on the town, he favored formal wear, including spats, an opera cape and an expensive silk homburg. He even affected a monocle, though he didn't really need it.[31]

Hiding just beneath this severe exterior, according to Moskowitz, was a sharp sense of humor: "The truth is that Gernsback socially is a man of almost rapier-like wit, with a mischievous gleam in his eyes and with the rare ability to joke about his own misfortunes."[32] Throughout his life, friends and colleagues noted Gernsback's relentless energy and the way that it seemed to sweep up everyone around him. Visions completely out of step with his surroundings seemed to fall out of him wherever he went. A distant relative recalls a 1910 visit from Gernsback on his way to Chicago to purchase new equipment for Electro Importing: When a ringing telephone interrupted one of his many stories of "robot doctors, retirement colonies on Mars, domed cities orbiting Earth," Gernsback (who had arrived in a horse-drawn carriage) reminded his seven-year-old niece as she ran toward the receiver, "Hildegarde, fix your hair. It won't be long before the caller can see your face over the telephone wires."[33]

Throughout his youth, Gernsback's parents never approved of his interest in electrics, and especially didn't see it as a viable career choice for him. But when his father Moritz died at age 57, Gernsback sensed that it was time to branch out on his own. In February 1904, at the age of nineteen, he emigrated to New York by himself, appearing in a photograph wearing an elaborate three-piece suit aboard the S.S. *Pennsylvania* on his way across the Atlantic.[34] Perhaps inspired by his first experiences with that Leclanché cell as a boy, Gernsback carried with him the design for a new kind of dry cell battery whose electrolytic paste could replace the inefficient liquid of wet cells like that of the Leclanché. Having been denied patents in both France and Germany for the battery, Gernsback decided to try his luck in the United States. A year later, he began publishing his ideas, with his first printed article appearing in *Scientific American,* again under that most American of names, "Huck" (**A New Interrupter**). He was able to sell his battery technology to the Packard Motor Car Company, who ended up using the device in their ignition systems. With the profits of his sale, Gernsback formed the Electro Importing Company, an importer of specialized electrical equipment from Europe and one of the first mail-order radio retailers in the country. Through their

29. Henzig, Lesch, and Letsch, *Hugo Gernsback,* 19.

30. O'Neil, "The Amazing Hugo Gernsback."

31. Daniel Stashower, "A Dreamer Who Made Us Fall in Love with the Future," *Smithsonian* 21, no. 5 (August 1990): 44.

32. Moskowitz, *Explorers of the Infinite,* 235.

33. Stashower, "A Dreamer Who Made Us Fall in Love with the Future."

34. Henzig, Lesch, and Letsch, *Hugo Gernsback,* 40.

FIGURE I.3.
Electro Importing
Company store, circa 1908.

catalog and retail store on the Lower West Side of Manhattan, the company provided access to specialized wireless and electrical equipment not found anywhere outside of Europe. Electro Importing catered to a diverse clientele, first manufacturing the Telimco in 1905 for their novice users, and providing their more advanced amateur experimenters with the first vacuum tube offered for sale to the general public in 1911.

After several issues of their mail-order catalog and a growing subscription list, Electro Importing began including features, editorials, and letters to the editor. Between 1906 and 1908, the catalog evolved into *Modern Electrics,* a monthly magazine for the wireless homebrewer. The transition from mail-order catalog to monthly magazine was smooth, evidenced by the fact that the third and fourth editions (1907 and 1908) of the Electro Importing catalog bear the title of the new full-format magazine, *Modern Electrics.* The offshoot Experimenter Publishing Company, founded in 1915, published expanded how-to manuals, pamphlets, and complete books like *The Wireless Telephone, One Thousand and One Formulas,* and *The Wireless Course.* While *Modern Electrics* still advertised the equipment Electro Importing offered for sale in a familiar grid layout with ordering instructions, it also included feature articles detailing the latest research into experimental media technologies in America, throughout Europe, and in Gernsback's own company offices. Each 36-page, 6 by 9.5 inch issue sold on newsstands for ten cents and contained regular reports from E. I. Co. employees like Harry Winfield Secor and René Homer, who would contribute to Gernsback titles for years to come. Some freelancers attributed their decision to pursue science as a profession to their experiences with *Modern Electrics,* as did Donald H. Menzel, later director of the Harvard Observatory, who earned money for college by writing for the magazine.[35] Lewis Mumford, the architectural critic, philosopher of technology, and author of the now-classic *Technics and Civilization,* published his first piece of writing at the age of 15 in the same issue of *Modern Electrics* that carried the first installment of Gernsback's *Ralph 124C 41+.* And computer scientist John McCarthy was inspired to pursue his profession by reading Gernsback's stand-alone how-to pamphlets.[36] Celebrity guest contributors like Lee De Forest, and later Thomas Edison and Nikola Tesla in the pages of *Electrical Experimenter,* raised the profile of the magazine among a readership hoping to emulate the success of these famous inventors.[37] Regular dispatches and photographs printed

FIGURE I.4. An Electro Importing Company Catalog bearing the title of *Modern Electrics,* 1907, with cover art by Louis Coggeshall. Courtesy collection of Jim and Felicia Kreuzer, Grand Island, NY.

35. O'Neil, "The Amazing Hugo Gernsback."

36. Silberman, *NeuroTribes.*

37. Tesla first published his autobiography in serial installments in *The Electrical Experimenter,* a book that was recently republished by Penguin Classics. Nikola Tesla and Samantha Hunt, *My Inventions and Other Writings* (Harmondsworth, UK: Penguin Classics, 2011).

on the magazine's quality stock from the unnamed Paris, Berlin, and Brussels correspondents kept readers informed on developments in television, wireless telephony, and the use of novel electrical apparatuses in film and theatrical productions, each of which would go into great technical detail.

But the hallmark of the magazine became its more speculative articles, those that were willing to extrapolate fantastic scenarios out of the technical details at hand. Gernsback and his contributors wrote as if the more detail a particular technology's description contained, the more plausible were the projections of its future possibilities. In a sense, the future stood as the horizon of technical description. As Mumford wrote in his autobiography, "In my youth, as a zealous reader of Hugo Gernsback's *Modern Electrics,* I shared my generation's pious belief in our future."[38] With **Signaling to Mars,** for example, Gernsback detailed the conditions that would have to be met in order for Earth to send messages via wireless telegraph to the red planet. The quantitative description of the transmitting apparatus in terms of its necessary output (a gargantuan 70,000 kilowatts) and best time of year to signal (summer) is only one aspect of this scenario. Gernsback goes on to take into account the nature of Martian intelligence that would be necessary for such a communicative circuit to be completed:

> we can only hope that the Martians are further advanced than we and may signal back to us, using a method new to us and possibly long discarded by them, when thousands of years ago they stopped signaling to us, and gave us up, as we did not have intelligence enough to understand.[39]

For readers of *Modern Electrics,* the technical context in which this highly speculative article appeared only lent credence to an idea as fanciful as the one that contact with an alien civilization was right around the corner. In the copy of this issue at Princeton University's Firestone Library, someone inserted a newspaper clipping (now a permanently affixed leaf within the bound volume) that tells of a new distance record for wireless signaling, from San Francisco to the Pacific Mail Line steamship *Korea* as it made its way across the ocean. Left there as if to vouch for the plausibility of the idea that we will soon be able to connect with our nearest planetary neighbor, the clipping provides a wonderful sense of how people read these magazines.[40] Although the Gernsback titles eventually became notable for some of their more outlandish claims—that electric current might clean us better than soap and water, that the success of a marriage can be predicted using gadgets assembled out of various household supplies—they were always presented through a lens of supposedly scientific rationality.

38. Quoted in Westfahl, *Hugo Gernsback and the Century of Science Fiction,* 62.

39. Continuing in the tradition of astronomers Percival Lowell and William Henry Pickering—the latter of whom offered a similar proposal on the front page of the *New York Times* to communicate with Mars using a series of mirrors—the projection of Martian technology (not to mention ecology) provided a topos upon which readers might assess the direction of its terrestrial analogues. See "PLANS MESSAGES TO MARS; Prof. Pickering Would Communicate by Series of Mirrors to Cost $10,000,000." *New York Times,* April 19, 1909.

40. While I haven't been able to determine the provenance of this particular clipping, other articles reporting on the *Korea*'s distance-signaling record were published in the *New York Times* on November 8, 1909, and the *Boston Evening Transcript,* November 6, 1909.

This frame affected the reception of the magazines by their readers, the design ethos that grew up around them, and the kind of fiction they eventually produced.

From the earliest of the Gernsback titles, we find "science" configured as an instrumentalized form of technological achievement that bore little relationship to what was happening in laboratories.[41] Contrary to the common division between the purely theoretical nature of the sciences and their application in the development of new technologies, Gernsback argues in a later editorial that *science* and *invention* are part of a continuum:

> The word *Science,* from the Latin *scientia,* meaning knowledge, is closely related to *Invention,* which, derived from the Latin *inventio,* means, finding out. There is little in Science that did not at one time require some inventive powers, while conversely most of the world's inventions are based upon one or more of the sciences. **(Science and Invention)**

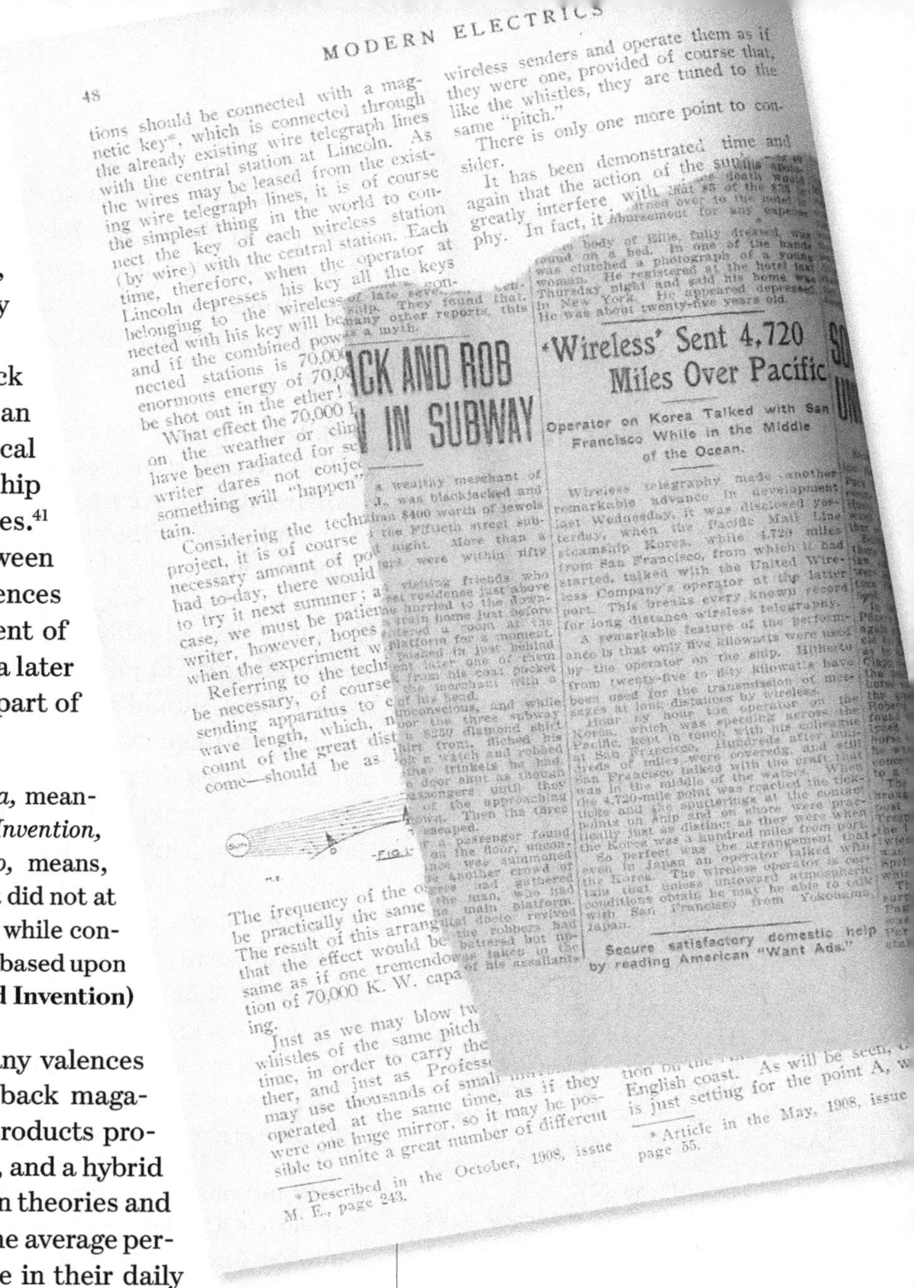
MODERN ELECTRICS

48

'Wireless' Sent 4,720 Miles Over Pacific

Operator on Korea Talked with San Francisco While in the Middle of the Ocean.

Secure satisfactory domestic help by reading American "Want Ads."

FIGURE I.5. Newspaper clipping from an unknown source left inside Princeton University's copy of the May 1909 *Modern Electrics.* Courtesy Firestone Library.

This article is a key of sorts for the many valences *science* can take throughout the Gernsback magazines. Science is the sum of its many products progressively connecting the modern world, and a hybrid ontology that saw no distinction between theories and their application. It is defined as what the average person understands of its growing presence in their daily life: "science no longer is the sombre book closed with seven seals. Quite the contrary, it is the public that popularizes science—not our scientists." It is even configured as a form of belief when Gernsback argues that skepticism is an entirely unscientific attitude:

> But our *real* scientists are as backward as in Galileo's times. The public applauds and instantly believes in anything new that is scientific, whereas the true scientist scoffs and jeers, just as he did in Galileo's times when that worthy stoutly maintained that the earth moved and did not stand still.

In many ways, this starry-eyed fanaticism for science as the sum of its progressive advance in the material world reflects how public discourse

41. For a theoretical overview of the relationship between science and technology, as well as an intellectual history of the intermediary term "applied sciences," see Jennifer Karns Alexander, "Thinking Again about Science in Technology," *Isis* 103, no. 3 (September 2012): 518–26.

was shifting on a larger scale as science entered mass-market news-stands, corporate research facilities, and public school classrooms. As John Rudolph has argued, it was during this period that popular conceptions of the "scientific method" emerged not from the practices of professional or academic researchers but in secondary school pedagogy:

> while the manner in which practicing scientists went about their work (the research strategies they used, their modes of inquiry, norms of argumentation, etc.) changed relatively little if at all from the 1880s to the 1920s, portrayals of the scientific method in American schools underwent a marked transformation.[42]

If science was a highly variable concept for Gernsback, conversation surrounding its application was just as muddled. The way we read these essays should also be complicated by the fact that *technology* was a word unknown to most English speakers at the time. So, for instance, when Gernsback writes in 1922 that "steam, electricity, and up-to-date technic have completely altered not only the face of the globe, but our very lives as well," the usage of the word *technic* where we might expect something like *technology* reflects an important terminological ambiguity at the time (**10,000 Years Hence**). In nineteenth-century English, according to Eric Schatzberg, *technology* referred to "a field of study concerned with the practical arts; except in anomalous usage, [it] did not refer to industrial processes or artifacts." Just as sociology names the study of society, technology was the science of technique, making, the useful arts. Somewhere around 1930, Schatzberg argues,

> new meanings derived primarily from the writings of American social scientists who imported elements of the German discourse of *Technik* into the English term technology, thus shifting the latter from its original definition as the science or study of the useful arts to a new one that embraced the industrial arts as a whole, including the material means of production.[43]

Science and technology for Gernsback are collapsed into one another, meeting somewhere in the middle as the thoughtful use of tools and making. This highly materialist understanding of science was foundational for Gernsback's later conceptualization of scientifiction.

Hoping to build on the success of *Modern Electrics,* Gernsback sold the magazine to the competing publisher of *Electrician and Mechanic* and launched a new title in May 1913: *The Electrical Experimenter.*[44] Continuing the work that began with *Modern Electrics*'s monthly radio set building contest and correspondence section, *The Electrical*

42. John L. Rudolph, "Epistemology for the Masses: The Origins of 'The Scientific Method' in American Schools," *History of Education Quarterly* 45, no. 3 (September 2005): 341–76.

43. This masterful history of the interrelated concepts of *Technik,* technology, and technique is highly recommended. Eric Schatzberg, "Technik Comes to America: Changing Meanings of Technology before 1930," *Technology and Culture* 47, no. 3 (2006): 486–512.

44. *Modern Electrics* continued under new ownership as *Modern Electrics and Mechanics* for two years before acquiring and taking on the name of *Popular Science Monthly* in April 1915, a magazine whose publishers were looking to update the format it had run since 1872: reprints of European science periodicals with little or no illustration. Under the editorship of Waldemar Kaempffert, the new *Popular Science* tried to increase its readership with short-form writing, dense photo spreads packed onto every page, and a far more generalist approach to "science." One would now find articles on developments in criminology, warfare, and motion pictures in a magazine that only a year earlier covered entomology, evolution, and pathology. It continues to be published today. For an overview of Kaempffert's editorial philosophy, see Waldemar Kaempffert, "The Vision of a Blind Man," *Popular Science Monthly* 88 (June 1916): vii.

Experimenter introduced several new ways for readers to participate. A section called "The Constructor" included tips for the home workshop, like a technique for tightly winding a spark coil using a hand-cranked drill. The "How-To-Make-It Department" offered prizes for designs submitted by readers that specifically "accomplished new things with old apparatus and old material," such as an alarm that activated a simple electric bell in the house when a mailbox's door was opened and closed. The "Electrical Magazine Review" provided a roundup of news recently published in other electrical magazines like *Electrical World* and London's *The Electrician,* while the "Patent Advice" column offered input on inventions that readers hoped to patent at the price of $1 per question. By May 1918, the magazine had reached a circulation of a hundred thousand copies per month, almost doubling *Modern Electrics*'s readership seven years previously.

To page through the print run of *The Electrical Experimenter* across the 1910s is to watch the activities of a quirky group of hobbyists grow into a mass cultural phenomenon. Over the course of its publication, Gernsback and his staff gradually widened their focus from the highly specialized electrical arts of *Modern Electrics* to a range of topics geared more toward the general public. As "recipes, wrinkles, and formulas" were joined by sensational depictions of future technologies, Gernsback changed the magazine's tagline from "The Electrical Magazine for the Experimenter" to "The Electrical Magazine for

FIGURE I.6. Science in the newsstands. New York City, June 1913, photographed by Lewis Hine. The larger 11 x 8.5 inch size of *The Electrical Experimenter* meant that it could be displayed alongside magazines like *Collier's* and *The Red Book.* Library of Congress, Prints & Photographs Division, National Child Labor Committee Collection, LC-DIG-nclc-03865.

Everybody." By the decade's end the title was completely rebranded as *Science and Invention,* a change that once again nearly doubled the magazine's circulation, to almost 200,000 copies per month by 1921.[45] The new look was accompanied by a significant change in editorial policy, proceeding from the idea that the image is the most effective way to communicate complicated scientific information to the public (**The "New" Science and Invention**).

Gernsback's visions were lent a shape and a color thanks to two new artists who were able to "illustrate difficult subjects in such a way that words really become superfluous," as Gernsback put it.[46] Though Howard V. Brown's career as a cover artist spanned a variety of subjects and techniques—his work appeared everywhere from *Scientific American* to *Argosy All-Story Weekly* to the children's magazine *St. Nicholas*—the signature of his Gernsback covers was the Rockwellesque way he was able to tell the story of an entirely speculative technology through facial expressions and dramatic gestures.[47] Seemingly familiar domestic scenes, detailed in vividly realistic brush strokes, were punctured by the presence of a strange apparatus and became the starting point for conversations on how a technology of the future might function. It was in no small part due to the vivid illustrations that readers were so drawn to these visions of the future. As a character in Robert A. Heinlein's 1940 short story "Requiem" says, he was one of those fans "who thought there was more romance in an issue of the *Electrical Experimenter* than in all the books Dumas ever wrote."[48]

The shape of media to come took on an iconography all its own through the illustrations of Frank R. Paul. Trained in Vienna, London, and Paris as an architectural draftsman, Paul is known for his exquisitely detailed cityscapes as well as the robots, aliens, and astronauts that would later adorn the covers of *Amazing Stories.*[49] Gernsback greatly respected the realist futurism of his artists, writing that their ability to fire the reader's imagination was one of the most important components of his magazines: "It is no easy matter to think out new things of the future and illustrating them adequately by means of expensive washdrawings or three-color cover illustrations. Indeed, there is nothing more difficult connected with the publication" (**Imagination versus Facts**).

As the magazines in which they appeared gained a wider following, these images began to circulate far beyond their original venues. Plans for the Osophone, a device Gernsback designed to replace headphones by transmitting sound through vibrations in the jawbone of the listener (**Hearing through Your Teeth**), were published and reviewed in the German journal *Der Radio-Amateur.*[50] Paul's sketch of a man using

45. *Science and Invention* continued for another decade before being bought out by the Chicago publisher of *Popular Mechanics* and merged into that title with the September 1931 issue. It is still published today. Mike Ashley, *The Gernsback Days: A Study of the Evolution of Modern Science Fiction from 1911 to 1936,* 1st Wildside Press edition (Holicong, Penn.: Wildside Press, 2004), 189–90.

46. Gernsback, **The "New" Science and Invention**.

47. Jon Gustafon et al., "Brown, Howard V.," in *The Encyclopedia of Science Fiction,* ed. John Clute et al. (London: Gollancz, 2015).

48. Quoted in Fred Erisman, "Stratemeyer Boys' Books and the Gernsback Milieu," *Extrapolation (Pre-2012)* 41, no. 3 (2000): 272–82.

49. Moskowitz, *Explorers of the Infinite,* 234.

50. Eugen Nesper, "Das Osophon von H. Gernsback," *Der Radio-Amateur* 2, no. 1 (1924): 10. For the history of *Der Radio-Amateur* magazine, see Heinz Sarkowski and Heinz Götze, *Springer-Verlag, Part 1: 1842–1945: Foundation, Maturation, Adversity* (New York: Springer Science & Business Media, 1996).

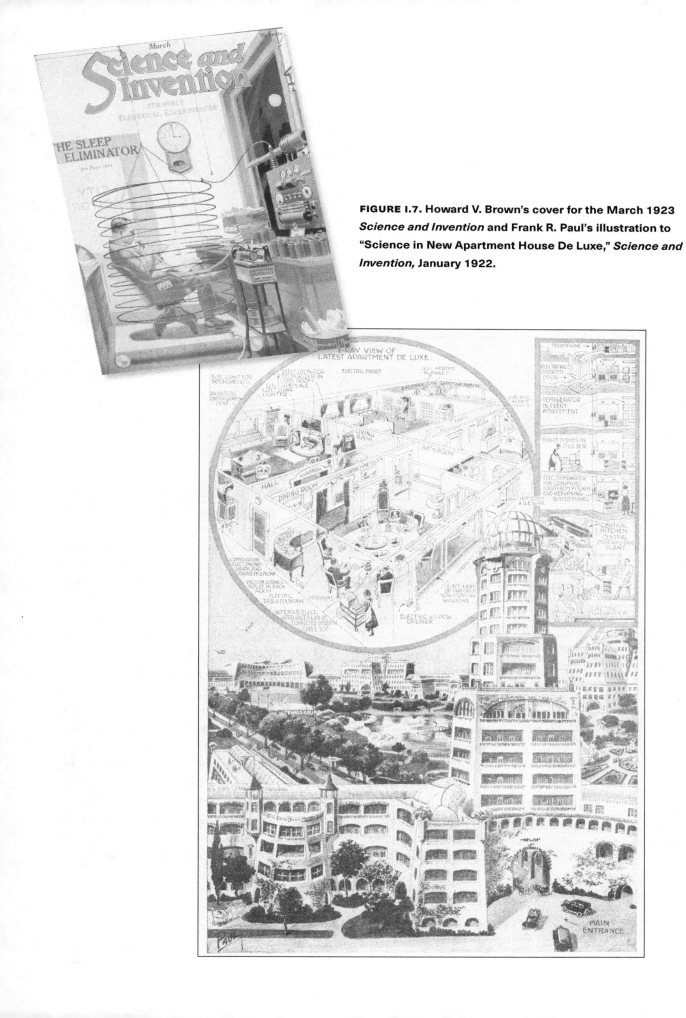

FIGURE I.7. Howard V. Brown's cover for the March 1923 *Science and Invention* and Frank R. Paul's illustration to "Science in New Apartment House De Luxe," *Science and Invention,* January 1922.

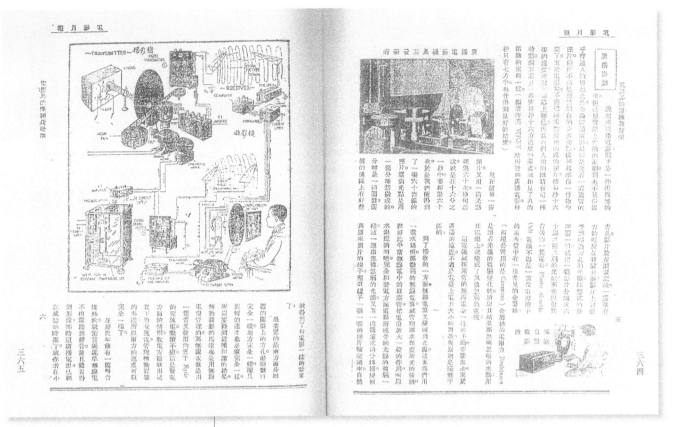

FIGURE I.8. *Science and Invention* illustrations republished in Chinese film journal *Yingxi zazhi.* Courtesy Princeton University Rare Books and Special Collections.

51. Weihong Bao, "Sympathetic Vibrations: Hypnotism, Wireless Cinema, and the Invention of Intermedia Spectatorship in 1920s China" (paper presented at *Media Histories: Epistemology, Materiality, Temporality,* Columbia University, New York, March 2011). Bao located Paul's illustration of the *Science and Invention* television receiver in Xiaose Shen, "Dianyingjie de Jiizhong Xin Faming [Several new inventions in the film world]," *Dianying yuebao* 9 (1929): 1–64. The term for television used in this article is "wuxian dianying" (wireless cinema, or, more literally, wireless electric shadow, or radio shadow). Paul's images were originally published as the accompaniment to Hugo Gernsback, "Radio Movie," *Science and Invention* 16, no. 7 (November 1928): 622–23.

52. "This Is How We Look in Arabic," *Electrical Experimenter* 7, no. 3 (July 1919): 287.

a tuning fork to calibrate the speed of the 1928 *Science and Invention* Nipkow disk television receiver was republished the following year in the Chinese film journal *Yingxi zazhi* (Shadow play magazine) as an illustration of recent research into television, what was referred to in the article as (directly translated) "wireless cinema."[51] Many more articles from *Electrical Experimenter* and *Science and Invention* were translated into French, German, Italian, Spanish, Japanese, Dutch, and Arabic, circulating widely along with their illustrations.[52] Gernsback even published a Spanish-language edition of *Radio News,* titled *Radio Internacional,* from March 1926 through April 1927.

It's the wide currency of these images then and now, perhaps especially as they are given new life as retrofuturist visions circulating on Twitter and Tumblr, that has resulted in the description of Gernsback for better or worse as a "prophet" of the future. Unfortunately, this approach tends to flatten the richness of his work into a list of the impressively early dates by which he had described the coming of technologies like in vitro fertilization, the transistor radio, atomic war, education by video, and telemedicine. Gernsback himself seemed to enjoy the continued notoriety these predictions brought him. In a sense, their sheer number is of course impressive, as it was for Arthur C. Clarke who dedicated his *Profiles of the Future: An Inquiry into the Limits of the Possible* "to Hugo Gernsback, who thought of everything."[53] Though many of these ideas were in the air at the time, Gernsback simply had a knack for tuning in like no other, for distilling the essence of a technosocial development from a cloud of diverse possibilities into a sleek, attractive form.

But descriptions of Gernsback as a prophet miss the fact that these imminent futures felt so close to hand thanks to a collective endeavor between contributing writers, assistant editors, illustrators, and readers. Throughout his writings, Gernsback relies on input from readers, quoting them sometimes at great length. As his magazines began to grow both in number and circulation, Gernsback increasingly delegated work to assistant editors like T. O'Conor Sloane, David A. Lasser, and Charles Hornig. While Gernsback wrote under a number of pseudonyms for *Modern Electrics* when he still didn't have many contributors, over time this situation was apparently reversed, with

53. Moskowitz lists the predictions in *Ralph 124C 41+* alone: "Fluorescent lighting, skywriting, automatic packaging machines, plastics, the radio directional range finder, juke boxes, liquid fertilizer, hydroponics, tape recorders, rustproof steel, loud speakers, night baseball, aquacades, microfilm, television, radio networks, vending machines dispensing hot and cold foods and liquids, flying saucers, a device for teaching while the user is asleep, solar energy for heat and power, fabrics from glass, synthetic materials such as nylon for wearing apparel, and, of course, space travel are but a few." Moskowitz, *Explorers of the Infinite*, 233.

FIGURE I.9. A Frank R. Paul caricature of *The Electrical Experimenter* newsroom from the April 1920 issue.

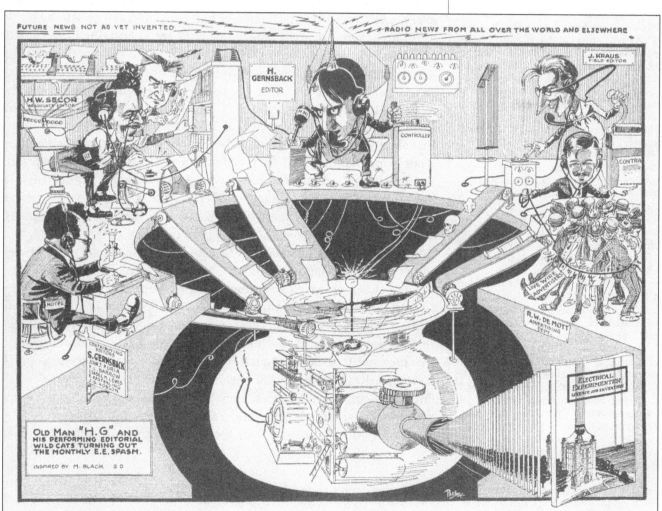

54. Justine Larbalestier, *The Battle of the Sexes in Science Fiction* (Middletown, Conn.: Wesleyan University Press, 2002), 17. Larbalestier cites Marshall B. Tymn and Michael Ashley, *Science Fiction, Fantasy, and Weird Fiction Magazines* (Westport, Conn.: Greenwood Press, 1985) and John Clute and Peter Nicholls, *The Encyclopedia of Science Fiction* (New York: St. Martin's Griffin, 1995).

55. Eric Leif Davin, "A Conversation with Charles D. Hornig," in *Pioneers of Wonder: Conversations with the Founders of Science Fiction* (Amherst, Mass.: Prometheus Books, 1999), 61–89. Hornig, who grew up in poverty, gained the attention of Gernsback as the editor of a fanzine and was hired as assistant editor at the age of seventeen. He remained with *Wonder Stories* until Gernsback sold it to Beacon Magazines in 1936. Out of a job and still just nineteen years old, Hornig "bought a large-sized paperback entitled *100 Ways to Make a Living*. In the thirties everybody wanted desperately to make a living, so it was a good seller. I thought I ought to be able to find one way that worked out of a hundred. The book had pictures and descriptions of all kinds—you could invent your own chemical formulae, invent products to sell, and whatnot."

56. Science fiction historian Mike Ashley challenges the notion that Gernsback was ever completely removed from editorial work bearing his name: "I'm not sure how accurate a word 'delegate' is. Clearly as his business grew Gernsback needed others to do much of the editorial work, and certainly the first reading of manuscripts, but Lasser, Hornig, and Lowndes all told me that Gernsback had to clear everything prior to publication. I think he initialed the various contributions to say he had agreed to them, and there were times he would send things back. He often corresponded directly with his regular writers, especially Keller and Ed Hamilton, so he kept his hand firmly on the tiller." Mike Ashley, e-mail to the author, December 10, 2015.

57. Larbalestier, *The Battle of the Sexes in Science Fiction*, 17.

"Gernsback" operating as a kind of collective editorial voice by the time of the science fiction magazines in the late 1920s. As Justine Larbalestier writes,

> Gernsback did not necessarily write all or even the majority of the editorial comments during the periods in which he was the publisher and editor of such magazines as *Amazing Stories* and *Wonder Stories* [1926–36]. However, such is the mythic force of Gernsback, the founding father, that in the majority of the work I have read on this period he is spoken of as though he wrote every word of editorial comment in the magazines he published. "Gernsback" has come to operate as synonymous with the magazines he founded, whether he was actively contributing to them or not.[54]

In one of the few existing interviews with Gernsback's editorial staff, Charles Hornig claimed in 1986 to have written the editorials while editor of *Wonder Stories,* as well as "handled 'The Reader Speaks' letters section, wrote all the blurbs and captions, and chose the work of all artists except for Frank Paul."[55] There is no telling how much of a hand Gernsback had in writing any of these later articles, from offering up his often clunky, nonnative prose for editorial revision, to suggesting the idea of a piece to one of his assistant editors, to simply allowing his staff full discretion.[56] Manuscript versions at the Hugo Gernsback Papers in Syracuse and in the possession of private collectors show that he did, in fact, author material for the technical magazines in longhand, with little revision before the articles went to print. But following Larbalestier and other science fiction critics, I think it important to continue "allowing 'Gernsback' to stand for the house editorial voice," not just as a convenient shorthand but as a means of tracing the many-voiced, overlapping perspectives on making, fiction, and science that emerged in these magazines over their thousands upon thousands of pages.[57] Whether Gernsback wrote every word of the later articles bearing his name or merely provided editorial guidance, we are left with the record of a new kind of conversation on technology. Gernsback, about whom we know little outside of the self-propagated myths, is thus best understood as an embodiment of this moment that falls between the gaps of literary and technological history, a distillation of what was in the air. To make Gernsback synonymous with his magazines, each of which was highly responsive to the interests and activities of its readership, is to follow the voice of a community that developed over the course of three decades.

"a perfect Babel of voices": Communities of Inquiry and Wireless Publics

"We are constantly being speeded up mentally," writes Gernsback in 1923 with the prescience of a Marshall McLuhan aphorism (**Are We Intelligent?**). Understanding media was a primary goal of his magazines, not just from the perspective of the expert experimenter but also from that of the bemused end user. "Everything we are doing now is too cumbersome. . . . Our lives are crowded to such an extent, that it is impossible to read as much as our grandfathers could." Sensory environments had become so packed, so distracting, that it was time to find a better medium to communicate complex information to the public. During this period, as Susan Douglas has shown, "the press's method of covering and interpreting technological change was developed. In other words, what scholars have identified as the functions of the mass media in the late twentieth century were being formulated and refined during the first twenty-three years of radio's history."[58] It is no mistake that Gernsback explicitly refers to the magazine as itself a "medium" capable of matching the new speeds at which people took in the world (**A New Sort of Magazine; The "New" Science and Invention**). His publications, his ideas, and his prose were purpose-built to move fast in a pulp environment with short deadlines and low overhead, and it was for this reason that they were uniquely suited to the analysis of rapidly shifting media conditions. While this speed sometimes results in seeming contradictions and an all-too-hasty embrace of the new, it also contributes to an inviting sense of openness for a diverse community of inquiry. A core set of his essays take up the question of where new media come from, as well as how ordinary people could participate in the conditions of their emergence rather than feeling like they had been swept up in their wake.

Many writings in this collection will leave the reader with a sense that media "evolve" according to their own logic. In **Is Radio at a Standstill?** Gernsback makes a striking media-historical analogy between the supposed threat that the rise of radio broadcasting posed to established phonograph manufacturers and the impact of "battery eliminator" radio sets (which could be plugged into wall outlets) to conventional battery manufacturers. In the model that Gernsback outlines, here, of the historical cycles that the technology industry goes through, competing formats do not replace but rather force one another to find their own unique attributes, simply as a matter of survival. In his media-historical writings, Gernsback often speaks of the "perfection" of an apparatus, a question of medium specificity he almost always raises through the lens of viable business

58. Susan J. Douglas, *Inventing American Broadcasting, 1899–1922*, Johns Hopkins Studies in the History of Technology (Baltimore: The Johns Hopkins University Press, 1987), xix.

models. Despite his penchant for projecting far-flung futures, Gernsback is often remarkably conservative when it comes to the specifics of precisely how long such perfection would take, and what it would mean for developments in emerging media like wireless and television. Given the small sample size of these very young histories, there was little the technologist had to go on when thinking about the stable, almost Platonic forms that might emerge. As he writes in **Edison and Radio:**

> The radio industry today is only five years old, and it may safely be predicted that when it becomes as old as the phonograph is today we shall hardly be able to recognize it as the same development. It is admitted that radio is not yet perfect. Neither is the phonograph, nor the automobile, or motion pictures, nor electric lights; nor, for that matter, a pair of shoes.

Not only was the concept of "wireless" itself—and later "television"—a moving target for its early adopters but also for legislators who scrambled to craft regulatory frameworks for information in the air. Gernsback's magazines avidly followed broadcast policy debates, and much of the power of their arguments against regulation derived from an understanding of media technologies as highly fluid entities that no legislation could hope to keep up with:

> Radio has always been able to take care of itself, and will continue to do so in the future. To be sure, we all want a radio law to straighten out some of our present tangles, but in the end radio engineering will make the best law obsolete.[59]

In the years leading up to World War I, when Gernsback was most active in organizing communities of amateur experimenters against proposed federal broadcast regulation, the magazine's staff had to rely on significant amounts of conjecture in reporting on developments that took place largely behind closed doors between the U.S. Department of Commerce and Labor and various private entities: "all of this maneuvering that could have so altered American communications and culture transpired behind the scrim of corporate confidentiality, not to be made public until scholarly investigation decades later."[60] Nevertheless, *Modern Electrics* was able to recommend best practices for wireless amateurs so that the community could avoid drawing the scrutiny of regulators (**The Wireless Joker**), play nicely with government stations whose apparatuses were "three years behind time" and whose manners were even worse (**The Roberts Wireless Bill**), as well as make technical recommendations on frequency allocation that were later incorporated into Congress's Wireless Act of 1912 (**The**

59. Hugo Gernsback, "The Broadcasting Situation," *Radio News* 8, no. 4 (October 1926): 323.

60. Tim Wu, *The Master Switch: The Rise and Fall of Information Empires* (New York: Knopf Doubleday Publishing Group, 2010), 80.

Alexander Wireless Bill). The Electro Importing Company began publishing its annual *Blue Book* in 1909, a telephone book of sorts that listed the names and call signs of amateur wireless operators around the country, and later, the world. Its presence was designed to encourage greater accountability for the content of wireless messages once the names of their senders were shared openly and freely (Signaling to Mars). In addition, Gernsback and his associates formed The Wireless Association of America in 1909, an education and outreach organization that ended up training many of the wireless operators that the Navy would need once the United States entered the Great War in 1917; one of these operators even developed a means of recording clandestine German U-boat commands that were being relayed through a New Jersey wireless station, unbeknownst to the American government (Sayville).[61] But with the outbreak of the war, the Navy outlawed all amateur wireless broadcasting activities and took sole control of the airwaves. *Electrical Experimenter* became a community forum for frustrations over this policy, as well as a drawing board for what broadcast regulation should look like once the war was over. While the positions taken up by contributors to and readers of the Gernsback magazines don't amount to a full-scale regulation policy, their arguments against hasty legislation that restricted the rights of everyday users were a series of reactions to conditions on the ground that are very much reminiscent of advocates for net neutrality and free and open-source software today (Wired versus Space Radio).

While at certain moments Gernsback evokes the perversity of media that seemed to evolve as if according to their own internal logic, at others he claims that user communities are driving these developments. While one article might obsess over the merits of a new detector, another might find that detector to be far less important than the new forms of connection it made possible. Differences in emphasis throughout Gernsback's writings don't necessarily happen over a certain period of time or unfold in any sort of linear fashion. While there were of course the signature themes that he returned to throughout his career, Gernsback's media theory does not constitute a rigorous, internally consistent philosophical system. Writing two, sometimes three or more, articles a month meant that he was firing off ideas as they came to him, ideas that were picked up and discarded as utility demanded. For this reason, it is not quite accurate to describe these variations as contradictions. Instead, they reflect the thinking of a tinkerer, comfortable with fragments and capable of applying any one of multiple perspectives to a single issue.

So when Gernsback's writings were not emphasizing material specificities as a means of understanding the shape a medium might take

61. See also Grant Wythoff, "The Invention of Wireless Cryptography," *The Appendix* 2, no. 3 (July 2014): 8–15.

in the future, they were describing ethereal meeting places and transcendent connections between people from around the globe. This was especially the case with his profiles of wireless. In "the most inspiring example of the triumph of mind over matter" (**A Treatise on Wireless Telegraphy**), simple minerals like zinc and copper could be harnessed to transmit the self live across the ocean. "We are now on the threshold of the wireless era, and just beginning to rub our intellectual eyes, as it were" (**The Wireless Association of America**). Wireless promised constant contact between friends, family, and complete strangers, regardless of location. "The radiophone," he wrote using another short-lived name for radio before the fact, "will link moving humanity with the stationary one" (**The Future of Wireless**).[62] In other words, developing "wireless tele-mechanics" meant that even people driving in automobiles could maintain a conversation with someone sitting at home. "In our big cities thousands of ears listen every minute of the day to what is going on in the vast ether-ocean" (**War and the Radio Amateur**). The announcement of the annual *Blue Book* not only promised more and more citizens of the ether popping up on the map with each passing year, it appeared alongside Gernsback's feature on sending and receiving wireless signals from Mars, making it seem as if soon we will even be able to look up our Martian friends in the universal wireless directory.

As Gernsback welcomed these newly minted citizens of the ether stepping "out into the star-lit night and myriads of voices" (**Amateur Radio Restored**), one wonders just how inclusive this global village of amateur experimenters was, in practice. Although Gernsback himself is remarkably silent on all questions regarding race, gender, and class, his magazines' advertisements for radio parts, books, and professional schools largely addressed a male audience. The most vocal participants in the technology magazines' correspondence sections, and thus their most immediately visible readers, were men roughly sixteen to twenty-four years old who often equated the pursuit of science with masculinity itself. As one reader commented, pursuing radio as a hobby "stills a thirst for scientific knowledge and hence stimulates manhood" (**Who Will Save the Radio Amateur?**). But this doesn't mean that other people weren't reading and using Gernsback's magazines. Eric Drown has convincingly argued that historians of science fiction have "repressed the working-class accents of pulp SF and been content to characterize it as a form of children's literature justly and rapidly replaced by a more mature form of 'speculative fiction.'"[63] Drawing on Michael Denning's scholarship on the complex relationship between writers and readers of nineteenth-century dime novels in which "mostly middle-class writers ventriloquized the working-class

62. Gernsback continues to use the term "radiophone" as late as 1923, signaling an understanding of radio as a point-to-point means of interpersonal communication even as the national broadcast networks were flickering to life. The term replaced the previously favored "wireless telephone."

63. Eric Drown, "Business Girls and Beset Men in Pulp Science Fiction and Science Fiction Fandom," *Femspec* 7, no. 1 (September 2006): 5–35 at 28.

accents of their readers to advance their views of the world," Drown reminds us that the common dismissal of these texts as "semi-literate" and "adolescent" constitutes a serious misunderstanding of the complex voicings of hourly laborers seeking the promise of secure, middle-class occupations through the "scientific" education these magazines provided.

> Readers were largely but not exclusively wage-earning people who describe themselves in letter columns as students, engineers, radio operators, amateur scientists, mill hands, office workers, salesmen, lathe operators, enlisted men, and government bureaucrats. While there was a significantly visible contingent of precocious mostly middle-class boys among the letter-writers, most readers were the adults who provided the routine intellectual, clerical, mechanical and physical labor that made the new mass production economy function.[64]

Further, this community may have been more heterogeneous than is apparent from the surface of the magazines. Steve Silberman identifies wireless amateurism as a haven for voices unable to find an audience anywhere else:

> The culture of wireless was also a strict meritocracy where no one cared about what you looked like or how gracefully you deported yourself in public. If you knew how to set up a rig and keep it running, you were welcome to join the party. . . . [It] also enabled shy introverts to study the protocols of personal engagement from a comfortable distance. "Through amateur radio . . . I've learned so much about communication between people. I've had the opportunity to observe and participate in the giving and getting process, which is what communication is all about," recalled Lenore Jensen, who co-founded the Young Ladies' Radio League in 1939 to encourage more women to join the conversation.[65]

Highly scripted conversational formats and the anonymity of code allowed for a neurodiversity and gender inclusion within this community that is difficult to recover from the pages of Gernsback's magazines alone. One has to read *Electrical Experimenter* against the grain to find the presence of women, for instance. But there are several notable examples in reports from contributors, including profiles of Maria Dolores Estrada, a refugee of the Mexican Revolution turned licensed wireless operator, and Graynella Packer, "the first woman wireless operator to serve aboard a steamship in a commercial capacity."[66] A woman identified as "Mrs. Alexander MacKenzie" used her son's wireless outfit in the service of the New York Women's Suffrage Party's twenty-four-hour election night demonstration in November 1915 to make speeches and relay messages "from various celebrities

64. Ibid., 8.
65. Silberman, *NeuroTribes*, 244–45.
66. "Mexican Girl Wins Radio Diploma," *Electrical Experimenter* 4, no. 10 (February 1917): 718.

and other suffragists."[67] And the "Autobiography of a Girl Amateur" offers an anonymous account of one woman's success in attaining a respected status among the wireless community by never revealing more about herself than her call letters, which she of course doesn't give up in the article.[68] This archive demands further digging.

Although it is indeed rare that women make more than a cameo appearance in the pages of these magazines, one *Science and Invention* editorial by Gernsback makes clear that they played a vital role at virtually every level of the magazine's production. A 1928 issue opens with a fanciful tale in which King Outis VII of Erehwon, a great fan of the magazine, visits the offices of Experimenter Publishing to witness how the magazine is made. He meets with editors, artists, advertising staff, linotypers, compositors, proofreaders, the binding department, and photographers. Following the supply chain even deeper, Gernsback introduces the King to the "chemical houses who furnished the chemicals for making the pulp and bleaching it," the coal mines that made possible the steam to make that sulfite pulp paper, even the hens that laid the eggs "used in the photo-engraving process to supply albumen." When the King finally asks "who all the pretty girls were that had clustered together in one aggregation,"

> We informed him immediately that they were secretaries, stenographers and typists, as well as editors and proofreaders and many others who had directly to do with the production of the magazine. The firm that did the composing had several dozen girls that performed various jobs in connection with the magazine. The printer had a number of girls who were either bookkeepers or stenographers, through whose hands passed the bills for the magazine, and the same was true of practically every other industry connected with the production of the magazine.[69]

Rarely made visible in the content of the magazines themselves, women were there at every stage in the process of their production, from the chemical supply houses to the newsdealers. For Gernsback, it was as if women were the very circuits that connected and regulated all of the magazine's material inputs and outputs.[70]

Given the potential diversity within the overlapping communities of wireless amateurism and later science fiction fandom, Gernsback's techno-utopianism seems somewhat empty. To what end was this future progressing, and who got to make decisions about the direction it took? In many ways, Gernsback's faith in the ceaseless advance of technological progress (he would write in **Wonders of the Machine Age** that "science fiction is based upon the progress of science; that is its very foundation. Without it, there could be no science fiction") makes his late flirtation with the short-lived political movement of

67. "The Feminine Wireless Amateur," *Electrical Experimenter* 4, no. 6 (October 1916): 396–97, 452.

68. "The Autobiography of a Girl Amateur," *Radio Amateur News* 1, no. 9 (March 1920): 490.

69. Hugo Gernsback, "A Magazine Wonder," *Science and Invention* 15, no. 12 (April 1928): 1073.

70. For a history of the influential role women played in periodical publishing throughout the nineteenth century, see Ellen Gruber Garvey, *Blue Pencils and Hidden Hands: Women Editing Periodicals, 1830–1910*, ed. Sharon M. Harris (Boston: Northeastern, 2004).

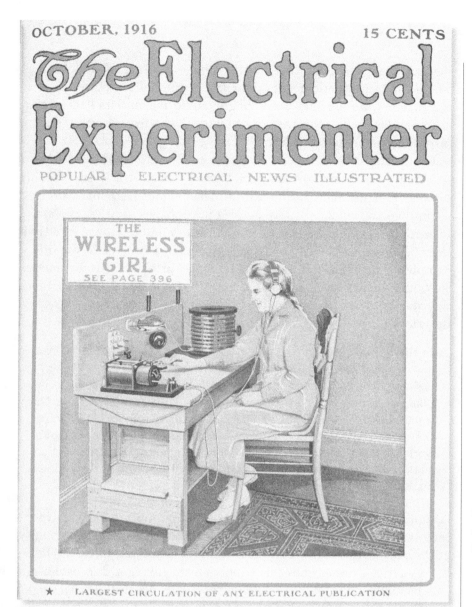

FIGURE I.10. Fifteen-year-old Kathleen Parkin, San Rafael, California, and the wireless set she constructed depicted in cover art by George Wall.

Technocracy seem inevitable. Technocracy blossomed in the United States during the Depression with its plan to put engineers and skilled technologists in charge of the government. While American socialists were reading Marx, the Technocrats took their inspiration from Thorstein Veblen, the American sociologist who advocated in *The Engineers and the Price System* (1921) for a more equitable form of price management through what he called a "Soviet of Technicians." These engineers and technically educated managers would be able to govern more effectively the production and equal distribution of energy, goods, and quality of life than capitalist bureaucrats or politicians were capable. His ideas were picked up by Howard Scott, a leading

71. Howard P. Segal, *Technological Utopianism in American Culture* (Syracuse, N.Y.: Syracuse University Press, 2005), 123. Harold Loeb, another leading proponent, headed up the rival Continental Committee on Technocracy and wrote the utopian road map *Life in a Technocracy: What It Might Be Like* (1933; repr., Syracuse, N.Y.: Syracuse University Press, 1996), a book that advocated a theory of value based on units of energy. For Howard Segal, Technocracy and other associated efforts to forge a better society through technology (e.g., the New Machine, the Technical Alliance, the Utopian Society of America) marked a shift in the 1920s and '30s in utopian thinking from the possible to the probable. For more on the Technocracy movement, see Daniel Bell, "Veblen and the Technocrats: On the Engineers and the Price System," in *The Winding Passage: Sociological Essays and Journeys,* 69–94 (1963; repr., New Brunswick, N.J.: Transaction Publishers, 1991); William E. Akin, *Technocracy and the American Dream: The Technocrat Movement, 1900–1941* (Berkeley: University of California Press, 1977).

72. The argument is developed in **Wonders of the Machine Age.** The inaugural issue of the *Technocracy Review* (February 1933) contained articles by Howard Scott ("Technocracy Speaks"), Paul Blanshard ("A Socialist Looks at Technocracy"), William Z. Foster ("Technocracy and Communism"), and Gernsback's assistant editor David Lasser ("Technocracy—Hero or Villain?").

73. Langdon Winner, *Autonomous Technology: Technics-Out-of-Control as a Theme in Political Thought* (Cambridge, Mass.: MIT Press, 1978), 146.

74. Daniel Czitrom, "Dialectical Tensions in the American Media Past and Present," in *Popular Culture in America,* ed. Paul Buhle, 7–20 (Minneapolis: University of Minnesota Press, 1987).

75. Eric Drown, "Usable Futures, Disposable Paper: Pulp Science Fiction and Modernization in America, 1908–1937," (PhD diss., University of Minnesota, 2001), 81–82.

figure in the movement who formed Technocracy Inc., an organization whose "militaristic demeanor, its rigid hierarchal structure, its special insignia, its special salute, its grey uniforms, and its fleet of grey automobiles" raised the eyebrows of more than a few observers during fascism's spread across Europe.[71]

Gernsback's short-lived *Technocracy Review* (1933) attempted to serve as a forum for the exchange of Technocratic ideas. This bimonthly magazine lasted only two issues, and Gernsback himself attempted to remain entirely neutral in its pages save for the argument that machines have throughout history created more jobs than they have taken.[72] But Gernsback's career-long project to provide a complete education in science and technical principles for the masses was fundamentally incompatible with Technocracy, in which there was absolutely no place for public participation. As Langdon Winner describes it,

> In the technocratic understanding, the real activity of governing can have no place for participation by the masses of men. All of the crucial decisions to be made, plans to be formulated, and actions to be taken are simply beyond their comprehension. Confusion and disorder would result if a democratic populace had a direct voice in determining the course the system would follow. Science and technics, in their own workings and in their utility for the polity, are not democratic, dealing as they do with truth on the one hand and optimal technical solutions on the other.[73]

Although Gernsback tried not to have an overt politics, and rarely formulated his positions in such terms, his gestures toward community participation, grassroots education, and social mobility were anything but apolitical. Sometimes this simply meant taking a step back from the bleeding edge of increasingly corporate innovation. Throughout **Radio for All**, a book designed to transform a novice readership into a polity of wirelessly connected citizens, Gernsback purposefully uses what were by then slightly outmoded (and thus simpler to understand) components in all of his examples. Daniel Czitrom describes this approach to educating a new public in the very technology that would constitute their awareness of themselves *as a public* as Gernsback's "hardware socialism."[74] At other times, this meant promising that anyone, even an immigrant with no professional credentials and little formal education like himself, could produce inventions that were beneficial to social progress for all. "Gernsback's electric futurism was a variation on progressive thought," writes Eric Drown.[75] Above all, Gernsback was determined, in the words of Andrew Ross, that *science* "would not be associated with exclusive rhetorical idioms or with obfuscatory accounts of the object world by over-accredited experts. For Gernsback, scientific language was a universal language of progress that ought to be accessible even to those without a college degree."[76]

Unfortunately, the rapid progress so valued by Gernsback in the electrical arts ended up rendering obsolete the circuits of amateur cooperation and frugal ingenuity he helped institute. By the 1920s, large corporations hoarded patents with the support of the federal government, research and development became an increasingly formalized institution, and hybrid public/private entities were able to draw on capital that no independent amateur could hope to compete with. In this regard, Gernsback's picture of popular scientific education and homespun innovation wasn't always able to adapt quickly enough. Falling back on hagiographic profiles of Edison, Marconi, and Tesla established unreal expectations for readers entering a new world of professionalized engineering. But the truth of what amateur experimenters were actually capable of is often more interesting than Gernsback's incongruous fantasy of socially contingent, lone genius inventors.

"'phone and code": Dynamophone, Radioson, and Other Emerging Media

The early twentieth century is commonly seen as a transitional period in American invention, from a reliance on the work of independent, almost mythologically brilliant inventors to corporate-based industrial research laboratories. For radio historian Hugh Aitken, the most meaningful development in the emergence of radio—the move from the spark gap to continuous wave transmission, which allowed the broadcasting of voice and audio rather than simple telegraphic code—happened at this "hazy boundary where radio stopped being a matter for visionary experimenters and started to become a hardheaded business capable of gaining and holding a commercial market."[77] But recent scholarship has explored the ways that the corporate model was not only preceded by older forms of collaborative invention but also accommodated independent inventors who worked well into the twentieth century. Historian of technology Eric S. Hintz argues that, despite the commonly accepted story, independent inventors continued to be a source of viable ideas well after the "heroic" era of inventors like Thomas Edison and Alexander Graham Bell:

> Lesser-known "post-heroic" inventors continued to contribute many important innovations throughout the twentieth century, among them [Samuel] Ruben's mercury battery, Edwin Land's Polaroid film, and

76. Andrew Ross, *Strange Weather: Culture, Science, and Technology in the Age of Limits* (London: Verso, 1991), 111.

77. Hugh G. J Aitken, *The Continuous Wave: Technology and American Radio, 1900–1932* (Princeton, N.J.: Princeton University Press, 1985), 9–10.

Chester Carlson's Xerox photocopying process. . . . There is certainly no denying the importance of science-based R&D at firms like GE, AT&T, and DuPont, but vertically integrated industrial research was not the only path to successful innovation. Small and medium-sized firms often pursued innovation strategies—licensing independent inventors, hiring consultants, and outsourcing inventions—that were much different from the ones followed by bigger firms.[78]

Muddling the distinctiveness of this transition from the other side is Paul Israel, who shows that many of the developments in telegraph technology throughout the nineteenth century were the product of a machine-shop style of collaborative problem solving "in which skilled operatives, superintendents, machinists, and manufacturers make up technological communities that draw on practical experience to design, build, and refine new technology."[79] In this shop culture,

few inventors actually worked alone and none worked in complete isolation. Conceptualization was often stimulated by access to new information. And the construction, testing, and redesigning of apparatus necessary for practical application almost invariably required an inventor to seek the assistance of others, whose own contributions often altered the original design.[80]

Much like the forums of questions and wrinkles shared by wireless amateurs, nineteenth-century craftsmen communicated through trade journals like *The Telegrapher: A Journal of Electrical Progress* (1864–1877). The communities making use of Gernsback's magazines are further evidence for a continuous tradition of collaborative development from the mid-nineteenth century on, calling into question whether there ever was anything like the heroic, lone inventor.

Falling somewhere in the middle of this collaborative–individualistic divide, Gernsback emphasized the virtues of amateurism in his writings on the development of new devices. A recurring argument in his editorials throughout the 1920s had it that the next great innovations, like television, would come not from corporate laboratories but from the avant garde of enterprising amateurs who could afford to take risks and try out wacky ideas (**Why the Radio Set Builder?**). But these amateurs were up against an establishment that was rapidly consolidating its power. After the federal government took control of the airwaves during World War I and assumed ownership of all wireless patents in order to aid the war effort, RCA was formed as the new steward of this amassed intellectual property. In a powerful partnership between the government, General Electric, and the Marconi Wireless Telegraph Company of America, the mission of RCA was to continue advancing radio technology in the interests of national security (**Silencing**

78. Eric S. Hintz, "Portable Power: Inventor Samuel Ruben and the Birth of Duracell," *Technology and Culture* 50, no. 1 (2009): 26–27, 56. For a chart showing that patents issued to independent inventors outnumbered corporate patents until 1933 and "still represented nearly 50 percent of total patents throughout the 1950s," see Eric S. Hintz, "The Post-Heroic Generation: American Independent Inventors, 1900–1950," *Enterprise & Society* 12, no. 4 (2011): 732–48.

79. Paul Israel, *From Machine Shop to Industrial Laboratory: Telegraphy and the Changing Context of American Invention, 1830–1920* (Baltimore: The Johns Hopkins University Press, 1992), 2.

80. Ibid., 57.

America's Wireless; Radio Enters into a New Phase).[81] "As a result," Andrew Ross writes, "the explosive age of industrial invention lay in the past, viewed as much too volatile in its effects for the scientifically regulated process of production favored by monopoly capitalism. The new emphasis on control, precision, uniformity, predictability, and standardization meant the extinction of the entrepreneur-inventor."[82] Radio amateurs could feel the effects immediately, with the number of radio-parts dealers in the United States (as opposed to consumer-ready, complete radio-set retailers) falling from an estimated 30,000 in 1922 to 2,500 in 1926.[83] The coming of continuous wave transmission and vacuum tube sets meant that radio was becoming increasingly complex for the average amateur. At the same time, radio was exploding in popularity among an American public who could comfortably listen in to nightly broadcasts. Seemingly overnight, radio was firmly cemented in American life as an everyday piece of household furniture. And after the so-called radio Christmas of 1924, when families around the country bought their first set, the legislative frameworks that would determine the structure of American broadcasting were finally in the process of being hammered out.[84] This was no longer "wireless," the province of technically savvy experimenters chatting about whatever came to mind. Radio was big entertainment.

This made it quite an interesting time for a new magazine to emerge as the voice of radio amateurism. The gradual shift we saw in *Electrical Experimenter* from a specialized companion magazine for tinkerers into the shiny, gold-covered *Science and Invention* tracked the growth of a reading public interested in a much wider array of developments across the sciences. But in order to maintain a relationship with his original audience of experimenters, Gernsback launched a new title, *Radio Amateur News* (later shortened to *Radio News*) in 1919. *Radio News* became the new home for hands-on readers, a magazine that took seriously the role of the independent experimenter throughout the 1920s, when, in Gernsback's words, "In all of the laboratories of our large industrial companies, research scientists are busy, day and night, in inventing and perfecting new devices."[85]

In a 1923 editorial for *Radio News,* we can see the prevalent anxieties over the amateur's role in a world of increasingly professionalized engineering and invention. Explicitly issuing a challenge to his readership, Gernsback argues that the amateur is nothing if he isn't visible to and valued by the public.

> Of what real use is the amateur of today? What does he really do to make the world a better place to live in? Of what use is he to the community at large? If the amateur will ask himself these questions, and search his heart, he will come to the conclusion that, indeed, his utility

81. Eric P. Wenaas, *Radiola: The Golden Age of RCA, 1919–1929* (Chandler, Ariz.: Sonoran Publishing, 2007), 29.

82. Ross, *Strange Weather,* 124.

83. Keith Massie and Stephen Perry, "Hugo Gernsback and Radio Magazines: An Influential Intersection in Broadcast History," *Journal of Radio Studies* 9, no. 2 (2002): 264–82.

84. "More than ten times the number of tube sets were manufactured in 1925 than just two years earlier, and four times as many speakers as in 1923." Susan J. Douglas, *Listening In: Radio and the American Imagination* (New York: Random House, 1999), 78.

85. Hugo Gernsback, "The Garrett Inventor," *Science and Invention* 15, no. 11 (March 1928): 977.

is microscopic. . . . *As far as the public is concerned, the radio amateur does not even exist.* (**Who Will Save the Radio Amateur?**)

The response from a passionate audience was swift. "The amateur's work is valuable," argues one reader with the conviction that tinkering is not play but important "work." Another recommends the approach of a "public-minded amateur," who hosts concerts, relays news, weather, and other useful information for free to small neighborhood gatherings. Still another suggests forming an Amateur Radio Research League whose sharing of expertise and ideas in "miniature laboratories" might be able to compete with the large corporations. I have seen many different editions of these magazines in various university and personal archives, and it is clear from the great number of issues containing underlining and annotations, scribbles in the margins from readers working out measurements, and relevant newspaper clippings slipped between the pages, that this was a highly active and responsive community. Some readers couldn't afford the components detailed in the experiments and diagrams published in *Radio News*. But as one writes in a letter to the editor, "You can better realize what these magazines mean to me when I say that I seldom, if ever, get a chance to have the material to experiment with, so I must content myself with reading matter, and I do like to read, especially such a wealth of material as is combined in the pages of your three publications."[86] These readers counted themselves as experimenters even if their lab equipment consisted solely of the magazines themselves.

Although Gernsback's readers and collaborators in many ways felt left out of the broadcast boom, *Radio News* remained a beacon for tinkerers around the world who knew that their homebrewed sets were far superior to the average RCA or Westinghouse radios sold in department stores. With the pace of their modifications and the channels in place for sharing these new developments, Gernsback argued that amateurs formed the true vanguard in the advance of the "radio science."

> The set builder naturally is well able to compete with the manufacturer, for two reasons. First, his time costs him little, and in price, therefore, he can compete easily with the factory-made set. Secondly, he has the jump on the manufactured set for the simple reason that, as like as not, his circuit is the latest out, and, therefore, will have improvements that the manufactured set can not boast for some months to come. (**Radio Enters into a New Phase**)

One incredibly fruitful area of experimentation for the amateurs was the short wave portion of the radio spectrum, initially deemed useless by government regulators and commercial operators. Amateurs

86. Richard J. Stephens, "The Opinion of One of Our Readers," *Radio News* 6, no. 8 (February 1925): 1434.

discovered in the early 1920s that short waves were easily reflected off the ionosphere and could travel great distances around the globe with small antennas and very little power **(The Short-Wave Era)**. Gernsback started an entirely new magazine devoted to this field in 1930, *Short Wave Craft*.[87] And as the previously speculative dreams of television came closer to fruition, Gernsback founded *Television News* in 1931. In his first editorial for this magazine, he emphasized that large corporations—who by necessity had to be depended on to produce complex and expensive prototypes—were actually stifling competition by "jealously guarding whatever improvements are made, because these large firms naturally wish to come out with a complete set that can be sold ready-made to the public." Continuing his argument that communities of amateur tinkerers who openly shared their results were a necessary element in the development of new media, Gernsback adds:

> Everyone knows that, the more people who are working on an art, the more rapid the progress will be in the end. Many improvements in radio have been due to experimenters who started in a small way and, later on, became outstanding figures in radio. . . . The more experimenters and the more television fans who become interested in the art, the quicker it will advance and the sooner it will be put on the stable basis which it deserves.[88]

Unlike *Modern Electrics* or *The Electrical Experimenter* with their devoted readership of wireless amateurs, *Television News* attempted to serve as the voice of a community of experimenters that had not yet come into being: "To this purpose I am dedicating this new publication, and the future will demonstrate the correctness of the assumption."[89] The idea that amateurs would eventually offer the world some of the most important contributions to the development of television was perhaps one of Gernsback's most far-flung science fictions, with many articles having to begin with a definition of what, precisely, television even was. Interestingly, one of Gernsback's articles borrows Edison's now-famous patent language for the phonograph, in saying that "television does for the eye what the telephone does for the ear."[90]

But by the end of the decade, the unfounded optimism of the idea that amateur experimenters were the engine of emerging media was laid bare. *Radio News* announced in January 1929 that it would end its monthly "constructional prize contest" due to a lack of quality submissions. In the initial announcement of the contest, Gernsback reminded readers that their ideas must be experimentally demonstrated: "It is not the idea, *but the putting of it into execution,* that makes a man an inventor. In addition to this, most purely theoretical

87. The changes in *Short Wave Craft*'s titles over the years reflects a history of technologies in which "craft" was seen to be applicable. It became *Short Wave and Television* in 1937, *Radio and Television* in 1938, and was finally folded into *Radio Craft* in 1941.

88. Hugo Gernsback, "The Television Art," *Television News* 1, no. 1 (April 1931): 7.

89. Ibid., 7.

90. For a reading of Edison's speculative description of cinema before the fact, see Tom Gunning, "Doing for the Eye What the Phonograph Does for the Ear," in *The Sounds of Early Cinema,* ed. Richard Abel and Rick Altman, 13–31 (Bloomington: Indiana University Press, 2001). Gernsback's entire definition reads: "To the layman who does not as yet know what television is, I may say that the term describes an electrical process, whereby it is possible to see at a distance and to view distant events as they are taking place. In this way television does for the eye what the telephone does for the ear. Your friend, using the telephone, talks to you from his office, while you are sitting in yours; while the television process is comparable in that you will see your friend as he is talking to you, and, vice versa, he will see you." Hugo Gernsback, "Television to the Front," *Radio News* 8, no. 12 (June 1927): 1419.

ideas are impracticable."[91] It seems that with the complexities of consumer radio by 1929, too few amateurs were capable of realizing their ideas in any material way:

> The rules of the contest stated very plainly that no one would be eligible for a prize unless some experimental work had been done and the practicability of the device had been demonstrated by the builder. Most of the entries consisted merely of ideas or suggestions, accompanied by the request that *Radio News* do the experimental work necessary for their full development.[92]

Perhaps sensing this problem, Gernsback had already begun focusing on older technologies in his recommendations for what to build. During a period of such rapid change in radio manufacture, new applications of older technologies became a form of dissent for tinkerers. Designs for cheaply made, easy-to-reproduce components like **The Radioson Detector** could rival the newer, industrially produced vacuum-tube sets: they were cheaper, easier to fix, and most importantly, easily understood by most home experimenters. In a move that is perhaps unexpected for a techno-futurist like Gernsback, he even begins to temper what he saw as a sometimes boosterish fervor over rapid developments in radio. Despite this seemingly overnight explosion, Gernsback writes in 1927, "there have not been any revolutionary improvements in radio and it is likely there will not be any." Instead, people should expect a long process of gradual refinement:

> It is this process of slow evolution that we may expect in the future, as well, and the old adage also holds true in radio: "Natura non facit saltum"—which, translated, says that Nature does not make jumps. In other words, all developments are part of a slow-moving plan of evolution. Even revolutionary inventions, when they do come along, will be found in the end to be not half as great a departure as they were thought to be at first.[93]

Taking the longer view allowed Gernsback to read current inventions in light of their not-so-distant precursors, profiling forgotten (and often quirky) paths not taken in the development of emerging media. Many of his editorials evoke the history of media not merely as a nostalgic trip back to the devices of yesteryear but as an archive of possibilities ripe for future experimentation. For instance, in another 1927 editorial, at a moment in which music and variety programs were flooding the country, he describes how the medium of radio had become a fixed idea in people's minds that papered over the inherent abilities of the underlying technology. Looking back on that strange trajectory in which the technology underlying wireless telegraphy

91. "Radio News Monthly Prize for Constructors—An Announcement," *Radio News* 10, no. 1 (July 1928): 46.

92. "End of Monthly Constructional Prize Contest," *Radio News* 10, no. 7 (January 1929): 651.

93. Hugo Gernsback, "Coming Developments in Radio," *Radio News* 8, no. 5 (November 1926): 465.

became "radio," a presence in virtually everyone's daily life, Gernsback writes, "the public at large is not aware of the fact that the art of radio is used for hundreds of different purposes aside from broadcasting and telegraphy. . . . There is hardly any industry today that cannot make use of radio instruments in some phase of its work."[94] His examples range from a force-field burglar alarm to automating the recording of lightning strikes, from measuring the miniscule weight of a fly to scanning factory workers for stolen metals. This wasn't merely a historiographical move, to remind us that the inherent features of wireless—information transmitted through the air—have now been scattered by the winds of media evolution in a way that influences countless other devices. In articles like **New Radio "Things" Wanted**, Gernsback encouraged amateurs to remix and retrofit old technologies in order to imagine what the next great advance might look like. Such pieces educated an increasingly interested public on precisely how the underlying technology of radio worked. But they also sought new paths forward that may have been overlooked.

One of the media-historical curiosities listed in this editorial is **The Dynamophone**, Gernsback's own invention from 1908. First published as a set of instructions in *Modern Electrics,* the Dynamophone consisted of a telephone mouthpiece connected to the transmitting end of a Telimco and an engine to the receiving end, so that the "power" of the human voice could be used to start that engine. "While it was at that time but a toy," Gernsback wrote in the 1927 retrospective, "the apparatus foreshadowed broadcasting, because this was before the days of the wireless telephone, and the human voice actually did create effects at the receiving end." This was a clever workaround: at a time when the voice couldn't be *transmitted* via wireless, perhaps it could at least be *translated* into a form of electrical or mechanical force. Just like the affective voice of a news bulletin or an operatic tenor transmitted over the airwaves in the late 1920s, the 1908 Dynamophone used the power of the human voice to "create effects." While Reginald Fessenden had already produced an elegant solution to the technical problem of using radio waves to produce frequencies in the audible range with his discovery of the heterodyne principle—in which two electrical waves of different frequencies can amplify one another—Gernsback's Dynamophone was a kind of metaphorical resolution to the same problem.[95] It was a trickster solution to one of the biggest stumbling blocks in the history of radio. This is why Gernsback, like many of the amateurs he represented, is best understood not as an "inventor" but a tinkerer whose idiosyncratic know-how lent his creations a quirkiness, a sense of humor.

Ever since the *Modern Electrics* days, Gernsback's activities as a

94. Hugo Gernsback, "Radio Steps Out," *Radio News* 8, no. 6 (April 1927): 1205.

95. For a discussion of heterodyning, see the annotations in **The Dynamophone.**

FIGURE I.11. Hugo Gernsback at work in the "special research laboratory on fifth floor," and H. W. Secor with Hugo's brother Sidney Gernsback in "general testing and research laboratory on second floor" of the Electro Importing Company offices. The offices were then located at 233 Fulton Street, now the site of the new World Trade Center Transportation Hub. From the January 1915 *Electrical Experimenter.*

tinkerer were conducted alongside his publishing endeavors. It began with the Electro Importing Company's main factory and office building at 233 Fulton Street, a location that now sits across from the new World Trade Center Transportation Hub. Alongside Louis Coggeshall, Harry Winfield Secor, and his brother Sidney Gernsback, Hugo manufactured not only traditional equipment like telephone parts and headphones but far more speculative devices.[96] The magazines often read like lab reports, detailing the progress made in these experiments in a way that opened them up to suggestions from the community. **The Physiophone** introduced a means of transforming music from a phonograph record into tactile, rhythmic pulses that could allow people with hearing impairments to enjoy music through touch. **The Detectorium** allowed for a much smoother wireless tuning action by linking components that previously had to be adjusted separately with multiple knobs. Other, more farfetched creations like **The Isolator,** an oxygen-filled, soundproof helmet that aided concentration, were interspersed with the more serious proposals. These efforts continued when Gernsback founded WRNY in 1925, a medium-range broadcast station operating out of New York that was used to debut new broadcast media and analyze the effects of various instruments and signal-processing techniques on the auditory perception of the station's listeners. Several articles profile these experiments, including a live radio concert that featured one of the earliest electronic keyboards in history, developed at the successor to the E. I. Company, Radio News Laboratories. **The "Pianorad"** looked like an automobile engine with a horn attached to the end, covered in radio tubes that looked like engine cylinders, with each tube controlling a single note on the keyboard. WRNY is also notable for conducting one of the earliest regularly scheduled television broadcasts on record, using a unique method of interleaving audio and visual signals on a single frequency **(Television and the Telephot).**[97]

While conducting these groundbreaking experiments at Electro Importing, Radio News Laboratories, and WRNY, Gernsback kept all developments open to the contributions and participation of his readers.[98] Often, a new device he profiled was offered for sale in subsequent issues of the same magazine. This often makes the mode of address confusing, as it is in **The Radioson Detector.** What begins as an objective description of a new wireless component eventually seems to boil down to a product pitch. Indeed, the Radioson would be advertised for sale from the Electro Importing Company in competing publications in the coming months. But the level of detail Gernsback goes into here is unique, making this essay not just a product announcement but a detailed discussion of how the device is constructed and

96. "The E. I. Co. Laboratory at New York," *Electrical Experimenter* 2, no. 9 · (January 1915): 139.

97. The *New York Times* printed an announcement that WRNY would begin regular television programming. During an otherwise normal radio broadcast, the image of the performer's face was sent out periodically throughout the show. "WRNY to Start Daily Television Broadcasts; Radio Audience Will See Studio Artists," *New York Times*, August 13, 1928: 13.

98. WRNY was short-lived. In 1927, under the guidance of then–Secretary of Commerce Herbert Hoover, the newly formed Federal Radio Commission included WRNY in its list of 164 smaller-scale stations who had to justify their existence or be forced to shut down. Tim Wu: "In effecting its program of clearing the airwaves, the agency relied on a new distinction between so-called general public service stations and propaganda stations. These were, in effect, synonyms for 'large' and 'small' respectively, but it was apparently easier to assault the underdog if one could label him a propagandist, even though the term's present pejorative sense would not take hold until used to describe communications in Nazi Germany. In any case, by whatever name, the FRC favored the large, networked stations affiliated with NBC (and later CBS). Because the large operators had superior equipment, and fuller and more varied schedules, the FRC could claim not implausibly that they better served the public." Wu, *The Master Switch,* 83.

FIGURE I.12. Gernsback delivering a lecture titled "The Future of Radio" from WJZ in Newark, New Jersey. From Hugo Gernsback, *Radio For All* (1922).

the decisions that went into its design. This is surprising, especially given that the intense commercial competition to improve on detector technology in the 1910s "meant that there often was a great deal of secrecy about technical details."[99] One wonders, then, who Gernsback is pitching this device to, and why. If the Radioson is such a valuable advance in radio technology, why would he share its detailed blueprints for the cost of a magazine issue?

I think this Gernsback device and others like it are best considered as a proof of concept for a growing community of amateur experimenters. The Telimco, for instance, was not exactly the revolutionary device that first brought radio to the masses, as Gernsback later liked to claim it was. Although its advertisements claimed the set was "guaranteed to work up to one mile," the Telimco was notoriously finicky.[100] As it was sold, the outfit had a range of merely 300–500 feet and could only receive signals from farther distances when a large antenna was hooked

99. D. P. C. Thackeray, "When Tubes Beat Crystals: Early Radio Detectors," *IEEE Spectrum* 20, no. 3 (March 1983): 64–69.

100. "Wireless Telegraph [Advertisement]," *Scientific American*, November 1905: 85.

up. Further, it was highly susceptible to any kind of electrical interference, such as the elevator motor in the Electro Importing Company building, which caused difficulties during in-store demonstrations of the apparatus. The Telimco's untuned circuits, which would produce a high degree of interference for any nearby radio station, would soon be outlawed by federal legislation. For these reasons, its metal filings coherers "had all but disappeared from commercial work in 1910," according to historian of early radio Thomas White.[101] From this perspective, the Telimco seems less a practical means of communication than the promise of its future iterations.

So while in essence the Telimco was little more than a gimmick, a parlor trick—press a button and a bell in another room would ring without the need for any intervening wires—it was also a rough prototype, an aggregate of handmade components that encouraged and enabled a conversation on what the wireless medium might look like in the future. The technical limitations of the device as it then stood didn't stop Gernsback from describing how thousands of perfectly synchronized home wireless sets would be capable of sending a message to the aliens that surely must be listening for us on Mars (**Signaling to Mars**) or, months later, from laying out in great technical detail how it would be possible to send and receive motion pictures over the airwaves using currently available equipment (**Television and the Telephot**). Gernsback's inventions blurred the lines between the real and the imaginary. While some were awarded patents and available for sale to consumers, others seem more like open invitations for a community of tinkerers to "fork" his hardware and continue development on an as-of-yet unrealizable idea. These amateurs and their work in the aggregate may not have been entirely responsible for the technologies Gernsback loved to claim he had predicted, but they were part of a collective endeavor to imagine the future by slowly, painstakingly, feeling their way toward it.

"certain future instrumentalities": The Mineral Proficiencies of Tinkering

Media theorists have been attempting to devise a conceptual language for an object that is collectively imagined before the conditions of its material possibility, from André Bazin's writings on the precinematic myth of total cinema ("The concept men had of it existed so to speak fully armed in their minds, as if in some Platonic heaven, and what strikes us most of all is the obstinate resistance of matter to ideas") to Friedrich Kittler's description of occult traditions and

101. Thomas H. White, "Pioneering Amateurs," *United States Early Radio History*, 1996, http://earlyradiohistory.us/1910ei.htm.

late eighteenth-century magic lantern shows as a "relay race from illusionists to engineers."[102] For his part, Gernsback was able so successfully to outline the material possibility of an invention before it existed because he understood tinkering as a practice that provided us glimpses of the future. In their interview, Thomas Edison dismissively responded to Gernsback's question about the future: "You know, Mr. Gernsback, I am an inventor, and as such I do not concern myself overmuch with philosophical research" (**Thomas A. Edison Speaks to You**). But for Gernsback, invention was every bit a matter of this so-called philosophy of the future.

Take for instance *The Electrical Experimenter*'s January 1916 "Wrinkles, Recipes, and Formulas" section. Amid exhaustive instructions for a homemade heliograph printing process, all the way down to a recipe for concocting the ink, there appears a curious piece by Gernsback that recommends the reader try biting a sewing needle tightly between his teeth while holding the sharp end to the groove of a phonograph record:

> With a little practise one will become proficient in moving the head at the same ratio of speed as the ordinary reproducer arm is moved from the outside of the record towards the inside. As soon as the needle touches the record with sufficient pressure, the inside of the head will be filled immediately with music exceedingly loud and clear. A curious result of the experiment is that a person standing near by can hear the music, the head acting as a reproducer in this case. (**Hearing through Your Teeth**)

The experiment may have seemed familiar to those who had just read the latest installment of Gernsback's serial novel *Baron Münchhausen's New Scientific Adventures* earlier that issue, which introduced a very strange Martian technology known as the Tos Rod that allows Martians to hear "not with their ears. They were listening with their brains!"

> The two reddish plates pressing against the bare temples are made of two metals unknown on earth, and the metals are distributed over the surface of the plate in honeycomb fashion without touching each other. Now if the two plates are pressed against the temples and when wireless waves are passing through them, the waves are translated into vibrations of a certain frequency. It has been found that if these vibrations reach the conscious sense of hearing which is located in the *Temporal Lobe* of the brain, sounds can be impressed upon the brain without requiring the ear and its auditory nerve. In other words, the sound is "heard" directly within the brain without the agency of the ear's mechanism.[103]

102. See André Bazin, "The Myth of Total Cinema," in *What Is Cinema?* vol. 1 (Berkeley: University of California Press, 2005), 17. Friedrich A Kittler, *Optical Media: Berlin Lectures 1999*, trans. Anthony Enns (Cambridge, UK: Polity Press, 2010), 101. Gunning, "Doing for the Eye What the Phonograph Does for the Ear." Geoffrey Batchen, *Burning with Desire: The Conception of Photography* (Cambridge, Mass.: MIT Press, 1999). In a similar vein, John Guillory writes that "this peculiar latency is characteristic of discourse about media from the sixteenth to the later nineteenth centuries. . . . The transposition of writing into print did not elicit at first a theoretical recognition of media as much as a reflection on the latency of print in the technology of writing itself. It was as though print were there, already, in the medium of writing. This is how Bacon understands print in the New Organon: 'the technique of printing certainly contains nothing which is not open and almost obvious.' Bacon explains that 'men went without this magnificent invention (which does so much for the spread of learning) for so many centuries' because they somehow failed to notice that letters and ink might be used for inscription with a different instrument than the pen, namely 'letter types' inked and pressed on paper, an ingenious form of endlessly reproducible writing." Guillory, "Genesis of the Media Concept," 324.

103. Hugo Gernsback, "Baron Münchhausen's New Scientific Adventures: Thought Transmission on Mars," *Electrical Experimenter* 3, no. 9 (January 1916): 475.

Martians in mind, the reader could now have a felt sense of the operative principle behind this alien technology thanks to the sewing needle experiment: that the ear itself is not "absolutely essential for hearing." This is just one of many examples of how, throughout the Gernsback magazines, tinkering with something as simple as available household materials became the starting point for new inventions, both functional and fictional. Seven years later Gernsback would patent a device he called the Osophone, a "small, compact and handy instrument which can be easily carried about and used without attracting undue attention." The Osophone employed the same principle as the *Electrical Experimenter* editorial, this time with a finished mouthpiece attached to a wire that provided a "simple and practical means by which hearing may be effected by sound vibrations transmitted directly to the osseous tissue of the body."[104] Projecting out from these available materials and describing their experience in the future was a means of aiding design practice. And it was through the perspective of the future that Gernsback developed his tinkerer's ethic.

What did it mean to be a tinkerer in the early twentieth century? The appellation was variable. Gernsback himself interchangeably used *tinkerer, experimenter,* and *amateur,* while Edison preferred *mucker.*[105] Today, we might refer to people who take pride in their ability to build seemingly complex technologies from scratch as *makers.* Not just a hobby, tinkering in the early twentieth century was a special form of intuition or creativity, a means of advancement for a self-educated working class, as well as a political position opposed to the increasingly complex and unreliable consumer technologies produced by corporations like RCA and AT&T. Throughout his editorials on technology, understood in the older, Aristotelian sense of "a reasoned state of capacity to make,"[106] Gernsback argued that the incremental modification associated with a term like *tinkering* involved just as much creative expression, expertise, and skill as the whole-cloth production of a new technology we might associate with *invention.*

Across Europe, political circumstances in the wake of World War I led some amateur radio groups to become highly politicized, as was the case with the *Arbeiterradio-Bewegung* (Workers' Radio Movement) in Germany. What began as "a loose association of amateurs who wanted to learn how to build their own apparatus so that they could transmit and receive radio messages themselves" soon grew into a labor movement operating in an environment of high unemployment and growing government repression. Associated with this movement and its publications like *Der Neue Rundfunk* [New Radio], *Arbeiter-Funk* [Workers' Radio], and *Unser Sender* [Our Radio Station] were

104. Hugo Gernsback, "Acoustic Apparatus," December 1924, http://www.google.com/patents/US1521287.

105. "'A mucker is an experimenter—a scientific experimenter. . . . Mr. Edison is a mucker himself. He calls all experimenters muckers.' It was explained that Mr. Edison had been president of a muckers club. So there you are, muckers!" "Edison Says You Are a 'Mucker'!" *Electrical Experimenter* 4, no. 1 (May 1916): 15.

106. Aristotle, *The Nicomachean Ethics,* rev. ed. (Oxford: Oxford University Press, 2009), 105.

prominent artists and intellectuals such as Walter Benjamin, Bertolt Brecht, Paul Hindemeith, and Kurt Weill. According to Siegfried Zielinski,

> They began to criticise the programmes of established radio broadcasting and presented their own suggestions for organising the medium according to their interests, organised evenings for collective listening (particularly to Soviet radio stations), and protested about censorship of programmes.[107]

While tinkering never evolved into a coherent political position in the United States, its practitioners nevertheless saw themselves as operating in opposition to the hoarding of intellectual property by corporate interests and as advocates of freely sharing their knowledge and skills. Forms of self-education during the period varied widely according to class. While middle-class Americans were now able to culturally enrich themselves during their leisure time through several new "great book" series geared toward mass audiences, like the Book of the Month Club and the Harvard Classics "Five Foot Shelf" collection, the urban working classes "were asked to invest their sense of self in hands-on technical knowledge," writes Eric Drown. Commercial technical institutes and correspondence schools promised upward mobility and job security within a number of growing fields like automobile repair and electrical engineering, in addition to radio.[108] Companies began selling products that allowed youths to learn electrical principles at home:

> Parents eager to set their children on the right path bought them chemistry sets with names like Chemcraft, produced by the Porter Chemical Company, which also sponsored Chemists Clubs. Young adult books on electricity and technology flourished, many of them written for boys and intended to instill the noble desire to invent. . . . This was certainly a far cry from the genteel instruction on natural philosophy to promote polite drawing room conversation on scientific theory published just a few decades earlier. It's interesting to consider just what a young Michael Faraday would make of . . . *Harper's Electricity Book for Boys* (1907) in which the author, Joseph H. Adams, wrote: "Theory is all very well, but there is nothing like mastering principles, and then applying them and working out results for one's self."[109]

This was the environment in which Gernsback began laying out his theories on the process by which "new things" come into the world and the special forms of creativity that emerge when an amateur draws equally upon his familiarity with science and the arts. In many ways, he was far ahead of his time in proposing several frameworks for understanding invention, even if they were underdeveloped. It would

107. Siegfried Zielinski, *Audiovisions: Cinema and Television as Entr'actes in History* (Amsterdam: Amsterdam University Press, 1999), 126–27.

108. Drown, "Usable Futures, Disposable Paper," 77–78. Joan Shelley Rubin, *The Making of Middlebrow Culture*, 1st ed. (Chapel Hill: University of North Carolina Press, 1992).

109. Henry Schlesinger, *The Battery: How Portable Power Sparked a Technological Revolution* (New York: HarperCollins, 2011), 200–201. Schlesinger points out that an amateur experimenter culture preceded Gernsback's magazines, noting in particular L. Frank Baum of *The Wonderful Wizard of Oz* fame and his novel *The Master Key: An Electrical Fairy Tale* (1901), in which a young man sets out on "an electrical adventure with all manner of gadgets and gizmos that benefit mankind." The book, as Baum noted in the preface, was "Founded Upon the Mysteries of Electricity and the Optimism of Its Devotees."

be many years before the social and historical study of technology would be accepted as a serious academic pursuit, as Hugh Aitken writes in the 1976 prologue to his magisterial, two-volume history of radio:

> If the "new things" of science, technology, and economic life often seem overwhelming and uncontrollable, the reason may well be that we have, in modern societies, created highly efficient structures for the generation of new knowledge without seriously attending to the processes by which new knowledge is put to use. . . . historians of the future will find it strange that so few social scientists of the twentieth century could bring themselves, in their work, to treat science as one of the great social institutions of the time.[110]

Although this is of course no longer the case, it is important to note that the questions raised by Gernsback in a pulp publishing context would not have sufficient answers even among academic historians and sociologists of technology for many decades to come.

Gernsback's views on these issues evolved over the years thanks in part to the debates, suggestions, and questions he and his staff printed in lively correspondence sections like "The Oracle" and "Patent Advice." Questions from readers ranged from the factual (How many feet come in a pound of No. 12 B.&S. triple-braided weather proof wire?) to the experimental (Might two lead plates submerged in sulphuric acid make a good storage battery?). It was important for knowledge to be freely shared among readers, contributors, and editors, since the construction of wireless sets involved many different forms of expertise. "A radio set builder," wrote Gernsback in a 1925 editorial, "must be a carpenter, an electrician, a metal worker, a tinsmith, and a radio engineer, all rolled into one."[111] Initially, Gernsback argued that such impressively skilled inventors were born, not made. While many could learn to be "mechanical" inventors, or those "who are suddenly confronted with a certain device that to their minds seems imperfect, whereupon they will bend their energies towards improving the existing device," true inventors were hard to come by (**The Born and the Mechanical Inventor**).

Eventually, perhaps in response to the variety and number of designs sent in by amateurs over the years, Gernsback began to shift his views on invention so that they were more in line with the educational goals of his magazines. In **Why the Radio Set Builder?** we begin to understand that the biggest contributions to this community were not necessarily made by its most broadly educated members. To the contrary, the most important things that an experimenter needed to know would emerge organically from the process of tinkering, in a

110. Hugh G. J. Aitken, *Syntony and Spark: The Origins of Radio* (New York: Wiley, 1976), 9. Aitken is citing here Robert K. Merton, "Priorities in Scientific Discovery: A Chapter in the Sociology of Science," *American Sociological Review* 22, no. 6 (December 1957): 635–59.

111. Hugo Gernsback, "Radio and the Student," *Science and Invention* 12, no. 12 (April 1925): 1177.

dialogue between a person and an inanimate object. Gernsback advocated making things by hand not just as a hobby, a means of diversion, or release after the working day is over, but as a complete education in tools, materials, and techniques:

> If, for instance, you are making an elaborate radio console for your living room, you will get more information on the subject than you could possibly get from the best text-books. You will, first of all, become familiar with the various tools necessary to fashion the wood, and if you own a woodworking lathe you will learn quite a good deal about the operation of wood-turning and the tools to be used for this particular purpose. You will learn what it means to sandpaper, you will learn to recognize the different kinds of wood, and you will know the difference between green and kiln-dried variety of woods. You will soon know how to use glue, and what kind. You will study the various fillers, and, last but not least, you will get a thorough education in varnishes and paints, and the use of all of these.[112]

For Gernsback, the know-how that emerges from making things is cascading: hands gradually feeling their way along an interlocking series of dependent skills and material properties. As we see in **The Perversity of Things,** objects only ever seem like frustrating "obstacles" in the way of a good idea when we don't fully understand the proper affordances of those objects. "It is not the things that are perverse, it is ourselves who make them seem perverse." With enough experience, barriers and constraints begin to look like creative opportunities. Gernsback refers to this capacity to see finished form in raw materials as a "knack" or an "intuition." It's what allows one reader to build an audion tube from scratch, that ultimate symbol of inscrutable corporate complexity, by learning glassblowing.[113] Or another to make the first-ever permanent recording of a radio broadcast by modifying his phonograph player so that it could register the signals from his wireless set.[114] The more experience an experimenter had with a wide variety of materials and techniques, the better attuned his intuition would be when it came to making new devices.

The intense material specificity of the technologies as they were described to the readers of Gernsback's magazines is impressive. In his writings, one is struck not only by the romance of communicating in private through secret codes and the intimacy of a headset but also by the weird materials that made this experience possible. Pieces like **The Dynamophone** and **A Treatise on Wireless Telegraphy** address a readership with an already impressive material awareness of their apparatus, and introduce new possibilities each month to the wireless medium's construction and operation. Thanks to these mineral proficiencies, a rural Midwestern reader would know that news from New

112. Hugo Gernsback, "Handicraft," *Science and Invention* 14, no. 10 (February 1927): 881.

113. Thomas Reed, "The Unterrified Amateur," *Electrical Experimenter* 4, no. 5 (September 1916): 325.

114. Charles E. Apgar, "The Amateur Radio Station which Aided Uncle Sam," *Electrical Experimenter* 3, no. 7 (November 1915): 337.

York may become audible if only he could find some molybdenite, or a supplier willing to ship him the nitric acid needed to try out a new electrolytic detector.[115] The sheer alchemy that went into the construction of these apparatuses—glass light bulbs filled with argon gas, impure silicon crudely fused into thin wafers, influential experiments conducted by an employee of AT&T with the contents of a "Minerals of Maine" souvenir box—makes the effects they were able to achieve seem all the more wonderful.[116]

Details such as these seem especially poignant in light of recent calls for a political awareness of the rare earth minerals used in the production of today's increasingly complex digital devices, as well as the environmental toll of the electronic waste these devices leave behind once we inevitably acquiesce to their fragility and replace them with the latest release.[117] In an article on the dust produced during the manufacture and disposal of hardware, Jussi Parikka writes of a "persistence that lingers across scales from minerals and chemical elements to the lungs and organic tissue."[118] Timothy Morton refers to these persistent objects, often produced and disposed of under appalling working conditions in China and countries across Africa, as "hyperobjects": manufactured materials or devices that "do not rot in our lifetimes. They do not burn without . . . releasing radiation, dioxins, and so on."[119] That gadgets like smartphones and tablets today are described as "magical" or like "touching the future" should come as no surprise.[120] Not only are they black boxes that with every passing year become increasingly difficult to modify, repair, or tinker with, but our use of them is blissfully disconnected from the labor practices and material conditions that go into producing them. We encounter a very different kind of moment through Gernsback's writings. Of course the advent of wireless was pure magic: it allowed people to skim disembodied voices from the air. But wireless was magical to Gernsback's readers not because they didn't understand how the trick worked but because they did. That elemental, raw materials could produce such effects was absolutely fantastic and provided an endless source of fascination.

"we exploit the future": Scientifiction's Debut

In many ways, Gernsback's new fiction magazines *Amazing Stories* (1926) and *Science Wonder Stories* (1929) contained nothing that had not already been introduced in one form or another by his technical titles. The future continued to be defended as a topic of serious

115. For more on molybdenite and how amateurs used it, see **A Treatise on Wireless Telegraphy.**

116. The "Minerals of Maine" box included iron pyrite, galena, chalcopyrite, and magnetite. See Alan Douglas, "The Crystal Detector," *IEEE Spectrum* 18, no. 4 (April 1981): 64–69; Thackeray, "Communications."

117. See especially Jennifer Gabrys, *Digital Rubbish: A Natural History of Electronics* (Ann Arbor: University of Michigan Press, 2011).

118. Jussi Parikka, "Dust and Exhaustion: The Labor of Media Materialism," *CTheory* (October 2013), http://www.ctheory.net/articles.aspx?id=726.

119. Timothy Morton, *The Ecological Thought* (Cambridge, Mass.: Harvard University Press, 2012), 130.

120. Lori Emerson writes on the rhetoric of magic as it is applied to contemporary screen interfaces in *Reading Writing Interfaces: From the Digital to the Bookbound* (Minneapolis: University of Minnesota Press, 2014), 1–46.

discussion (**Our Cover**). Fantastic fiction served as a means of describing and explaining present-day technologies. The letter column hosted lively debates among readers, the importance of self-instruction was emphasized throughout, and Gernsback invited "plots" from readers in a manner similar to the "wrinkles, recipes, and formulas" solicited by the technology magazines (**What to Invent**). In the November 1929 issue of *Science Wonder Stories,* for instance, he offers a prize of $150 in gold for story treatments: "the more interesting, the more exciting, and *the more scientifically probable* you can make it, the better. Remember, *anyone* can participate in this contest."[121] The approach taken by *Modern Electrics, Electrical Experimenter, Science and Invention,* and *Radio News* served as the blueprint for narrative works appearing in *Amazing Stories* and *Science Wonder Stories*: extrapolate a technology of the future from given materials and describe its engineering specifications. The "story" itself was almost of secondary concern. In A. Hyatt Verrill's "The Man Who Could Vanish," for instance, the narrator prepares the reader for the amount of technical detail that will follow in a tale of invisibility technology:

> To many readers much of this matter will, no doubt, prove rather dry, and, if I were writing fiction, I would omit all those portions of the tale which deal with the scientific side and the preliminaries. But both Dr. Unsinn and myself feel that to omit such matters would be a great mistake, and that as the story is of as much interest and importance to the scientific world as to the layman, nothing should be left untold. Moreover, we feel that unless such matters were included my story would be considered as purely fictitious. And at any rate the reader is at liberty to skip such portions of my narrative as the appreciative reader may find to be lacking in real and genuine interest.[122]

In stories by Abraham Merritt, Murray Leinster, and Philip Francis Nowlan, characters recede into the background as the pretense for setting a parade of gadgets in motion, endowing a field of inanimate objects with agency and conflict.[123] The character in a work of scientifiction, in the words of Samuel R. Delany, is a "subject-that-doesn't-even-exist-without-objects."[124] It is for this reason that Gernsback, as late as the 1960s when science fiction was taking on far more socially and culturally complex topics, referred to Clement Fezandié as a "titan of science fiction."[125] Fezandié, author of Doctor Hackensaw's Secrets, a dry series of lectures masquerading as short stories, simply posits a technology and describes in minute detail how it works.

Gernsback's own short fiction perhaps best exemplifies this form. **The Killing Flash** (1929) explores the idea that, with news, operas,

121. *Science Wonder Stories* 1, no. 6 (November 1929): 485.

122. A. Hyatt Verrill, "The Man Who Could Vanish," *Amazing Stories* 1, no. 6 (January 1927): 900–913.

123. According to Sam Moskowitz, Gernsback's most important finds as an editor included Dr. David H. Keller, author of a ten-volume *Sexual Education Series* later published by Gernsback and "the pioneer in what later was to become popular as 'psychological' science fiction stories; Edward E. Smith, the creator of the super-science tale; John W. Campbell, Jr.; Philip Francis Nowlan, whose stories of Anthony (Buck) Rogers ran in *Amazing Stories* some time before they appeared as a comic strip; Stanton A. Coblentz, poet and, in his fiction, magazine science fiction's best satirist in the tradition of Jonathan Swift; A. Hyatt Verrill, outstanding archaeologist; Fletcher Pratt, later celebrated naval writer; Harl Vincent, Bob Olsen, Miles J. Breuer, M.D., and Jack Williamson." Moskowitz, *Explorers of the Infinite,* 237. Other notable writers published by Gernsback include Francis Flagg, Claire Winger Harris, the first woman to be regularly published in the specialized SF magazines, R. F. Starzl, Edmond Hamilton, Abraham Merritt, and Murray Leinster.

124. Delany, *Silent Interviews,* 200.

125. Hugo Gernsback, "Guest Editorial," *Amazing Stories* 35, no. 4 (April 1961): 5–7, 93.

and dramas now sent over the airwaves, one might now be able to conduct a wireless murder. A fictional apprentice to Nikola Tesla named Why Sparks devises a means of harmlessly disabling German weapons in **The Magnetic Storm** (1918), providing an imaginary resolution to the tensions of the Great War at its height. And in the October 1915 installment of **Baron Münchhausen's New Scientific Adventures,** the Baron explains how he set up the very communications relay between Mars and the Earth that transmits the story to readers. Much like the Baron, stringing electrical components together in order to tell a story, Gernsback was inventing the form of scientifiction as he wrote, altering the narrative's setting, tone, and emphasis over the course of his serial novels.[126] It was sometime during the summer of 1911, for instance, that Gernsback first discovered Mark Wicks's utopia *To Mars via the Moon: An Astronomical Story* (1911). According to Gary Westfahl, this encounter radically changed the course of *Ralph 124C 41+,* which had been running since the beginning of that year as a guided tour of the brilliant Ralph's amazing inventions and the marvelous future city in which he lived. By the November 1911 installment, the narrative shifts drastically to an interplanetary space opera that included many more illustrations and footnotes, elements borrowed from Wicks.[127] Gernsback borrowed freely and intermittently from other generic influences as well, including the stock romantic intrigue of melodrama, "the colorful storylines of villainy, pursuit, and resurrection" from contemporary popular fiction, and "his own emphasis on imagining and explaining at length new scientific achievements."[128]

So when we say that Hugo Gernsback was one of the first science fiction critics, we have to understand the word "critic" in a very specific sense. Gernsback was primarily a technologist. He "almost certainly never read any contemporary book reviews, literary magazines, prefaces to literary works, or critical studies, and he did not research such materials from previous eras."[129] Unlike the insularity of many literary critical debates, his arguments about the nature of scientifiction were honed through engagement with readers. As John Cheng has shown, the publication of readers' letters in the pages of *Amazing Stories* allowed writers, editors, and readers to engage in a dialogue not only on scientifiction but on a form of science itself that was not merely a passive spectator sport. These dialogues built a popular consensus on what counted as scientific fact:

> As they redefined the character of participation that publishers sought to create for them, readers' exchanges also reshaped the character of science within their writers' fiction and their editors' features, reading

126. Scientifiction was utilitarian not just in content but in form. One story has it that Gernsback's decision to write *Ralph 124C 41+* was an editorial stopgap in order to fill space in a *Modern Electrics* issue that was running short of its page quota. Moskowitz, *Explorers of the Infinite,* 232. Westfahl, *Hugo Gernsback and the Century of Science Fiction,* 98–99.

127. Westfahl argues that the anonymous review of Wicks's book in the September 1911 issue of *Modern Electrics* was probably written by Gernsback. The review refers to the Wicks text as a "scientific novel," according to Westfahl, "the first Gernsback-magazine use of a term for science fiction." Westfahl, *Hugo Gernsback and the Century of Science Fiction,* 101.

128. Ibid., 105.

129. Ibid., 18.

it with unanticipated intent and unintended consequence. Science fiction's conversations not only fostered a community of readership, they imagined and reimagined science and fiction within the social dynamics and historical circumstances of the 1920s and 1930s.[130]

One of these readers (and later contributor of the famous short story "The Man from the Atom"), G. Peyton Wertenbaker, offered perhaps one of the most eloquent of these letters. Gernsback quotes Wertenbaker's letter in the editorial **Fiction versus Facts** and attempts to provide an address to his warning about clinical objectivity. Wertenbaker writes that if a writer attempts to capture the sublime through too-rigorously objective a lens (scientifiction in his view being the only modern literature truly capable of approaching the sublime), then scientifiction could lose much of its power:

> Scientifiction goes out into the remote vistas of the universe, where there is still mystery and so still beauty. For that reason scientifiction seems to me to be the true literature of the future. The danger that may lie before *Amazing Stories* is that of becoming too scientific and not sufficiently literary. It is yet too early to be sure, but not too early for a warning to be issued amicably and frankly.

Ironically, Gernsback responds with a completely mechanical formula: "If we may voice our own opinion we should say the ideal proportion of a scientifiction story should be seventy-five per cent literature interwoven with twenty-five per cent science." Nevertheless, one can see here that the genre began forming itself around a problem to be solved. As if anticipating Darko Suvin's highly influential definition of science fiction as a negotiation between "cognition," or empirical verifiability, and degrees of "estrangement" from the author's given world, pulp practitioners were already thinking in their own way about how to marry technical complexity with imagination.[131] Technocratic solutions to this problem of poetics were of course favored, as we can see from the 1928 *Amazing Stories* contest to design a logo for scientifiction. The image that resulted—an amalgamation of suggestions from several readers—depicts the dual cogs of "fact" and "theory" working in lockstep to automatically ink the genre's name. Gernsback describes the final image:

> It was our aim to incorporate as much science as possible in the design, so the frame of the design, representing structural steel, suggests more machinery. The flashes in the central wheel represent Electricity. The top of the fountain pen is a test tube, which stands for Chemistry; while the background with the moon and stars and planet, give us the science of Astronomy.[132]

130. John Cheng, *Astounding Wonder: Imagining Science and Science Fiction in Interwar America* (Philadelphia: University of Pennsylvania Press, 2012), 78. Justine Larbalestier agrees on this point: "In the early years of pulp science fiction magazines in the 1920s and 1930s there are not many letters that discuss what science fiction or scientifiction is. Everyone knew what 'it' was—it had something to do with 'science'—but everyone seemed to have a different notion of what constituted 'science,' what does and does not belong to the universe of science fiction and who has the rights to make these decisions. What science fiction is, the genre, is taken for granted in the letters, but the texts, the instances of this genre, are not so defined, so that attempts to limit the field become a series of test cases about which there is not always agreement." Larbalestier, *The Battle of the Sexes in Science Fiction*, 33.

131. Darko Suvin, "On the Poetics of the Science Fiction Genre," *College English* 34, no. 3 (December 1972): 372–82.

132. Hugo Gernsback, "Results of $300 Scientifiction Prize Contest," *Amazing Stories* 3, no. 6 (September 1928): 519–21.

What precisely counts as "fact" and "theory" in Gernsback's most programmatic definitions of the science fiction genre (the latter category is also referred to as "fiction," "imagination," and "fantasy") changes with time, and he mixes these elements in different proportions (**A New Sort of Magazine; The Lure of Scientifiction; How to Write "Science" Stories**). Gernsback of course demanded that the works of scientifiction he published be sufficiently plausible (**Reasonableness in Science Fiction**), a criterion that fans referred to as the "Gernsback Delusion."[133] But he had an unusually high tolerance for what constituted plausibility. Receiving criticism as early as *Electrical Experimenter* for lending "so much space to the exploitation of the future," Gernsback argues that progress makes anything possible with time. "We are fully aware of the fact that some of the imaginary articles which we publish are wildly extravagant—now. But are we so sure that they will be extravagant fifty years from now? It is never safe in these days of rapid progress to call any one thing impossible or even improbable" (**Imagination versus Facts**).[134] This simultaneous faith in technological progress and sense of satisfaction that the present day has fulfilled the utopian dreams of the past leads to a strange idea embedded in the installment of **Baron Münchhausen's New Scientific Adventures** included here, that there is no fiction:

> If, as often—no, always—has been proved that the most violent fiction at some time or other invariably comes true, then by all proceeds of modern logic, there cannot be such thing as fiction. It simply does not exist. This brings us face to face with the startling result that if fiction always comes true some time or other, why then, bless their dear souls, all fiction writers must be prophets!!

All fiction is in this sense "true," whether it be entirely fantastic or a simulation of the most mundane day-to-day occurrences. This is no small point, and evokes the argument of several contemporary literary theorists who see all fiction to be in fact a subgenre of science fiction. If science fiction deals in degrees of difference from the "real" world, realistic or literary fiction then exists at a specific point on the spectrum of variation from the real.[135]

Gernsback follows a similar line of thought in **Science Fiction versus Science Faction**, where he outlines a spectrum of the fantastic that includes the "probable, possible, and near-impossible." A given story's horizon of predictability is dependent on the technical expertise of the writer, who must be an "up-to-date scientist" (**10,000 Years Hence**). Crafting an important work of scientifiction thus involves more of the mind of a technologist than of a literary writer, as we see in a piece he wrote for *Writer's Digest*, **How to**

133. Eney and Speer, *Fancyclopedia 2*. Available online at http://fancyclopedia .wikidot.com/gernsback-delusion.

134. Readers responded favorably to Gernsback's defense. As one put it, "I have been more or less narrow-minded myself. I could not see, for the life of me, why it was that the editor continued the story of the 'Baron Münchhausen's Scientific Adventures,' but after reading a few installments I began to look at improbabilities with the far distant view of probability." E. A. Norstadt, "Letter," *Electrical Experimenter* 4, no. 5 (September 1916): 331.

135. Carl Freedman writes: "As we argue that the qualities that govern texts universally agreed to be science fiction can be found to govern other texts as well, it may be difficult to see just where the argument will stop. It may even begin to appear that ultimately nearly all fiction—perhaps even including realism itself—will be found to be science fiction. Does not that conclusion preclude success in defining science fiction as a recognizable *kind* of fiction? In fact, I do believe that all fiction is, in a sense, science fiction. It is even salutary, I think, sometimes to put the matter in more deliberately provocative, paradoxical form, and to maintain that fiction is a subcategory of science fiction rather than the other way around." Freedman, *Critical Theory and Science Fiction*, 16. Seo-Young Chu also touches on this idea in *Do Metaphors Dream of Literal Sleep? A Science-Fictional Theory of Representation* (Cambridge, Mass.: Harvard University Press, 2010), 7.

136. "I'd say you saw a semiotic ghost. All these contactee stories, for instance, are framed in a kind of sci-fi imagery that permeates our culture. I could buy aliens, but not aliens that look like Fifties' comic art. They're semiotic phantoms, bits of deep cultural imagery that have split off and taken on a life of their own, like the Jules Verne airships that those old Kansas farmers were always seeing. But you saw a different kind of ghost, that's all. That plane was part of the mass unconscious, once. You picked up on that, somehow. The important thing is not to worry about it." William Gibson, "The Gernsback Continuum," in *Universe 11*, ed. Terry Carr, 81–90 (New York: Doubleday, 1981).

137. "The Moon Hoax," *Amazing Stories* 1, no. 6 (September 1926): 556.

Write "Science" Stories. Of paramount importance is "description" and making that description interesting for the reader: "What description of clouds and sunsets was to the old novelist, description of scientific apparatus and methods is to the modern" scientifiction writer. The very best works in the genre are "prophetic" and slide along the continuum of plausibility as its real-world referents change. These works mutate with time, shifting in parallax with historical developments and gradually aging from works of science fiction into what he calls "science faction," losing their imaginative charge as they become simple lists of facts. Much like the decaying futures haunting the protagonist of William Gibson's short story "The Gernsback Continuum," one of scientifiction's signature characteristics is that its works have a half-life.[136] There is some particular quality that weakens over time from prophetic vision to what SF writers today refer to as the realism that exists outside of the genre: "mundane fiction."

Given the high degree of technical detail scientifiction writers engaged in, Gernsback often referred to imagination as a form of invention. In this sense, Luis Senarens literally "invented" the submarine when he described it in a nineteenth-century story (**An American Jules Verne**). So too did Leonardo da Vinci "invent" the breech-loading gun and Roger Bacon the telescope (**The Lure of Scientifiction**). Likewise, the automobile was invented by eighteenth-century approximators who constructed steam-powered prototypes of what would later evolve into the modern car: "inventors over 150 years back *knew* the automobile" (**Predicting Future Inventions**). But "modern" scientifiction is distinguished from these older forms of technical prophecy by the speed with which a fantastic idea could be proven through communications media. Among the fiction in the September 1926 issue of *Amazing Stories* was "The Moon Hoax," a reprint of articles published by the *New York Sun* throughout August 1835 that erroneously claimed the astronomer Sir John Herschel had recently discovered vegetation, animals, and an intelligent species that built temples on the moon. The editorial introduction to the story states that it was due to a lack of modern media that the hoax spread so easily to a credulous public:

> At that time, when there were no cables and no radio, and communication was slow, it was a simple matter to spring such a hoax, where today it would not last twenty-four hours, because verification or denial would speedily be brought about.[137]

"The Moon Hoax," a "charming story that will live for ever . . . [and] one of the finest pieces of imagination that has ever appeared," nevertheless by Gernsback's standards for modern scientifiction fell short not only because it did not end up becoming fact but also because it could not participate in the economy by which prophetic scientifiction is translated into fact: debated, tested, proven, or disproven by a networked community of tinkerers and aficionados.

The nineteenth century nevertheless provided an endless source of inspiration for Gernsback and his writers. As Gary Westfahl has persuasively argued, Gernsback not only set up the conditions in which a genre of science fiction could flourish, he also referenced and republished the work of nineteenth-century writers in an attempt to establish a tradition upon which it could build. In addition to his appreciation for Luis Senarens (**An American Jules Verne**) and Clement Fezandié (**Predicting Future Inventions**), Gernsback reprinted stories by Richard Adams Locke and Fitz James O'Brien. Other nineteenth-century authors mentioned in the pages of *Amazing Stories* included H. Rider Haggard, Edward Bellamy, Garrett P. Serviss, M. P. Shiel, and Arthur Conan Doyle. Contrary to later histories of proto–science fiction that emphasized texts commenting on the social or political context from which they emerged, Gernsback's nineteenth-century canon revolved around prognostication. Westfahl:

> In associating writers whose careers began in the 19th century with SF, Gernsback, unlike later historians, did not attribute their work to larger events in that era; they were rather persons ahead of their time, "prophets" who anticipated both the value of scientific progress and the value of literature about scientific progress. All on his own, Poe "conceive[d] the idea of a scientific story." Thus, according to Gernsback, 19th-century SF was simply the product of isolated individual geniuses.[138]

Using this definition of scientifiction as a "prophetic vision" of future machines and inventions "that were only to materialize centuries later," Gernsback argues that scientifiction can be traced all the way back to Leonardo da Vinci and Francis Bacon (**The Lure of Scientifiction**). Contrary to later claims by science fiction critics that the genre definitively began in 1926 with the introduction of *Amazing Stories,* these pieces make clear that a very sophisticated understanding of the genre's roots was already in place by the 1910s. For instance, in a 1915 satirical patent application for a wearable apparatus for bookworms who couldn't stop reading both day and night, the illustration depicts an insatiable reader holding a copy of *Deadwood Dick.*

138. Gary Westfahl, "'The Jules Verne, H. G. Wells, and Edgar Allan Poe Type of Story': Hugo Gernsback's History of Science Fiction," *Science Fiction Studies* 19, no. 3 (November 1992): 340–53 at 342–43, citing **The Lure of Scientifiction.**

This suggests that Gernsback and his colleagues were aware of an even broader tradition upon which they built (**Phoney Patent Offizz**). *Deadwood Dick* was a series of dime novel westerns written by Edward L. Wheeler from 1877 to 1885 and other ghostwriters from 1886 to 1897.[139] Dime novels serve as an important precursor to magazine-era science fiction both materially—printed on very cheap paper and "sent to the army in the field by cords, like unsaved firewood" during the American Civil War—and thematically—featuring stories of marvels, lost races, and inventions like Edward S. Ellis's *Steam Man of the Prairies.*[140]

And though dime novels and the popular pulp magazines that followed were the product of "a revolution in industrial mass production that capitalized on North America's vastly increased immigrant population, its decline in illiteracy rates, its rural postal delivery, and its increased urbanization," their thinly drawn "enemies" were almost always signaled through ethnic markers.[141] Everett F. Bleiler, in his entry on the dime novel in the *Encyclopedia of Science Fiction,* writes:

> Stress was on iron technology, with little or no science; narratives contained random, thrilling incidents, often presented in a disjointed and puerile way. Typical social patterns were: a conscious attempt to capitalize on age conflict, with boy inventors outdoing their elders (Edisonade); aggressive, exploitative capitalism, particularly at the expense of "primitive" peoples; the frontier mentality, with slaughter of "primitives" (in the first Frank Reade, Jr. story Frank kills about 250 Native Americans, to say nothing of destroying an inhabited village); strong elements of sadism; ethnic rancor focused on Native Americans, blacks, Irish and, later, Mexicans and Jews.[142]

Four decades later, what had scientifiction learned from the example of dime novels? John Rieder has persuasively argued in *Colonialism and the Emergence of Science Fiction* that interplanetary invasion narratives popular in the 1920s and '30s invoked the specters of colonial encounters between different cultures at the time. In a mirror image of colonialism on Earth, stories confronted white male heroes with civilizations and individuals far more "advanced" than anything they have seen before. Within these stories

> lurks the possibility of finding oneself reduced by someone else's progress to the helplessness of those who are unable to inhabit the present fully, and whose continued existence on any terms other than those of the conquerors has been rendered an archaism and anomaly. Colonial invasion is the dark counter-image of technological revolution. In relation to technology, as in other contexts, the history, ideology, and discourses of colonialism dovetail with the crucial, double perspective

139. Wheeler "wore a Stetson hat, and is said to have greeted even strangers as 'pard.'" Albert Johannsen, *The House of Beadle and Adams and Its Dime and Nickel Novels; the Story of a Vanished Literature,* 1st ed. (Norman: University of Oklahoma Press, 1950), available online at Northern Illinois University Libraries's Beadle and Adams Dime Novel Digitization Project, http://www.ulib.niu.edu/badndp/.

140. See George C. Jencks, "Dime Novel Makers," *The Bookman* 20 (October 1904): 108–14; E. F. Bleiler, "From the Newark Steam Man to Tom Swift," *Extrapolation* 30, no. 2 (1989): 101–16.

141. Ross, *Strange Weather,* 106

142. E. F. Bleiler, "Dime-Novel SF," in *The Encyclopedia of Science Fiction,* ed. John Clute et al. (London: Gollancz, 2014); updated February 10, 2016, http://www.sf-encyclopedia.com/entry /dime-novel_sf.

FIGURE I.14. *Deadwood Dick Jr.; or, the Sign of the Crimson Crescent* (1886), one of the weekly dime novels published by Beadle and Adams. Courtesy of E. M. Sanchez-Saavedra.

that runs throughout the genre: on one hand, the wondrous exploration of the new and the marvelous encounter with the strange, but on the other, the post-apocalyptic vision of a world gone disastrously wrong.[143]

For Andrew Ross, the language of technical wizardry "functioned as a mark of the heroes' superiority in coping with conditions in exotic localities like Mars or Venus, whose climates were simply displaced from the popular action-adventure regions of the arid West and tropical Africa respectively."[144] The futures imagined in the pages of *Amazing Stories* were, perhaps unsurprisingly, overwhelmingly white. The participation of women in these futures was very specifically circumscribed as well. At one point, Gernsback defended the title *Amazing Stories* over the originally proposed *Scientifiction* that was favored by many readers by arguing that foregrounding "science" on the cover might frighten off women readers. For him, the title of *Amazing*

143. John Rieder, *Colonialism and the Emergence of Science Fiction* (Middletown, Conn.: Wesleyan University Press, 2008), 32–33.

144. Ross, *Strange Weather*, 111.

Stories was "the best one to influence the masses, because anything that smacks of science seems to be too 'deep' for the average type of reader. . . . we are certain that if the name of the magazine had been *Scientifiction*, they [women] would not have been attracted to it at a newsstand" (**Editorially Speaking**).

A similar debate over the social identity of scientifiction broke out among readers when *Science Wonder Stories* simplified its name to *Wonder Stories* in June 1930. Just as Gernsback does in his comments above, the conversation insidiously conflated genre and gender. Describing the change in title, John Cheng writes that

> readers' responses were swift, heated, and divided. Those readers interested in science fiction's science were appalled. "Your aim, I take it, is to make the title more 'catchy,'" wrote Lloyd E. Foltz of Indianapolis, Indiana, "to that class of magazine addicts who are already reading, 'Sappy Stories,' 'Slushy Romances,' and so on, ad nauseam." He argued that the changes betrayed science fiction's premise. "I believe this is a mistake," he said. "It will attract a type of reader to whom S. A. means sex appeal and not scientific adventure." Other readers disagreed. "Now for the burden that some of the readers seem to carry—romance," wrote Darrel Richards, [. . .] "Why shouldn't 'females' . . . be included in the science fiction stories?" he asked. "Remember, 'Love makes the world go around.'"[145]

Justine Larbalestier points out that in many conversations on genre among readers, "the terms 'love interest,' 'romance,' 'sex,' and 'women' are used interchangeably. They become one and the same. On the whole this equivalence between 'women' and 'love interest' disqualifies women from the field of science fiction, since love belongs to the field of romance."[146] An argument for minimizing the presence of romance elements in scientifiction thus becomes—in some cases explicitly—about marginalizing women.[147] When a young Isaac Asimov wrote, "When we want science-fiction, we don't want swooning dames, and that goes double. . . . Come on, men, make yourself heard in favor of less love mixed with our science!" Mary Byers of Springfield, Ohio, responded: "To his plea for less hooey I give my whole-hearted support, but less hooey does not mean less women; it means a difference in the way they are introduced into the story and the part they play."[148] Larbalestier argues that Byers's retort echoes later feminist science fiction critics like Joanna Russ. Given the communal, conversational nature of scientifiction, the very structure of the genre actually ended up giving women a way into an arena they were formerly excluded from. And the rules of that supposedly masculine genre, which demanded a rigorous consideration of radically

145. Cheng, *Astounding Wonder*, 116.
146. Larbalestier, *The Battle of the Sexes in Science Fiction*, 10.
147. See ibid., 117–35. Larbalestier provides readings of women SF writers as well, including Claire Winger Harris and Leslie F. Stone, whose story "The Conquest of Gola" in the April 1931 issue of *Wonder Stories* thematized the "battle of the sexes."
148. Mary Byers, "In Other Words, It Isn't What You Say, It's the Way You Say It," *Astounding Science Fiction* (December 1938): 160–61.

alternate worlds, allowed it to call into question the hard and fixed gender norms enforced by many readers.[149]

"Science," wrote Gernsback in the inaugural issue of *Amazing Stories,*

> through its various branches of mechanics, astronomy, etc., enters so intimately into all our lives today, and we are so much immersed in this science, that we have become rather prone to take new inventions and discoveries for granted. Our entire mode of living has been changed with the present progress, and it is little wonder, therefore, that many fantastic situations [. . .] are brought about today. It is in these situations that the new romancers find their great inspiration. (**A New Sort of Magazine**)

Gernsback envisioned scientifiction as a form of popular education in which the reader might not even be aware that she was learning about the science that went into the construction of her everyday experience. Because "the average man or woman does not wish to laboriously wander through miles of scientific facts . . . we have tried to reduce all scientific matter to entertainment instead of study" (**The "New" Science and Invention**). What Gernsback could not have anticipated was how the genre itself would end up learning from the diverse makeup of its community. Today, this is a powerful reminder during a moment in which the Hugo Awards celebrate voices that have not only gone unrecognized in science fiction for far too long, but are also the source of some of the genre's most exciting new ideas.[150] From its humble beginnings, scientifiction would actually become expansive enough to address "all our lives," if only through the gentle reminder that there is an entire universe of radically different vistas to experience right here at home.

149. For more on women in early science fiction, see Jane Donawerth, *Frankenstein's Daughters: Women Writing Science Fiction* (Syracuse, N.Y: Syracuse University Press, 1997); Robin Roberts, *A New Species: Gender and Science in Science Fiction* (Urbana: University of Illinois Press, 1993); Lisa Yaszek, *Galactic Suburbia: Recovering Women's Science Fiction* (Columbus: Ohio State University Press, 2008); Lisa Yaszek and Patrick B. Sharp, eds., *Sisters of Tomorrow: The First Women of Science Fiction* (Middletown, Conn.: Wesleyan University Press, 2016).

150. A movement calling themselves the Sad Puppies attempted to influence the 2015 Hugo Awards by creating a campaign to nominate more white male writers, whom they felt had been excluded from recent ballots. For a profile of the controversy that resulted and the eventual withdrawal of authors from the ballot who didn't want to be associated with these views, see Amy Wallace, "Sci-Fi's Hugo Awards and the Battle for Pop Culture's Soul," *WIRED*, October 2015, http://www.wired.com/2015/10/hugo-awards-controversy/.

A New Interrupter

Scientific American, July 29, 1905

Experimenting with different magnetic and electric interrupters, the idea occurred to me that it might be possible to construct an interrupter whose chief functions would be based upon the expansion and contraction of mercury, when heated, by passing a current through it.[1]

After many fruitless experiments I succeeded in making such an interrupter, and the definite form that proved most satisfactory is explained in the following lines:

A barometric glass tube of about 15 centimeters length, with a central opening of 3 millimeters, is heated in an oxy-hydrogen flame and drawn into the shape, as shown in the accompanying drawing. This is by no means easy, as the tube, C, which represents the main part of the interrupter, must be so attenuated as to leave a capillary bore within, its minute diameter not surpassing ⅛ of a millimeter.

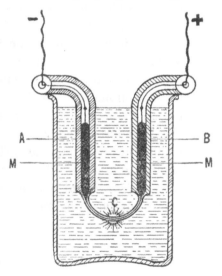

FIGURE 1.

Heat the middle part of the tube over the flame by constantly rolling the ends between three fingers of each hand, till it is red hot and soft. Take the tube quickly out of the flame, and draw it straight out, till it is thin enough; then bend it into the right shape, and let it cool slowly. Of course, these manipulations have to be done quickly, because the glass will not remain soft very long in the open air, and it is nearly impossible to draw the capillary tube when the flame touches it. The tube has to be filled then with chemically-pure mercury, which is easily done by placing the end of the open column, A, in a receptacle containing the quicksilver. By drawing the air out of B, the mercury will quickly mount in A, then pass through C, and rise up in B.[2] It is well to only half fill both columns. The apparatus will generally work satisfactorily, when the whole arrangement can be placed in any desired position without the mercury flowing out of it. This is a sign that the capillary tube, C, is sufficiently attenuated.

1. This short article, attributed to Huck Gernsback, was the twenty-year-old's first piece published after immigrating to the United States. Written in the concise prose of an electrical engineer, it details his design for an interrupter, a device that repeatedly interrupts a low-voltage power supply and transforms it into higher-voltage pulses. This setup used rising bubbles in a chamber of heated mercury—an electrically conductive liquid, or electrolyte—to interrupt current at a very high frequency. It was designed to either replace or supplement the induction coil, which used delicate mechanical contacts. The inaugural issue of Gernsback's first magazine, *Modern Electrics,* contains an article on constructing a very similar apparatus. Hugo Gernsback, "How to Make a Mercury Interrupter," *Modern Electrics* 1, no. 1 (April 1908): 22. Twenty years later, Hugo's brother Sidney described an updated version of this design using sulfuric acid as the electrolyte in his encyclopedia entry for "electrolytic interrupter." Sidney Gernsback, *S. Gernsback's 1927 Radio Encyclopedia* (Palos Verdes Peninsula, Calif.: Vintage Radio, 1974), 93–94.

2. The intellectual heritage of mercury as an electrolytic solution is interesting, as the method was first produced not in relation to wireless telegraphy and the transmission of audio signals, but during mid-nineteenth-century attempts to arrest and quantify the speed of light. Electrochemical discharges like those described by Gernsback were first discovered by Léon Foucault and Hippolyte Fizeau as an accidental by-product of their 1844 research into the formation of light spectra using various sources of illumination. For more on this discovery, see Rolf Wüthrich and Philippe Mandin, "Electrochemical Discharges—Discovery and Early Applications," *Electrochimica Acta* 54, no. 16 (June 2009): 4031–35. Roughly ten years later, Foucault constructed a system of rotating mirrors described by the English inventor Charles Wheatstone to produce the first accurate measurement of the speed of light. For more on Foucault, who began his career as a microscopy assistant, was a daguerreotype enthusiast, and conceptualized the problem of measuring light as the "observation of a moving image as a fixed image," see Canales, *A Tenth of a Second* (Chicago: University of Chicago Press, 2009), 158–59.

Two thin platinum wires are introduced into A and B till they dip in the mercury. The apparatus is put into a vessel containing water, which serves to constantly cool C, which part would break in the open air. Connect the two wires with two small storage batteries, and the interrupter will start instantly. In the middle of C there will be a bright bluish-green spark, and a high-pitched tone will emanate from the interrupter, indicating that the interruptions are of high frequency.

I found that this interrupter works most satisfactorily with 4 to 6 volts; it will consume, when made according to directions, from ¼ to ½ ampere, and run as long as desired. By making the part, C, of a larger cross-section, the voltage may be higher and more current will be absorbed, but the interruptions will be very unsteady and irregular, and will very often give out entirely.

The instrument, I believe, cannot be used with high tension currents, as it is too delicate, but it will work satisfactorily in connection with small induction coils, for instance, although a condenser will be required.[3]

The explanation as to how this interrupter works is as follows:

The instant the current is closed, the mercury at the smallest cross-section in C will become so heated that it commences to boil, and the force of the resulting bubbles, falling against each other, will be sufficient to make a momentary rupture in the thin mercury column.[4] There will be a little shock, and the expanding quicksilver will rise in A and B. Of course, a vacuum will be created at the place where the rupture occurred; and as the tube is immersed in water, the mercury will stop boiling; it cools instantly, then contracts, and the atmospheric pressure, combined with the weight of the quicksilver columns in A and B, will help to bring the metal in contact again, after which the same play commences as described.[5]

3. A condenser, today known as capacitor, is a component that can store an electrical charge. The Leyden jar, invented in the 1740s, was the first experimental demonstration of this technology.

4. The high frequency with which bubbles in the mercury interrupted a current was a method that could in theory replace the slower and more delicate mechanical contacts that then were common, such as an opening and closing magnetic arm. Gernsback's interrupter was one of many proposed around the turn of the century for use in high-current applications such as wireless telegraphy, due to the fact that the high currents necessitated by early spark-gap radio transmitters would often destroy any physical contacts.

5. A reader with the Minnesceonga Supply Co. in Haverstraw, New York, reported in the October 1914 issue of *The Electrical Experimenter* to have some trouble building the Gernsback electrolytic interrupter: "in a half hour run, the glass jar cracked from the heat, and it did not interrupt properly. What is the reason for this?" An Electro Importing employee and assistant editor to Gernsback, Harry Winfield Secor, replies that the component was designed for use in wireless telegraph applications almost exclusively. Due to the nature of telegraphic signaling, "this service is invariably intermittent, and the interrupter is never left in circuit continuously, for more than one to two minutes, without opening the circuit." If the interrupter is to be used in other applications ("such as for X-ray work, etc."), Secor recommends the reader "arrange special means for cooling the solution, etc." Minnesceonga Supply Co., Haverstraw, N.Y., "Gernsback Electrolytic Interrupters," *Electrical Experimenter* 1, no. 9 (January 1914): 142.

The Dynamophone

Modern Electrics, vol. 1 no. 2, May 1908

hile conducting some experiments in wireless telephony I made the discovery of quite an interesting combination which, to my knowledge of the art, has not been tried up to this date.

To produce mechanical effects directly, man is obliged to exert his muscular forces, by bringing his muscles in contact with the object to be moved, or through an indirect way, namely, by interposing a tool between the muscle and the object.[1] This tool might be a lever or it might be a telegraph key, which latter, if desired, will move or disturb the far-off object through another medium, electricity.

To apply power to an object the hand is used more than any other part of the human body; the foot probably ranges next. The whole body, mostly applied as a lever, follows. The head is practically never used at all. The lungs are used very little comparatively, for instance in glass blowing, etc.

The voice has never been used, no case being on record that a motor or a dynamo was started solely by talking, to or through a medium. I am, of course, well aware that by talking in a transmitter the telephone diaphragm at the other end will vibrate, but that can hardly be called power, it being proved that in most cases the vibrations of a receiver diaphragm measure less than one-five-thousandths of an inch.[2] To provide a contact on the diaphragm in order that the vibrations should close a certain circuit, which in turn could be relayed, to transmit or start power, has always proved a failure on account of the vibrations of the diaphragm, created by the human voice, being exceedingly weak.

My arrangement, as described below, to transmit or start power, etc., simply by talking into a transmitter, will therefore be found novel, especially if it is considered that the transmitter is not connected with the receiving station whatsoever. The transmission is made by means of wireless electric radiations.

No new apparatus being needed in my arrangement, any amateur can easily perform the experiment.

Referring to plan, M represents the common transmitter as used on most telephones. R is a fairly sensitive pony relay of seventy-five ohms. C is a condenser to absorb excessive sparking; I, induction coil; S, oscillator balls.

1. A sentence linking human "muscular forces" to "mechanical effects" in a metaphor that isn't signaled as such becomes possible thanks to the rise of the "human motor" as a dominant form of political, social, technical, and scientific thought from the late nineteenth century through the interwar period. Anson Rabinbach has shown how labor is transformed into "labor power" over this period, "a concept emphasizing the expenditure and deployment of energy as opposed to human will, moral purpose, or even technical skill." Predicated on a utopian ideal of a body without fatigue, discourse on the human motor sought to marry the motion of the body with that of the machine: "European scientists devised sophisticated techniques to measure the expenditure of mental and physical energy during mechanical work—not only of the worker, but also of the student, and even of the philosopher. If the working body was a motor, some scientists reasoned, it might even be possible to eliminate the stubborn resistance to perpetual work that distinguished the human body from a machine. If fatigue, the endemic disorder of industrial society, could be analyzed and overcome, the last obstacle to progress would be eliminated." Anson Rabinbach, *The Human Motor: Energy, Fatigue, and the Origins of Modernity* (Berkeley: University of California Press, 1992), 2.

2. This is the problem that first led Reginald Fessenden to discover the

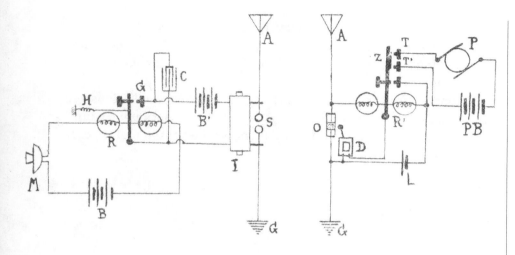

FIGURE 1.

The tension of the adjusting spring H on the relay should be just sufficient to keep the armature away from contact G. With a little experimenting the right tension will be ascertained.

By talking medium loud in the transmitter, the resistance in same will be varied accordingly and the armature will close the circuit at G, through battery B and coil I, which may be a common one-inch spark-coil not deprived of its vibrator.

Every time the circuit is closed at G a series of sparks will jump across the balls at S, creating oscillations. These oscillations, traveling through the ether, arrive at the receiving station, where they impinge on the antennae A and operate the coherer O through relay R'. The decoherer D is also shown. Relay, coherer and decoherer are all operated by a single dry cell L.[3] This is the same circuit as in my

3. The coherer, first designed by the physicist and inventor Édouard Branly (1844–1940) in 1891, is one of the earliest forms of detector, the most important component of a wireless set. The detector is responsible for demodulating radio frequency signals into an audio frequency current, ready to be piped through the listener's headphones. The idea behind the coherer is that when a radio frequency signal is applied to a glass vial filled with metal powder, the metal filings will "cohere" and an electrical current will pass through the device. Each telegraphic "dot" or "dash" causes the filings to cohere, after which some form of "decoherer" mechanically taps the vial to loosen them once again.

It was left for Marconi to, in the words of electrical engineer Thomas H. Lee, combine Hertz's spark-gap transmitter with Branly's coherer and "tinker like crazy" to produce the first wireless telegraph set. It is worth noting that the physical principle behind the coherer's operation is still not understood today. It remains a device with practical applications whose inner workings are nevertheless not understood. Thomas H. Lee, "A Nonlinear History of Radio," in *The Design of CMOS Radio-Frequency Integrated Circuits* (Cambridge: Cambridge University Press, 2004), 3.

heterodyne principle in 1901: two electrical waves of different frequencies can amplify one another, even to the point of producing frequencies in the audible range. Hugh Aitken: "The word [heterodyne], now part of every radio engineer's vocabulary, was Fessenden's coinage, reflecting his early training in Greek: to heterodyne meant to mix two different forces—in this case, two waves of different frequency." Fessenden was exploring the question of why early detectors like the liquid barretter couldn't produce a signal powerful enough to produce sound through a telephone earpiece, or "move the diaphragm . . . at audio frequencies": "Fessenden's answer was to feed two currents into the earpiece: one the antenna current, a train of oscillations at the signal frequency, and the other a train of oscillations generated either locally at the receiver or at a second transmitter, with the two frequencies differing slightly from each other. The two frequencies would mix or 'beat' against each other, and if the difference between them were correctly chosen, the result would be a wave train that the metal diaphragm of the earpiece could follow and that the human ear could hear." Much like the sympathetic vibrations of a piano tuner or undamped guitar strings, these two currents resonate with one another. Hugh G. J Aitken, *The Continuous Wave: Technology and American Radio, 1900–1932* (Princeton, N.J.: Princeton University Press, 1985), 58.

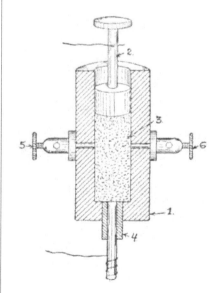

Branly coherer, from Archie Frederick Collins, *Wireless Telegraphy* (1905).

"Telimco" wireless system.[4] Relay R' has in addition two stationary contacts T and T', which, when the armature Z closes, complete another circuit, as, for instance, through a small motor P, an incandescent light, etc.

As long as words are spoken in the transmitter M, oscillations will be set up in S and the receiving station will work continuously until the voice at M stops. Motor P will, of course, be kept in motion only as long as the voice talks into M.[5]

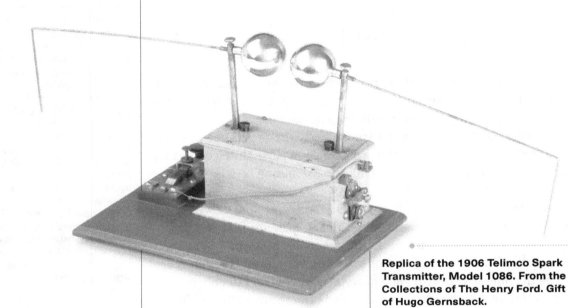

Replica of the 1906 Telimco Spark Transmitter, Model 1086. From the Collections of The Henry Ford. Gift of Hugo Gernsback.

4. The Telimco—a portmanteau of "The Electro Importing Company" and their flagship product—was the first fully assembled radio set ever sold to the American public, with transmitting and receiving models made available. For the Dynamophone described here, imagine a Telimco transmitter retrofitted with a telephone mouthpiece, and a Telimco receiver hooked up to some form of motor. As long as the user speaks into that mouthpiece on the transmitter, the motor on the receiving end would run.

Blueprints for such jury-rigged apparatuses were typical of *Modern Electrics* at the time and were presented as if they were easily replicable by readers who could stock their home workshops with parts from the *Electro Importing Catalog*. This issue, for instance, also contains instructions by Gernsback for building an alarm clock ("An Electric Sun Alarm") connected to light-sensitive selenium cells (used in early television prototypes—see **Television and the Telephot** in this book) that would set the alarm ringing when hit by the morning sunlight. This issue also contains the first winner of a regular "Laboratory Contest" section. Reader Irving Kimball impressed with the submission of his "vitaphone," a unique device that synced a motion picture projector with sound reproduction.

5. Although the coherer wasn't as sensitive as later electrolytic and crystal detectors, it had the merit of being the only one to work in mobile applications until the late 1910s. In a later editorial on what he refers to as "radio kinetics," Gernsback details the shortcomings and unique benefits of the coherer. Hugo Gernsback, "Radio Kinetics," *Radio Amateur News* 1, no. 12 (June 1920): 677.

The Aerophone Number

Modern Electrics, vol. 1 no. 7, October 1908

October 1908 cover of *Modern Electrics*.

It affords the Editor great pleasure to present to his readers herewith the first "Aerophone" number.

As is most likely the case a good many readers were surprised at this title, and an explanation is due.

We have grown so accustomed to the word "telephone" that we use it over and over without being conscious that it really means "far voice." You will say: "I shall telephone you," but nobody would think to say: "I shall far-voice you."

A short word has long been needed to express what is known now under name of "wireless telephone."

It sounds decidedly odd to say: "I shall wireless telephone you," or "I shall telephone you wirelessly."

The word "radio-telephone" expresses the idea a good deal better, but still it sounds strange if we say: "I shall radio-telephone you." Better would be the shorter word "radiophone." But it does not seem to sound quite right when we say: "I shall radiophone you," or: "I have received a radiophonic message."[1]

Somehow or other it sounds harsh. The Editor suggests the word "Aerophone," which not alone sounds well, and is easily remembered, but expresses the idea correctly. Translated it means: "air-voice." In other words, talking through the air, while telephony stands for talking over the wire. The word radiophone does not convey the idea that no wire is used, while Aerophone does.

The words, Aerophone, Aerophony, Aerophonic, sound good, and are to the point.

As will be seen by perusing this issue, the new word has been used almost throughout and the Editor shall continue to use it until a better one is found, or until another word is universally adopted.

The Editor shall furthermore be grateful if *every* reader would have the kindness to drop him a postal card stating which word he desires to become universal.

Results will be published in next issue.

Modern Electrics claiming several records, with this issue adds a new one to its list. No magazine heretofore issued an "Aerophone Number," or a "Wireless Telephone Number," the honor belonging entirely to *Modern Electrics,* leading, as usual.

1. The medium that Gernsback attempts to name here is best understood as an early conceptualization of radio. While the voice had been transmitted wirelessly as early as 1900, most notably by the Canadian inventor Reginald Fessenden using an electrolytic detector (see **The Radioson Detector** in this book for more), a reliable means for accurately sending sound, voice, and music signals was still a ways off. Because the broadcast model of radio we are familiar with today had yet to be imagined, most projections of wireless voice transmission involved a point-to-point model resembling a telephone conversation.

"Aerophone" never took off as a name, and other contenders like "radiophone" and "wireless telephone" were eventually simplified to "wireless." While the term "radio" replaced all of these by the 1920s, today there is a resurgence in the use of "wireless" thanks to new applications of the technology in WiFi, near field communication, Bluetooth, and cellular systems like GSM and CDMA. But it is interesting to note how the cultural form of broadcast radio—only one application of wireless, or the physical transmission of electromagnetic signals—was for a time synonymous with the general technology.

The Wireless Joker

Modern Electrics, vol. 1 no. 8, November 1908

The Editor is in receipt of several communications from the government and commercial wireless stations, all of whom file complaints against wireless amateurs, who annoy the large stations throughout the country.

By publishing the call letters and names of the wireless stations in the U.S., we simply wished to keep our readers informed, so that when any of them "caught" a message they would be in a position to know from whence the message was dispatched.

Modern Electrics being the leading wireless telegraph magazine, naturally was appealed to by the large stations. It seems that a number of experimenters are in the habit of calling up government and commercial stations, thus interfering with regular work and annoying them a great deal. Usually the large stations do not know the location of the sender and they are thus quite powerless to stop the mischief.[1]

This state of affairs can naturally not go on, and we most earnestly request all those who are in the habit of calling up the large stations to refrain in future from doing so.

We say this in behalf of all those interested in the wireless art, as several large companies are now endeavoring to have a law passed licensing all wireless stations in the U.S. This, of course, would be the end of the amateur and experimental wireless stations, as under the new law heavy licenses would be imposed.

In view that the art as yet is in its infancy, such a law would be deplorable, from the standpoint of the amateur, and the Editor earnestly hopes that the mischief will come to a speedy end.

Nobody cares how many messages the amateur catches, as long as he keeps in the dark and does not "talk back."

While we talk of wireless mischief making, we must mention the wireless "joker" (?). This pest located in New York, Chicago or anywhere will send a plausible message, calling up an ocean liner, stating that the machinery of the ship is damaged, or some other plausible yarn. He signs off, giving the name of a large ocean boat, and does of course not forget to state position of his ship which—in the message—is about 2,000 miles from land.

Of course the result is that some stations in the vicinity of the "joker's" one, catch the message and next day the owners go bragging about that they were able by means of a "new connection" to hear

1. A user skimming the airwaves in 1908 would have found a cacophony of code flashing in multiple directions, including: government stations sending out time signals; maritime communications going ship to ship and ship to shore with information like weather conditions, iceberg reports, and ship position and speed; personal "Marconigrams" or "Aerograms" sent from or to these ships; experimenters sending messages across the street or, at most for the time, across town; and commercial stations conducting experiments and sending messages for paying clients. In addition, there was no standard for wireless communication, which meant that one user might have identified himself by name while another might have made up a call sign, which were not standardized until 1927.

such and such boat 2,000 or 3,000 miles away, far out in the ocean. Naturally this nonsense only serves to mislead others, as it cannot but create wrong impressions in the minds of students who do not receive messages from more than 100 miles away and who perhaps are trying hard to improve or invent new instruments, etc.

We earnestly hope that those in the habit of sending these "fake" messages will see the harm they are doing and that they will turn their efforts to more fruitful directions.

The Wireless Association of America

Modern Electrics, vol. 1, no. 10, January 1909

It affords the Editor genuine pleasure to offer to the thousands of wireless enthusiasts an appropriate New Year's present in form of the Wireless Association of America. Every reader perusing the announcement of the association, found elsewhere in this issue, will agree that such an organization has been needed for a long time.

Wireless Telegraphy may now be said to have left the experimental stage. It has become an everyday necessity, and already rivals the wire and cable telegraph. Between sea and land wireless is of the utmost importance and to-day forms as necessary an adjunct to every up-to-date vessel, as its coal and lifeboats.

Only a few days ago Wireless demonstrated anew its utility. Several Italian warships were at sea, when the disaster occurred in southern Italy, necessitating prompt assistance by outside help. The warships were at once recalled by means of wireless, and through it have saved the lives of hundreds of sufferers and brought the much needed relief.[1]

Not alone on sea, or between land and sea, but on the mainland also, Wireless is of high importance. Only a few weeks ago two young Wireless experimenters saved several hundred lives in southern France in a dramatic manner. One of the young men, living on a hill near a dam, discovered that it was giving way, which meant the destruction of a small village in the valley below and the drowning of hundreds. He promptly "called up" his friend living in the village below who gave the alarm. Over two hundred people reached the mountain in safety, just in time to avoid the flood. Without Wireless they would have been surely drowned, as telephone and telegraph are unknown in that locality and the person in charge of the dam was sick on account of a recent accident.

The greatest aim of the Wireless Association of America is to bring young experimenters together—not in clubs, but in practical work.[2] The Editor would like to see every reader of "Modern Electrics" in possession of a wireless station; there is really nothing more instructive and entertaining than a wireless between two isolated, or for that matter, distant points in a city. The necessary instruments, thanks to competition are now so cheap that they are in reach of everybody,

1. The 7.1 magnitude Messina earthquake struck Sicily on December 28, 1908, triggering a tsunami that claimed 123,000 lives. Three years later, the central struggle of Howard Roger Garis's novel *Tom Swift and His Wireless Message* (published under the *nom de plume* Victor Appleton) was an escape from Earthquake Island using a spark gap wireless set:

> "But I think I can summon help."
> "How?" demanded Mr. Hosbrook. "Have you managed to discover some cable line running past the island, and have you tapped it?"
> "Not exactly," was Tom's calm answer, "but I have succeeded, with the help of Mr. Damon and Mr. Fenwick, in building an apparatus that will send out wireless messages!"
> "Wireless messages!" gasped the millionaire. "Are you sure?"
> "Wireless messages!" exclaimed Mr. Jenks.
> "I'll give—" he paused, clasped his hands on his belt, and turned away.
> "Oh, Tom!" cried Mrs. Nestor, and she went up to the lad, threw her arms about his neck, and kissed him; whereat Tom blushed.

Victor Appleton, *Tom Swift and His Wireless Message* (New York: Grosset & Dunlap, 1911), 168.

2. In addition to numerous local clubs around the country, rival publications attempted to build their own national organizations of wireless operators. *Electrician and Mechanic* started a "Wireless Club" the previous year, garnering a modest 114 members as of September 1908. A monthly section of the magazine, "Wireless Club," published short letters from its members: "experiences, discoveries, and suggestions, which may be helpful to all interested." The first installment is available at Thomas H. White's *United States Early Radio History* site: http://earlyradiohistory.us/1908wc2.htm. *The Wireless Age* formed a National Amateur Wireless Association in 1915 with a greater emphasis on citizen preparedness for potential war. Major J. Andrew White, editor of *The Wireless Age* and announcer at the famous 1921 experimental radio broadcast of the Jack Dempsey v. Georges Carpentier fight, wrote that he hoped this organization would help boys from the ages of twelve to eighteen "learn what military discipline means, accustom themselves to the rigors of camping and forced marches, perfect themselves in the operation of portable wireless equipment

and become trained soldiers in embryo." J. Andrew White, "A Word with You," *The Wireless Age* (December 1915): 184–88.

Hiram Percy Maxim proposed a different idea for a wireless organization in 1914: one that would explicitly work toward relaying messages across long distances, in effect forming a communications network of small, volunteer broadcast stations. This American Radio Relay League formed its own magazine, *QST,* and though it tried to collaborate with the Wireless Association of America, *Electrical Experimenter* refused to publish advertisements for the ARRL, writing that it was "distinctly competitive to *The Electrical Experimenter.*" A secretary of the AARL wrote to Milton Hymes, *The Electrical Experimenter*'s advertising manager, trying to convince the then-largest wireless publication in the country that "We regard the *Electrical Experimenter* as a companion magazine; a magazine of an entirely different type. Your paper is devoted to articles which would interest an experimenter and one who wished to keep abreast of the electrical news. Our paper is to keep each amateur in touch with the other regarding the operations of their stations."

But *The Electrical Experimenter* held firm. *QST* published this correspondence in its July 1916 issue, and a bitter rivalry ensued between the two organizations. See Thomas H. White, "Pioneering Amateurs," *United States Early Radio History*, 1996, http://earlyradiohistory.us/1910ei.htm. The letters exchanged between *QST* and *The Electrical Experimenter* are available at http://earlyradiohistory.us/1916QST.htm.

3. By 1912, the Wireless Association of America boasted 22,300 members.

Wireless Association of America member pin. Courtesy Collection of Jim and Felicia Kreuzer, Grand Island, New York.

4. Lee de Forest, an inventor of the vacuum tube (which he called the Audion), had a troubled relationship with

and we have as yet to hear of the person regretting an investment in wireless instruments.

The Association's button will also be of great help to bring young experimenters, unknown to each other, together. As only a few thousand buttons were ordered, prospective members should send at once for same, as it will take several weeks to obtain a new supply.[3]

WIRELESS ASSOCIATION OF AMERICA
Under Auspices of "Modern Electrics"

BOARD OF DIRECTORS
Dr. Lee de Forest, President
John S. Stone, Vice-President
Wm. Maver, Jr., Secretary
Hugo Gernsback, Chairman & Business Manager

The Wireless Association of America has been founded with the sole object of furthering the interests of wireless telegraphy and aerophony in America.

We are now on the threshold of the wireless era, and just beginning to rub our intellectual eyes, as it were. Sometimes we look over the wall of our barred knowledge in amazement, wondering what lays beyond the wall, as yet covered with a dense haze.

However, young America, up to the occasion, is wide awake as usual.

Foreign wireless experts, invariable exclaim in wonder when viewing the photographs appearing each month in the "Wireless Contest" of this magazine. They cannot grasp the idea that boys 14 years old actually operate wireless stations successfully every day in the year under all conditions, but they are all of the undivided opinion that Young America leads the rest of the world wirelessly.

Even Dr. Lee de Forest, America's foremost wireless authority, confessed himself surprised that so many young men in this country should be in the possession of such well constructed and well managed wireless stations, which is only another proof that the clear headed young men of this country are unusually advanced in the youngest branch of electrical science.[4]

So far America has lead [sic] in the race. The next thing is to stay in the front, and let the others follow. In fact he would be a bold prophet who would dare hint at the wonders to come during the next decade.

The boy experimenting in an attic today may be an authority tomorrow. However, not even the cleverest inventors or experimenters always have opportunity of making themselves known to the world,

THE PERVERSITY OF THINGS

and it is right here that we are confronted with a mystery so far unsolved. Out of 100 per cent of young wireless experimenters, 90 per cent are extremely bashful. Why this should be so is a mystery.

As stated before the new Wireless Association's sole aim is to further the interests of experimental wireless telegraphy and aerophony in this country. Headed by America's foremost wireless men, it is not a money-making institution. There are no membership fees, and no contributions required to become a member.

There are two conditions only. Each member of the Association *must be an American citizen* and MUST OWN A WIRELESS STATION, either for sending or for receiving or both.

The Association furnishes a membership button as per our illustration. This button is sold at actual cost and will be mailed to each member on receipt of 15 cents (no stamps nor checks).

This button is made of bronze, triple silver-plated. The flashes from the wireless pole are laid in in hard red enamel, which makes the button quite distinctive. The button furthermore has the usual screw back making it easy to fasten to buttonhole. The lettering itself is laid in in black hard enamel. Size exactly as cut.

On account of the heavy plating it will last for years and is guaranteed not to wear "brassy."

Its diameter is ¾ inch. This is a trifle larger than usual, the purpose being to show the button off so that it can be readily seen from a distance. The reason is obvious. Suppose you are a wireless experimenter and you live in a fairly large town. If you see a stranger with the Association button, you, of course, would not be backward talking to the wearer and in this manner become acquainted with those having a common object in mind, which is the successful development of "wireless."

The Association furthermore wishes to be of assistance to experimenters and inventors of wireless appliances and apparatus, if the owners are not capable to market or work out their inventions. Such information and advice will be given free.

Somebody suggested that Wireless Clubs should be formed in various towns, and while this idea is of course feasible in the larger towns, it is fallacious in smaller towns where at best only two or three wireless experimenters can be found.

Most experimenters would rather spend their money in maintaining and enlarging their wireless stations, instead of contributing fees to maintain clubs or meeting rooms, etc., etc.

The Board of Directors of this Association earnestly request every wireless experimenter and owner of a station to apply for membership in the Association by submitting his name, address, location,

Gernsback. They began as close associates, with De Forest contributing to some of the earliest issues of *Modern Electrics,* serving as President of the Wireless Association of America's Board of Directors, and presenting alongside Gernsback at industry banquets. See "Wireless Banquet," *Electrical Experimenter* 2, no. 1 (April 1909): 24. But De Forest later sued Gernsback for patent infringement, claiming that the Electro Importing Company was selling devices that used his Audion technology without paying royalties. The $4.00 Electro Audion, sold from 1911 to 1913 in the *Catalog,* is most likely one of the products he had in mind. Although the court sided with Electro Importing, arguing that the devices they manufactured contained significant enough improvements on the original Audion design, De Forest continued to contribute to the Gernsback magazines for years to come. Hugo Gernsback, "De Forest vs. The Electrical Experimenter," *Electrical Experimenter* 4, no. 11 (March 1917): 808.

instruments used, etc., etc., to the business manager. There is no charge or fee whatever connected with this.

Each member will be recorded and all members will be classified by town and State.

After February 1st, 1909, members are at liberty to inquire from the Association if other wireless experimenters within their locality have registered. Such information will be furnished free if stamped return envelope is forwarded with inquiry.

To organize the Association as quickly as possible it is necessary that prospective members make their application at once, and without delay.

If you are eligible fill out application sheet and state particulars as follows: Full name; town; state; age; system and apparatus used; full description of aerial.

In order to facilitate quick classification, please be brief and keep application sheet separate from your letter.

[Editorials]

Modern Electrics, vol. 1, no. 11, February 1909

A t last wireless telegraphy has had its real christening. For the first time in history, due directly to wireless telegraphy a terrible disaster was averted and close to 500 human beings are now alive instead of resting on the bottom of the ocean, like so many others, before the days of wireless.

Only a few weeks ago there were many people who doubted the practicability and usefulness of wireless. In fact the public at large rather had an idea that wireless was a scientific toy, and on a ship nothing but a diversion for passengers, who had too much money to spend by sending telegrams, "which never arrived at their destination, anyway."

All this has been changed in less than a week. Wireless has at last become public property, a position which it was denied for years. It has stood the fire test and has emerged from it gloriously.

Accidents such as the one just witnessed when the steamer Florida collided with the ill-fated Republic will soon become a thing of the past, thanks to wireless. Already the U. S. Government has taken steps to make wireless compulsory on ocean ships and other Governments will undoubtedly follow soon, as the tremendous importance of the subject has at last been realized.[1]

However, when talking of the Republic, let us not forget its gallant wireless operator, John Robinson Binns, the now famous "C. Q. D." man. It was due chiefly to his efforts that the passengers of the Republic were saved so promptly without the loss of a single life. He has set an unforgetable [sic] example to all wireless ship operators and has shown us what the duties of the operator are when his ship is singing. His story will be found elsewhere.

The announcement of the Wireless Association of America in the January number has brought such an avalanche of mail that it is impossible at this date of writing to do the entire correspondence justice. Members and prospective members will please have patience, all letters will be answered as fast as it is possible to write them. Apologies are also due new members who ordered the association button after January 15. As stated in the January issue only 1,000 buttons were ordered; this entire amount has been mailed about the 15th, and the second thousand could only be sent out by January 26th. The delay in

---

1. The RMS *Republic* was a luxury steamship that collided with the SS *Florida* on January 23, 1909. The ship was equipped with a Marconi-designed wireless set and became the first ever to send a "CQD" distress signal, effectively saving all 1,200 passengers aboard. "Official Story of the Florida," *New York Times,* January 27, 1909. The operator of the wireless aboard the Republic was Jack Binns, who became a hero among the amateurs and was later hired as a correspondent for the *New York Tribune* to report on the activities of local tinkerers. For more on Jack Binns and the collision, visit http://www.jackbinns.org/the_republic -florida_collision.

shipping was unavoidable, as the great demand has not been anticipated, and the editor trusts that the new members will understand the situation now.

The editor is truly amazed at the tremendous interest shown in the association, and as up to this writing over 3,200 members have applied, it is safe to say that the Wireless Association of America is the largest official wireless association in the world.

A membership application card has been issued which will be sent free of charge to any prospective member.

The demand for the association button is so tremendous that we will possibly run short again. Every mail brings dozens of orders, and if you do not receive your button at once, please do not get impatient. The buttons are shipped just as quick as the factory can turn them out and just as quick as our mailing department can address and ship them.

Members are requested to be kind enough to refrain from asking for information as to members near their localities at least till the 15th or 20th of this month.

On account of the great amount of work necessitated in classifying and assorting the membership cards and the mail, it will be an impossibility to furnish any information before that date.

The Editor, in behalf of the Wireless Association of America, wishes to thank all members for the interest shown and trusts that by next month the amount of members will have doubled.

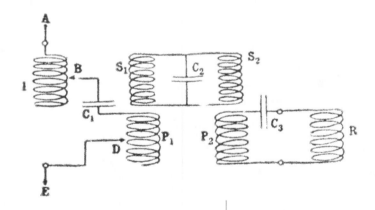

FIGURE 1.

2. Tuning wireless sets so that they did not transmit and receive on one wavelength only was still a great challenge at this point.

Tuned Wireless Telegraphy

At the recent exhibition of the Physical Society in London, the Marconi Wireless Telegraph Company showed a multiple tuner for tuning a wireless telegraph receiver so as to render it immune to interference from other stations. It can also be used for measuring the lengths of the transmitted and received waves and for estimating the distance of a known station. The instrument has been designed so as to stand several climates and comparatively rough usage, and is suitable for all wave lengths from 300 ft. to 8,000 ft. Its general principle is shown in illustration where A represents the aerial, E the earth, R the receiver or detector.[2]

[. . .]

Wireless on Mars

By Our Martian Correspondent

Mr. Spif Marseroni, the great national wireless scientist, has scored another great triumph.[3] As will be recalled, Martians have been for a long time in the habit of receiving and sending telephone messages, no matter if they were walking in the street or gliding in an aeroplane. In fact, this system is now so popular that the "Interplanetarian Remembering Co." has found no trouble whatsoever in getting over 60 million subscribers to their new system. The system is simple enough.

Suppose you are a busy man. During the night you suddenly recall that you must see a certain party to-morrow afternoon, 4 p.m. While you stay in bed you call up the Remembering Co. and tell the operator to call you to-morrow afternoon at 3.30 p.m.

Next day at 3.30 p.m. the little buzzer which you carry in your vest pocket suddenly "goes off" and when you put your pocket phone to the ear a young lady will tell you in a silver voice, that you have to meet a party at 4 p.m. The service of the new company is so efficient that it does not make much difference *where* you are. The Remembering Co. will locate you, whether you are taking a bath, or whether you are napping in a Morris chair in the lobby of an airship.

It will also be recalled that Mr. Marseroni is the inventor of the Telewirltransport. This as will be known, has been considered up to a few days ago, his greatest invention. By means of his system Martians may ride on electric motor rollers, the energy being supplied from a central station wirelessly through the ether. The power on all Martian airships and aeroplanes is furnished wirelessly to them from the same central station to which the users must be subscribers.

Now Mr. Marseroni has succeeded by conveying food through the ether wirelessly for unlimited distances. Already a large syndicate has been formed under name of "Interplanetarian Wireless Food Co." to exploit the invention. If you are a subscriber and you are walking in the street, and if it is 12 o'clock noon, your call buzzer suddenly rings. You put the phone to your ear and this may be what you hear:

"Luncheon ready, please. What will you have?"

"Ham sandwich and a glass of milk," you call back.

You then draw your silver case out of your pocket and connect its terminals with your antenna, fastened on your hat. Two seconds later and a ham sandwich has "materialized" in the silver case. The milk is received in the same manner. In fact, Mr. Marseroni has succeeded to send almost anything now from champagne down to lobster salad. The only thing he does not send are onions, as the odor is

3. This section was included as part of the regular "Wireless Screech" humor section and written under the pseudonym Mohammed Ullyses Fips, a name that Gernsback used for all of his tongue-in-cheek columns throughout the years.

lost in transmission and an onion without smell is like a river without water.

The process of sending food by wireless is not as difficult as might be thought at first.

The food is passed through "puffers," which blow it to atoms step by step. It is finally reduced so much that its consistency is brought in "balance" with the ether. It is then passed through a system of Leyden jars and sent out in form of ether waves, carrying the infinite minute food particles. The receiving apparatus condenses these particles again and the food appears in its original condition, only far more palatable.

Of course the sending operator must be careful so as not to "mix" things, as in the beginning it happened a few times that a subscriber got hot when he received coffee mixed with chopped pickles or buckwheat cakes soaked in Worcestershire sauce.

However, by using tuning separators this defect has been entirely overcome now.

Signaling to Mars

Modern Electrics, vol. 2, no. 2, May 1909

Every time our neighbor Mars comes in opposition to the earth a host of inventors and others begin to turn their attention to the great up-to-date problem, "signaling to Mars."[1]

It is safe to say that all the proposed, possible as well as impossible, projects would fill a good-sized volume, especially the ones invented in this country, which by far leads the world in number of projects and "inventions."

So far only one feasible plan has been worked out. The writer refers to Professor Pickering's mirror arrangement, now being discussed all over the world. But even Professor Pickering is skeptical, as he apparently does not like to take the responsibility of spending ten million dollars on a mere idea which might prove fallacious.[2]

Wireless telegraphy has been talked of much lately as a probable solution to the problem. The writer wishes to show why it is not possible at the present stage of development to use wave telegraphy between the two planets, but at the same time he would like to present a few new ideas how it could be done in the near future.[3]

Take the average present-day wireless station having an output of about 2 K. W. On good nights and under favorable circumstances such a station may cover 1,000 miles. Very frequently, however, only about 800 miles can be spanned.

40,000,000 miles would have to be more than a quarter of a mile of glass a single mirror would not be practicable. We would have to use a number of mirrors.

"These mirrors would all have to be attached to one great axis parallel to the axis of the earth, run by motors, and so timed as to make a complete revolution every twenty-four hours, thus carrying the reflecting surface around with the axis once a day and obviating the necessity of continually readjusting it to allow for the movement of the planets." "Plans Messages to Mars: Prof. Pickering Would Communicate by Series of Mirrors to Cost $10,000,000," *New York Times*, April 19, 1909.

Signaling to Mars did not mean that Pickering believed there to be life on Mars. A few months later, *Popular Mechanics* magazine noted that Pickering "is also among those who seriously doubt that there are any living beings upon Mars, although he has due respect for the theories of those opposed to him, but he does believe that his scheme of sending messages is the one practical way of finding out, once for all, whether there are such beings, although he admits that if no answering signals were made, it would not disprove the theories that Martians exist." "The Scheme to Signal Mars: Prof. Pickering's Practical Plan," *Popular Mechanics* 12, no. 1 (July 1909): 10.

A version of Pickering's plan was actually carried out in 1924, when Swiss astronomers "mounted a heliograph in the Alps to flash signals to Mars. The U.S. Navy maintained radio silence for three days to listen to messages from the Martians." Markley, *Dying Planet*, 158.

1. In his book on the intertwined histories of astronomical observations and literary depictions of Mars, Robert Markley succinctly describes the red planet's orbital opposition, "For a few weeks every twenty-six months, Mars and the Earth are aligned on the same side of the sun in their elliptical orbits. During these periods of opposition, Mars is visible through comparatively small telescopes, and, since the mid-seventeenth century, scientific observations of the planet's surface and atmosphere have clustered during these periods." Robert Markley, *Dying Planet: Mars in Science and the Imagination* (Durham, N.C.: Duke University Press, 2005), 33.

The occasion for this article, the orbital opposition of 1909, was also the first time Percival Lowell took a successful series of telescopic photographs of Mars in an attempt to prove his theory that a vast infrastructure of canals was built across the Martian surface. For more, see William Graves Hoyt, *Lowell and Mars* (Tucson: University of Arizona Press, 1996); Oliver Morton, *Mapping Mars: Science, Imagination, and the Birth of a World* (New York: Macmillan, 2002); Robert Crossley, *Imagining Mars: A Literary History* (Middletown, Conn.: Wesleyan University Press, 2011).

2. William Henry Pickering was an astronomer with the Harvard Observatory known for discovering Saturn's moon Phoebe, developing new techniques in telescopic photography, and advancing popular knowledge of the surface of Mars. In April 1909, Pickering proposed a plan to communicate with Mars using a massive heliograph. Pickering described the system in a front-page article in the *New York Times*: "My plan of communication would necessitate the use of a series of mirrors so arranged as to present a single reflecting surface toward the planet. Of course one mirror would do as well, but as the area necessary for reflecting the sunlight over

3. William Preece, engineer-in-chief of the British Post Office and radio experimenter, suggested as early as 1898 that wireless could be used to contact Mars: "If any of the planets be populated with beings like ourselves, then if they could oscillate immense stores of electrical energy to and fro in telegraphic order, it would be possible for us to hold commune by telephone with the people of Mars." William Preece, "Ethereal Telegraphy," *Review of Reviews* 18 (December 1898): 715; quoted in Susan J. Douglas, "Amateur Operators and American Broadcasting: Shaping the Future of Radio," in *Imagining Tomorrow: History, Technology, and the American Future*, ed. Joseph J. Corn (Cambridge, Mass.: MIT Press, 1986). See also Thomas Waller, "Can We Radio a Message to Mars?" *Illustrated World* 33 (April 1920): 242.

Next summer for a few days Mars will be nearer to us than for many years to come. The distance between the two planets will then be about 35 million miles.

If we base transmission between the earth and Mars at the same figure as transmission over the earth, a simple calculation will reveal that we must have the enormous power of 70,000 K. W. to our disposition in order to reach Mars.

Now it would be absolutely out of the question to build a single station with that output. Even a station of 700 K. W. would be a monster and a rather dangerous affair to meddle with. This may be understood better when considering that none, even the most powerful stations, to-day have 100 K. W. at their disposal.

A solution, however, presents itself to the writer's mind. As it is impossible and impracticable to build and operate a single station with an output of 70,000 K. W., let us divide the 70,000 K. W. in small stations of, say, 2, 10, or 50 K. W. Neither do we have to build these stations for the sole purpose of using them to signal Mars; they are already being erected by the governments and commercial stations at the rate of about 150 per month.

At the present time of writing the entire output of the U.S. Government, commercial and ship wireless stations combined, is about 2,500 K. W. By adding the stations of private individuals, ranging in power from ¼ to 2 K. W., the total sum is brought up to about 3,500 K. W., as far as the writer is able to ascertain from the latest reports.

If, however, the art progresses as it has during the past four years, it is safe to predict that in 20 or possibly 15 years the United States, Canada, and Mexico will reach the combined output of 70,000 K. W.

It will then be comparatively easy to seriously undertake to signal to Mars, and the writer proposes the following plan, which has the great advantage that the experiments can be made at practically no cost, against Professor Pickering's project, involving the enormous expense of ten million dollars.

The idea is simple enough. A central point of the continent, such as Lincoln, Nebr., should be selected preferably. By previous arrangement all wireless stations on the continent should be informed that on certain days their stations should be connected with a magnetic key,[4] which is connected through the already existing wire telegraph lines with the central station at Lincoln. As the wires may be leased from the existing wire telegraph lines, it is of course the simplest thing in the world to connect the key of each wireless station (by wire) with the central station. Each time, therefore, when the operator at Lincoln depresses his key all the keys belonging to the wireless stations connected with his key will be pressed down, and if the combined power

4. Gernsback: "Described in the October, 1908, issue of M. E., page 243."

THE PERVERSITY OF THINGS

of the connected stations is 70,000 K. W., the enormous energy of 70,000 K. W. will be shot out in the ether!

What effect the 70,000 K. W. will have on the weather or climate after they have been radiated for several hours the writer dares not conjecture, but that something will "happen" is almost certain.

Considering the technical side of the project, it is of course feasible. If the necessary amount of power was to be had to-day, there would be no difficulty to try it next summer; as this is not the case, we must be patient and wait; the writer, however, hopes to see the day when the experiment will be tried.

Referring to the technical side, it will be necessary, of course, to tune all the sending apparatus to exactly the same wave length, which, naturally—on account of the great distance to be overcome—should be as long as possible. The frequency of the oscillations should be practically the same for all stations. The result of this arrangement would be that the effect would be practically the same as if one tremendously large station of 70,000 K. W. capacity was sending.

Just as we may blow two or a dozen whistles of the same pitch at the same time, in order to carry the sound further, and just as Professor Pickering may use thousands of small mirrors all operated at the same time, as if they were one huge mirror, so it may be possible to unite a great number of different wireless senders and operate them as if they were one, provided of course that, like the whistles, they are tuned to the same "pitch."

There is only one more point to consider.

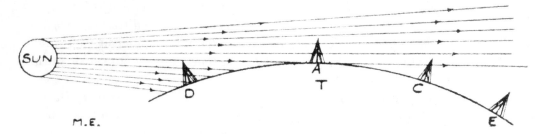

It has been demonstrated time and again that the action of the sun's rays greatly interfere with wireless telegraphy. In fact, it is possible to send twice as far over water during the night than during the day. This may be better understood by quoting Mr. Marconi's views:

FIGURE 1.

> Messages can now be transmitted across the Atlantic by day as well as by night, but there exists certain periods, fortunately of short duration, when transmission across the Atlantic is difficult and at times ineffective unless an amount of energy greater than that used during what I might call normal conditions is employed.

Thus in the morning and in the evening when, due to the difference in longitude, daylight extends only part of the way across the Atlantic, the received signals are weak and sometimes cease altogether.[5]

Mr. Marconi's explanation is that illuminated space possesses for electric waves a different refractive index to dark space and that in consequence the electric waves may be refracted and reflected in passing from one medium to another.[6]

The writer wishes to offer a different explanation, which seems far more plausible.

Referring to Fig. 2, let T represent a section of the earth. Let A be a station on the American, E a station of the English coast. As will be seen, the sun is just setting for the point A, while E has night already (no sun rays reach E).

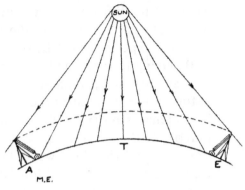

FIGURE 2.

When A is sending the waves are shot out *parallel* with the sun's rays and *carried with the rays*. The action of the sun's rays is so strong that most of the electric waves are carried along, and therefore never reach E at all. Only by using more powerful waves can this effect be overcome. This action is not surprising at all. Electromagnetic waves are closely related to light rays. As Svante Arrhenius has also shown us, the rays of the sun exert a certain amount of pressure on all encountered objects.[7] It is therefore easy to prove that considering the close relationship of light rays and Hertzian waves, the latter *will be carried in the direction away from the sun* under favorable circumstances. Again considering Fig. 1, such favorable circumstances would be reached during sunset or during sunrise.

That this explanation is not a mere theory is best proved by the fact that a point D and C will communicate with each other best during sunrise and sunset, the signals received being the strongest. The electric waves during these two periods *travel parallel with the sun's rays, following the line of least resistance.*

During the day (Fig. 2), let A again represent the American, E the English coast station. Now it will be easily understood why messages can be sent almost twice as far during the night as during the day. In this instance the electric waves must *cut directly through the vast field of light,* and are being "held down" to a certain degree by the pressure of the light.

Now let us turn our attention to Fig. 3. This represents the earth

5. Gernsback: "Article in the May, 1908, issue M. E., page 55."

6. Oliver Heaviside proposed the existence of an electrically charged layer in the atmosphere in 1902, but not until 1927 was the existence of the ionosphere confirmed. While Gernsback is correct here that the presence of sunlight affects the transmission of radio waves, it is not the sun's rays themselves but rather their interaction with the ionosphere that causes interference. He could not have known that in 1909, but this was another area where technological achievements outpaced their scientific explanations. Marconi's first successful transatlantic wireless message in 1901, sent *around* the curvature of the earth, was possible because it bounced off the then-undiscovered ionosphere.

7. Gernsback: "'Worlds in the Making.' See Panspermie [sic]." This note refers to Svante Arrhenius, *Worlds in the Making: the Evolution of the Universe* (London: Harper & Brothers, 1908). Panspermia is the theory that some form of microscopic life is spread just as evenly throughout the universe as matter itself. Arrhenius, a 1903 Nobel Prize winner in chemistry, writes, "According to this theory life-giving seeds are drifting about in space. They encounter the planets and fill their surfaces with life as soon as the necessary conditions for the existence of organic beings are established" (217). This would imply that "all organic beings in the whole universe should be related to one another and should consist of cells which are built up of carbon, hydrogen, oxygen, and nitrogen" (229).

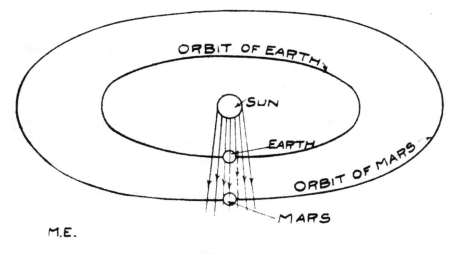

ORBIT OF EARTH

SUN

EARTH

ORBIT OF MARS

MARS

M.E.

FIGURE 3.

and Mars in opposition. It will be seen immediately that the earth has a great advantage over Mars as far as wireless is concerned.

Messages sent from the earth during the opposition will go in the same direction as the sun's rays, and the writer is of the opinion that instead of the theoretical amount of power—that is, 70,000 K. W.— possibly only one-half or one-quarter will be required to signal to Mars, as the electric waves are undoubtedly assisted toward Mars by the rays of the sun.

On the other hand, Mars will find it difficult to signal back, especially during opposition, when his "operators" would have to work directly against the sun's rays.

However, we can only hope that the Martians are further advanced than we and may signal back to us, using a method new to us and possibly long discarded by them, when thousands of years ago they stopped signaling to us, and gave us up, as we did not have intelligence enough to understand.

Editorial

We call our readers' especial attention to the first annual official blue book which we are publishing and which is ready now.

This book contains all the information the student of wireless telegraphy is interested in. It enables him to tell where and from whom a message is sent and thereby increases and stimulates the interest in the art to a great extent.[8]

A contemporary publication a few months ago made the statement that it did not consider it wise to publish a list of Government and commercial stations, with their call letters, etc., as this would only

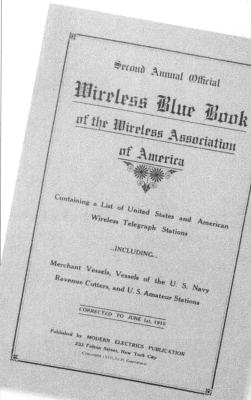

Second annual *Wireless Blue Book.*
Courtesy of the collection of Jim and Felicia Kreuzer, Grand Island, New York.

8. The *Blue Book* was the culmination of Gernsback's calls for greater accountability in **The Wireless Joker** in this book and one of the first radio directories in the country.

tend to put the experimenter in a position where he could successfully annoy the large stations. This statement is not very flattering to the mentioned publication's readers. It seems to have very little confidence in its own readers and their honesty.

The world has a right to be enlightened. There can be no progress without enlightenment. Any publication keeping valuable and enlightening news from its readers is not working in the readers' interests.

The wireless situation as it stands today, considered from the experimenter's standpoint, may be understood better by an analogy.

Take a blind man who only understands English. Let him walk over a dangerous road full of deep holes. The Italian laborer may vainly shout in Italian: "Look out, you will fall!" but the unfortunate blind man, while he hears the shout, of course, pays no heed—and falls in the first hole he encounters. *He heard, but did not understand.*

The same with the wireless experimenter. The large stations may vainly shout at him to stop sending. He hears but does not understand. He doesn't even know *who* shouts at him. He is far worse off than the blind man. If he knew *who* was doing the shouting he would more likely pay heed and respect the station whom he disturbed.

And still some people wish to keep the student in the dark, to refrain him from doing mischief! Sancta Simplicitas!

Television and the Telephot

Modern Electrics, vol. 2, no. 9, December 1909

Every now and then we see newspaper reports that Mr. So and So has discovered the real secret of television, only to be told again a few weeks afterwards that it has not been realized after all.[1]

For 25 years almost, inventors all over the world have been working strenuously to solve the problem, but so far none succeeded, apparently because they all seem to work along wrong lines.

The principle of television may be briefly stated thus: A simple instrument should be invented which should reproduce objects placed in front of a similar instrument (called a Telephot) at the other end of the line. In simple language, it should be possible to connect two mirrors electrically, so that one would show whatever object is placed before the other one and vice versa.

As in a mirror, the objects must be reproduced in motion (at the far-off station). The theory further requires that both instruments (one at each end) must be reversible, that is, each instrument must receive as well as transmit.[2]

A good parallel of this requirement is found in the ordinary Bell telephone receiver. As is known, the Bell receiver (without the use of a microphone transmitter) will receive as well as transmit, that is, one

1. This article contains the first mention of television in the Gernsback magazines, a favorite topic of his writers over the next few decades. Surveying the latest technical approaches to transmitting images across a distance, Gernsback introduces a system by Berlin-based technologist Ernst Ruhmer, one which bears more of a technical resemblance to today's liquid crystal displays (LCDs) than the electromechanical Nipkow disk scanners common at the time. In the photograph, Ruhmer displays a crude prototype with a 5x5 pixel cross transmitted from one display to another. Ruhmer's system used light-sensitive selenium cells arranged in a mosaic, transmitting differences in light intensity through variable current strengths. For more on Ruhmer, see Archie Frederick Collins, *Wireless Telegraphy: Its History, Theory, and Practice* (New York: McGraw Publishing Company, 1905), which contains a section on Ruhmer's "photo-electric telephone."

2. Although many point to this article as the first either to coin the term "television" or explain the concept to the layman (for instance Paul O'Neil, "The Amazing Hugo Gernsback, Prophet of Science, Barnum of the Space Age," *LIFE Magazine* 55, no. 4 [July 1963]: 63–68), the technology had been envisioned much earlier and explained in similarly generalist terms. Most historians date the earliest depiction of television *avant la lettre* to a George du Maurier cartoon in the December 9, 1878, issue of *Punch*. The cartoon, "Edison's Telephonoscope" was a prediction for the following year, part of *Punch*'s "Almanack for 1879." Friedrich Kittler dates the coinage of a name for this medium to Raphael Eduard Liesegang's 1891 book *Beiträge zum Problem des elektrischen Fernsehen* [Contributions to the problem of electrical television]. For more on the nineteenth-century origins of electric image-scanning techniques, see Friedrich A. Kittler, *Optical Media: Berlin Lectures 1999*, trans. Anthony Enns (Cambridge, U.K.: Polity Press, 2010), 208–12. Further explications of the principle of television for a popular audience in the German context can be found in the writings of the technologist Eugen Nesper, *Das elektrische Fernsehen und das*

**TELEVISION
—and—
THE TELEPHOT**

SEE PAGE 12

May 1918 cover of *Electrical
Experimenter.*

can talk in a receiver and also hear the other party, using one and the same instrument.

In the Telephot it should be possible to see the party at the other end while that party should see you, both through the medium of your Telephot.

Unlike the mirror, however, you should not be able to see your own picture in your own Telephot. In this the Telephot differs from the mirror analogy.

From this it will be seen immediately how difficult the problem becomes, as if you could see yourself in your own Telephot, as well as the picture of your friend, it is obvious that there would be a "mix-up" of personalities, the consequence being that you could not recognize your friend nor yourself, while your friend at the other end could of course not recognize you nor himself.

In the telephone the case is not so difficult, as it is absolutely necessary that one party talks while the other listens; if both talk and listen, none can understand, as the voices mix up.

In the Telephot this parallel does not hold good, as there is nothing to restrain you from looking at your friend at the same time he is looking at you.

Of course the problem can be simplified by getting the true parallel of the telephone, thus: When you wish to see A you keep in the dark, while A stands in full light. If A wants to see you he turns off his light while you switch on yours.

"Every evening, before going to bed, Pater- and Materfamilias set up an electric camera-obscura over their bedroom mantel-piece, and gladden their eyes with the sight of their children at the Antipodes, and converse gaily with them through the wire." Cartoon by George du Maurier, from *Punch,* December 9, 1878.

Telehor (Berlin: Verlag von M. Krayn, 1923). Nesper was a frequent contributor to the magazine *Der Radio-Amateur* and reviewed several of Gernsback's devices in this publication.

EDISON'S TELEPHONOSCOPE (TRANSMITS LIGHT AS WELL AS SOUND).
(*Every evening, before going to bed, Pater- and Materfamilias set up an electric camera-obscura over their bedroom mantel-piece, and converse gaily with them through the wire, and gladden their eyes with the sight of their Children at the Antipodes, and converse gaily with them through the wire.*)
Paterfamilias (in Wilton Place). "BEATRICE, COME CLOSER, I WANT TO WHISPER." *Beatrice (from Ceylon).* "YES, PAPA DEAR."
Paterfamilias. "WHO IS THAT CHARMING YOUNG LADY PLAYING ON CHARLIE'S SIDE?"
Beatrice. "SHE'S JUST COME OVER FROM ENGLAND, PAPA. I'LL INTRODUCE YOU TO HER AS SOON AS THE GAME'S OVER?"

However, this would be impracticable and is not the true solution of the Telephot.

So far most inventors seem to think that the problem can only be solved by means of the selenium cell, which being sensitive to light, can send out electrical impulses in the same ratio as the light falling upon the cell. Thus, if a strong light is thrown on a selenium cell a strong electric impulse is sent over the line which when operating a light relay (described below) can be made to throw a strong light upon a screen.

As a picture is made up of nothing but light and dark points it is easily to be seen that if several thousand very small selenium cells were arranged in a plane and just as many light relays at the other end, a good picture could be projected upon a screen—in theory. The trouble is that it is practically impossible to make two selenium cells with equal sensitiveness and just this is the most important part, as if one is not as sensitive as the other, it will of course not transmit the same impulse as the former. It can be imagined easily what kind of a picture a station would transmit having several thousand selenium cells, all of a different sensitiveness!

Then the next trouble is that each cell at best requires one wire (the ground might be used as return).[3] Think of two stations which, in order to work, require 3,000 to 5,000 separate wires! This seems to be as bad or worse than Sömmering's first telegraph (in 1809), which required 27 wires to operate. In Morse's subsequent telegraph only one wire is required, which unquestionably will be the case with the perfect Telephot.

Another great trouble with the selenium cell is that it works sluggishly, that is, its resistance will not drop instantaneously from the highest value to the lowest, which is a bad feature, as it would necessarily blur the picture at the other end. Furthermore, to work anywhere satisfactorily the selenium cell requires strong light.

The writer does not wish to throw cold water on selenium and selenium cells, as it is quite possible that the latter may be improved to such an extent as to do entirely away with the shortcomings mentioned above, although the greatest difficulty, the one that each cell requires at least one wire, is and will be the far greatest stumbling block.

Many different systems have been proposed in the past to solve the problem by means of selenium cells and although the list is quite long only a few will be mentioned in this article, as all systems are more or less on the same lines.

A. Knothe proposes to solve the problem as follows: C (Fig. 2), represents a camera into which the lines coming from the batteries enter. The space between each pair of wires is bridged by a selenium cell S.

3. The greatest benefit of Ruhmer's set-up is that the entire image is transmitted across a single wire, as opposed to one wire for each individual pixel, referred to in the article as "raster." This multiplexing was a unique solution to the problem of translating a two-dimensional image into a one-dimensional electrical current, which was seen as the biggest obstacle to television at that point.

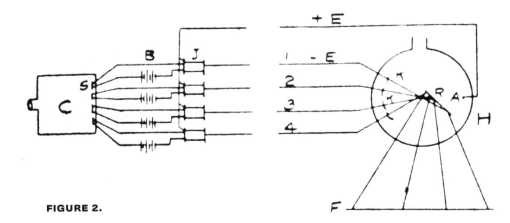

FIGURE 2.

4. The majority of Ernst Ruhmer's gorgeously illustrated book, *Wireless Telephony, in Theory and Practice*, trans. James Erskine-Murray (London: C. Lockwood; Sons, 1908), deals with the technical requirements, limitations, and possibilities of *image* transmission, rather than audio.

5. Dada artist Raoul Hausmann, in a letter to the art historian Yves Poupard-Lieussou, mentions that a 1901 book by Ruhmer on electro-acoustic phenomena was influential to his experiments with phonetic poetry using the magnetophone in the 1910s (Raoul Hausmann to Yves Poupard-Lieussou, November 25, 1959, Yves Poupard-Lieussou correspondence and collected papers on Dada and Surrealism, 1905–1984, Box 9, Folder 13, Getty Research Institute). Hausmann's most famous work, the "Mechanical Head [The Spirit of Our Time]" (circa 1920), shows a subject whose perception is determined by the mechanical apparatuses affixed to it.

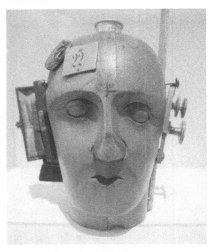

Raoul Hausmann's "Mechanical Head" (1920). Photograph by Pablo Ibañez. Reprinted under a Creative Commons Attribution 2.0 Generic license.

If now light enters the camera it falls on S (and all the other cells), and closes the current which operates the spark coils J. This furnishes a discharge as a single ray in the X-Ray pipe H at the receiving station E. This single ray is thrown as a single point on the fluorescent screen F.

It is understood that several hundred cells, spark coils and parabolic mirrors K are necessary to transmit a picture. The X-Ray tube would therefore necessarily be of monstrous dimensions. All the wires, 1, 2, 3, 4, up to several hundred, must of course, be well insulated, so that no sparking occurs between them. All these requirements make the arrangement almost impossible and quite impracticable.

The latest "Telephot" has been designed by Mr. Ruhmer, the well-known Berlin expert. Last June Mr. Ruhmer demonstrated a working model, which although it did not transmit pictures, served well to demonstrate the usefulness of the selenium cell for certain purposes.[4]

Fig. 2 shows the model clearly. The principle is as follows:

The transmitter has 25 squares, each containing a selenium cell. If any one of the 25 cells is exposed to light, it operates a sensitive relay, which sends an alternating current of a certain frequency over the line.

At the receiving end one resonating relay is stationed for each selenium cell at the sending station. The impulse sent from the selenium cell therefore operates only that relay having the right frequency.

Each relay operates an incandescent lamp which is placed in the same square at the receiver as the selenium cell at the transmitter.

If several cells are exposed to light at the transmitter, several alternating currents, but all of different frequencies are sent over the line. These currents do not mix, but operate only the relays for which they are intended.

These in turn operate the lamps in the various squares, assigned to them.

Mr. Ruhmer has perfected the selenium cell a good deal, and as his model worked very rapidly, it will be seen that the sluggishness of the cells has been overcome to a certain extent.[5]

Simple geometric figures were transmitted quite successfully as can be seen in the photograph, where 9 squares at the transmitter were lighted and the same amount, in the same position were reproduced at the receiver. This is quite remarkable if it is remembered that only *one wire* is used between transmitter and receiver.

Mr. Ruhmer intends to build a transmitter containing 10,000 cells, to reproduce pictures at the Brussels international exposition in 1910. The cost will be over one and a quarter million dollars and the writer is of the opinion that it is almost impossible to operate such a model on account of the 10,000 different frequencies necessary to accomplish the result.[6]

A simpler way could be brought about by the idea proposed by the writer some eight years ago.

Fig. 3 represents the well-known electrical harmonica, which for the sake of those not knowing the instrument, is described herewith:

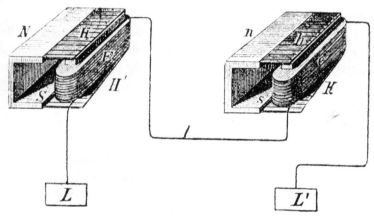

FIGURE 3.

A musical steel harp H is fastened to a permanent magnet NS. If any one of the steel harp-prongs is touched it will swing back and forward, at the same time sending an induced current through the windings of the electromagnet E. If we connect a similar instrument H through the line l, and ground LL' to H' it is evident that if we touch any of the steel prongs of one of the instruments the same steel prong on the other will be made to swing. If we have 12 prongs on each instrument and we touch prongs No. 1, 6, 9, 12 of H, all at the same moment, prongs No. 1, 6, 9, 12 of H' will be made to sound at the same time too, and so on.

Suppose we build such a harmonica having, say, 500 prongs P, Fig. 4, each responding readily at an extremely light touch.

Exactly over each of the 500 prongs we place a minute electromagnet E, 500 in all (only 6 shown in illustration), so when one of the small

6. A reporter for *Scientific American* visited Ruhmer's laboratory earlier that year, detailing his preparations for the Brussels Exposition: "Mr. Ernest Ruhmer, of Berlin, well known for his inventions in the field of wireless telephony and telegraphy, has succeeded in perfecting what is probably the first demonstration apparatus which may be said actually to solve the problem of tele-vision. [. . .] In fact, a complete and definite tele-vision apparatus, costing the trifling sum of one and a quarter million dollars, is to be the *clou* of this exposition." Daniel C. Schlenoff, "50, 100, and 150 Years Ago: Innovation and Discovery as Chronicled in Scientific American," *Scientific American* 301 (August 2009): 13.

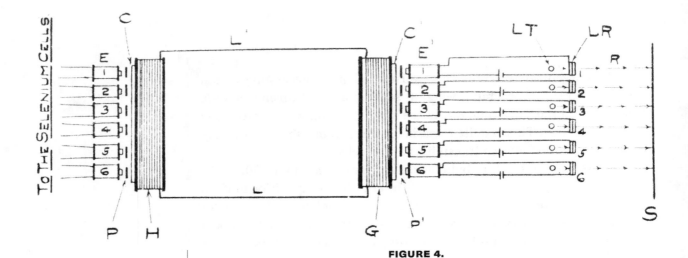

FIGURE 4.

electromagnets is actioned by means of a weak alternating current flowing through same it will cause the prong underneath it to swing as long as current flows through the electromagnet.

Now each of the small electromagnets is connected to a selenium cell of which 500 are placed in a plane.

It will be easily seen that if one or more of the selenium cells are acted upon by light, one or more of the small electromagnets is acted upon AND AS A PROPORTIONATE AMOUNT OF CURRENT IN PROPORTION TO THE INTENSITY OF LIGHT at the selenium cell flows through the small electromagnet, or electromagnets, it will cause the prong or prongs to vibrate IN THE SAME PROPORTION OF INTENSITY as the light falling on the selenium cell.

Thus if cell No. 1 is lighted with 10 C. P., assume that the small electromagnet connected to it causes its prong to swing through the distance of one millimeter. Then if cell No. 50 is only lighted with the intensity of 1 C. P., prong No. 50 will of course only swing $\frac{1}{10}$ millimeter and so on. Thus each prong will be caused to swing in exactly the same proportion as the amount of light falling upon the selenium cell, to which it belongs.

As each prong swings it sends a current over the line L' L". If now No. 1 and No. 6 of the electromagnets are energised through the selenium cells both cause their prongs to swing and send impulses over the line. At the receiving station G, prongs 1 and 6 must swing IN THE SAME PROPORTIONATE INTENSITY as the prongs at the sender H, consequently electromagnets E'1 and E'6 are energised (the prong acting as a telephone diaphragm, the electromagnets having for core a permanent magnet). E'1 and E'6 now operate the light relays LRl and LR6.

Now, then, if the selenium cell connected to El is lighted with, say, 10 C. P., a proportionate amount of energy—call it 10 energies— are received at LR1. The light relay therefore passes 10 energies of the small tungsten lamp TL through its opening, and 10 energies are projected on the screen S. If the cell connected to E6 receives the light of 100 C. P. it is evident that LR6 receives 100 energies and the screen is lighted with 100 energies and so on.[7]

Thus it will be seen that if we have enough selenium cells at H and enough light relays at G any picture in motion will be transmitted correctly and reproduced in its true phases on the screen S. It is only a matter of building the apparatus and instruments with sufficient precision.

The light-relay used in some of the writer's experiments is described herewith:

A light-relay is an instrument which has the purpose to utilize very weak electric impulses to throw a beam of light on a screen, which in intensity is proportionate to the strength of the electric impulses. In other words, if the impulses are strong, a large amount of light is caused to fall on the screen; if the impulses are weak, a small amount of light falls on the screen, and so on.

Fig. 5 shows the instrument in the perspective. Between the poles of a strong electromagnet NS, two extremely fine metal wires A and A'. are stretched. The wires may be stretched more or less by the regulating crews o and o'.

The two poles N and S, are each provided with a hole O and O', through which light rays are sent in the direction p, p'. On the two wires, A and A', a very light piece of aluminum foil B is attached in such a way that no light can pass from O to O' normally.

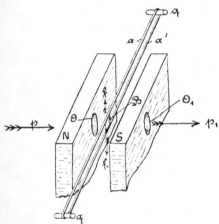

FIGURE 5.

If, however, a weak current passes from a to a', the aluminum foil is deviated in the direction f or f', as the case may be. In order to obtain very exact motions of the foil the thin wires are best replaced by fine metal bands 0.01 millimeters thick, 0.25 millimeters wide and about 6 centimeters long. The resistance of each band is about 7.5 ohms.

As far as the writer knows his plan so far is the only feasible one which can be used to transmit objects in motion over a single wire and at the same time receive a proportionate amount of energy at the receiving end to that received by the selenium cells at the transmitter.

No patents were taken out on this invention by the writer, as he considers the device too complicated for general use. He shall, however, consider himself happy if it will be the means to bring us nearer to the practical Telephot.

7. In a 1918 article on the telephot, Gernsback describes how successful two-way moving image communication would require both speakers' faces to be intensely lit in order for "light-relays" to be able to sense variations in the depicted subject: "In order that the distant person may see the speaker's face, it is of course necessary that the latter's face be illuminated. For it goes without saying that if the speaker was in the dark, his friend could not possibly see him on the other side because no light impulses would be thrown on the 'sending' lens. For this reason it will be necessary to provide a lamp R at the top of the Telephot, which lamp throws its rays on the speaker's face; from here the light rays are thrown onto the lens, thence to be transmitted to the distant station." Hugo Gernsback, "Television and the Telephot," *Electrical Experimenter* 6, no. 1 (May 1918). Just five years earlier, Gernsback had patented a "luminous electric mirror" whose design mimicked that of his speculative prototypes for the telephot. Hugo Gernsback, "Luminous Electric Mirror," April 1913, http://www.google.com/patents/US1057820.

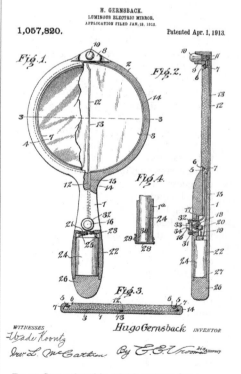

From Gernsback's 1913 patent for a "luminous electric mirror."

The Roberts Wireless Bill

Modern Electrics, vol. 2, no. 10, January 1910

The long threatened wireless bill has made its appearance at last.

The resolution, as introduced by Representative Roberts of Massachusetts, is reprinted in detail on the following page and should be carefully read by every one who has wireless progress at heart.[1] Personally, the Editor believes that there is no need of a wireless telegraph board.

It is of no practical value whatsoever, un-American, and will keep down the progress of a young and useful art, which in time may develop into an as yet undreamed-of asset of the nations' power.

Wireless telegraphy and telephony, in a country of such vast distances as America is a very valuable means for cheap transmission of intelligence, and it is the duty of the Government to encourage it, and not to pass a resolution to throttle it like England and Germany have done, in which two countries the art is almost unknown.[2]

It would be deplorable indeed to see Representative Roberts' resolution passed. The farmer, who three years hence will be in a position to own his wireless telephone to call up his next neighbor fifty miles distant from him, will much rather install his private wireless 'phone, rather than be forced to subscribe to an exorbitant rent of an instrument owned and controlled by the United Wireless Company or some other wireless trust, to which trusts such a resolution would give full swing to extort high rates.[3]

At first sight Representative Roberts' resolution appears very tame and gentle, but men acquainted with modern methods at Washington know full well what the "recommendations (!) to Congress" mean, with the big wireless interests dictating the "recommendations."

Despite the present telephone interests the farmer is allowed to put up his personal telephone line from his house to that of his neighbor's. If the national wireless board comes into power, the same farmer would undoubtedly not be allowed to operate a private wireless telephone between his and his neighbor's house.

As far as wireless telegraphy is concerned, it is ridiculous to maintain now that the amateur can interfere with the business of commercial stations.

With the present efficient weeding out tuners, loose couplers, variable condensers, etc., the amateur can no more interfere with the

1. Congressman Ernest W. Roberts of Massachusetts's 7th District introduced his legislation on wireless regulation as House Joint Resolution 95. Although Roberts's proposals did not go far, aspects of this Resolution were incorporated into the Radio Acts of 1910 and 1912.

2. For more on the development of wireless telegraphy in Great Britain and Germany (where it was largely controlled by the private company Telefunken), see Jonathan Reed Winkler, "Information Warfare in World War I," *Journal of Military History* 73, no. 3 (2009): 845–67; Heidi Evans, "'The Path to Freedom'? Transocean and Wireless Telegraphy, 1914–1922," *Historical Social Research* 35, no. 1 (2010): 209–36.

3. Jonathan Sterne writes on the postwar politics of rural access to a later medium—broadcast television—which encountered significant difficulties in particular geographic locations: "Once potential audience members saw an existing system and its limits, they demanded access on the principle that television was a kind of infrastructure, like roads or utilities. Issues of access and entitlement to television thus grew in importance precisely at the areas where profitability was most limited" (47). Gernsback's emphatic declaration of wireless as a "utility" is thus an accurate depiction of how Americans would later identify themselves within a national network of communications relays. Jonathan Sterne, "Television under Construction: American Television and the Problem of Distribution, 1926–1962," in *Television: Critical Concepts in Media and Cultural Studies*, vol. 1, ed. Toby Miller (New York; London: Routledge, 2003).

commercial or government stations than the transatlantic liners—equipped with powerful apparatus—can interfere with the messages flashed from coast to coast.

The trouble is, that the majority of commercial and government stations have antiquated instruments, and do not care to acquire new ones. Their operators are almost entirely wire telegraph men who have not the slightest idea of wireless, nor are they interested in it. The Editor, who is personally acquainted with over twentyfive [sic] such operators was amazed to find that *not four of them could draw a diagram how* their instruments were connected up.[4]

All their shortcomings are blamed on the innocent amateur, whose weak spark cannot be heard half a mile, as a rule, and the manager of the station of course takes the word of the operator every time.

There are to-day over sixty thousand experimental and amateur wireless stations in the United States alone.

That means that over sixty thousand young aspiring men stay at home evenings, enjoying an innocent sport, instead of dissipating outside in a questionable pastime.

We have as yet to find the father who objects to his son's "wireless." He knows it keeps the boy at home, away from mischief.

The Editor sounds a general call, and asks everyone to whom wireless is at heart, to send him at once a letter of protest against the wireless resolution. State in your letter, before all, the UTILITY of your wireless. These letters, in mass, will be presented in Washington, to the proper officials.

All letters must be received not later than January 25th. Act at once!

National Wireless Telegraph Board Proposed

Representative Roberts, of Massachusetts, has introduced a resolution in the House at Washington providing for the creation of a wireless telegraph board. Mr. Roberts said that there is the greatest need for such control, as he has information from the Navy Department, the revenue cutter service and the commercial wireless companies that the effect of the activities of amateur operators has been such as not only to make necessary a change from "C Q D" as the distress signal, but to interfere seriously with the operations of all Governmental and private services. As a result of these reports, Mr. Roberts, who is a member of the House Committee on Naval Affairs, considers it high time to take cognizance of the situation.

The perfection of wireless apparatus has reached such a stage,

4. A letter from a reader published in the June 1909 issue of *Modern Electrics* explains that it is no surprise for government stations to experience interference when they are using apparatuses that are "three years behind time." It goes on to say, "The experienced amateurs claim that the interference is generally caused by carelessness of the Government operators or by inexperienced amateurs. Some of these amateurs, when told to 'get out' by the navy operators, stay out for a time, but the average, after waiting half an hour, generally gets impatient and starts sending again. I hear more forcible conversations take place between them. The navy or army operators sometimes press down their keys for several minutes and drown out all sending which is being done in that range. Here the amateur steps in and says that if the Government operators cannot overcome amateur interference what will become of them in time of war when high powered stations of the enemy put the naval apparatus out of commission by simply holding down their keys?" John Crockford, "Amateur Defense of Interference," *Modern Electrics* 2, no. 3 (June 1909): 113.

he said, that if the service is to be permitted to grow unchecked it is absolutely essential that the Government take steps in the matter. The simplest solution of the matter lies in the passage of the resolution presented, or a measure of similar character, placing in the hands of a wireless board the control of wireless plants afloat and ashore. It has been brought to his attention in an official way that the wireless service of the navy has been rendered practically useless at times by amateur operators, who send meaningless and oftentimes vile and unmentionable language through the air from their instruments.

Mr. Roberts' resolution authorizes the appointment of a board of seven members, "one expert each from the War, Navy and Treasury departments, three experts representing the commercial wireless-telegraph and wireless-telephone interests, and one scientist well versed in the art of electric wave telegraphy and telephony."

The duties of the board, according to the resolution, shall be "to prepare a comprehensive system of regulations to govern the operation of all wireless plants afloat and ashore which come under the cognizance of the United States, with due regard alike for Government and commercial interests."

It is provided that within 30 days of the organization of the board it shall submit its report and recommendations to Congress. To defray the expenses of the board $2,000 is appropriated.[5]

W. A. O. A.

The Wireless Association of America was founded solely to advance wireless. IT IS NOT A MONEY MAKING ORGANIZATION. Congress threatens to pass a law to license all wireless stations. The W. A. O. A. already has over 3,000 members—the largest wireless organization in the world. When the time for action arrives, the thousands of members will exert a powerful pressure to oppose the "wireless license" bill. This is *one* of the purposes of the W. A. O. A. There are more.

5. The following month's editorial contains a roundup of the overwhelmingly positive responses Gernsback received, including "9,000 protest letters from experimenters and amateurs all over the United States," and sympathetic editorials published in *New York Evening World*, *New York American*, *New York Sun*, *Boston Transcript*, *New York Independent*, and *New York World*, the latter of which adds that "wireless communication itself is the invention of an amateur." By the April issue, which marked the third year of *Modern Electrics*'s run and a readership of 30,000, Gernsback was able to endorse a new proposal for wireless registration, the Burke Bill: "Representative [James F.] Burke's bill does not impose any license fees for operating a station, as the Roberts bill. It simply calls for registration of stations by the Secretary of Commerce—a law which when passed will certainly be just and fair to all. It will also do away with the 'Wireless Tramp,' and mischief maker, who should be banished from the map with all possible speed." The entire text of the bill was published on the next page.

FIGURE 1. "How to Make a Wireless Bicycle Outfit," submitted to the September 1910 issue by reader William Dettmer.

From *The Wireless Telephone*

1911

A Treatise on the Low Power Wireless Telephone, Describing all the Present Systems and Inventions of the New Art. Written for the Student and Experimenter and those engaged in Research Work in Wireless Telephony.

Preface

The present little volume is intended for the experimenter doing research work in wireless telephony and the student who wishes to keep abreast with the youngest branch of the new wireless art.[1]

The author realizes that the future use of the wireless telephone will be confined to the low power or battery system, as the present instruments, necessitating 220 and 550 volts for their successful operation, are not desirable nor practical enough for every day use.

The wireless telephone of the future must be as flexible as the wire telephone of today.

Every farmer will be able to operate his wireless telephone, when the sending and receiving instruments will be housed in a box a foot square, without depending on the lighting current for its operation. The author predicts that in less than 10 years this stage will have been reached as it is bound to come sooner or later.

Quite a little new matter will be found in these pages and while some old matter has necessarily appeared for the sake of completeness of the book, the author trusts that the necessity of reviewing such matter will be apparent.

The author shall feel happy if this little volume will be the cause to advance the new art if ever so little, and he will be pleased to bear honest criticism and suggestions as to the contents of the book.

H. GERNSBACK.
FEBRUARY 1910

Advertisement for *The Wireless Telephone* in the January 1911 issue of *Modern Electrics*.

1. This text is excerpted from a slim, seventy-three-page book published by Gernsback's Modern Publishing Company. At a time in which the wireless transmission of audio (referred to here as "the wireless telephone") was still workable only in limited experimental demonstrations, here was a book that introduced a popular audience to a speculative medium of the future by encouraging them to build it.

Chapter III: Early Experiments

Wireless telephony by electromagnetic induction is not a new invention. It has been known for over 50 years, but its use is very limited.

One of the best and simplest devices which may be used to transmit speech over small distances up to 50 feet (through stone or wood walls) can be made as follows:

How to Make a Simple Wireless Telephone

On the wall or on a large table tack 6 or 12 nails in form of a circle of 5 feet diameter, fig. 16.

Around the nails wind 80 turns of No. 28 B. & S. wire, enameled wire preferable.[2] Bring out two leads as shown. When completed, wind tape around the coil so the turns will not unwind. Two layers of tape are advisable as the finished coil will be somewhat stiffer then.

This coil is the secondary.

On the same nails wind 40 turns of No. 18 B. & S. wire and finish it with tape the same as secondary.

FIGURE 16.

This coil is the primary.

Two portable wood stands should then be built of wood laths as per dimensions shown in fig. 17. A shelf is also provided on which to place the instruments.

Each wire coil is attached to stand as shown in dotted lines.

Fig. 18 shows the connections.

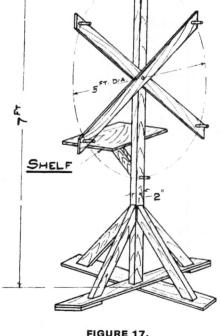

FIGURE 17.

2. B. & S. refers to the size of the wire. The Brown & Sharpe metal gauge favored by early electrical engineers and contractors to measure the diameter of a wire is still in use today.

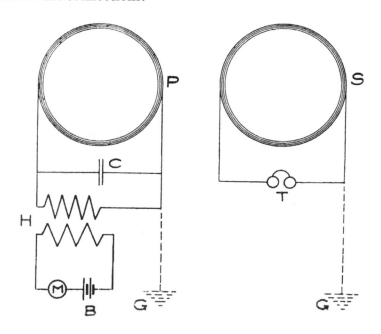

FIGURE 18.

P is the primary coil, M a common microphone transmitter, B 6-10 cells, C a condenser composed of 50 sheets of tinfoil 5x3 inches with parafined paper between the sheets.[3]

H is a medium size medical coil (without vibrator). The primary is connected with M-B, the secondary with C-P.

S is the secondary, to the ends of which is connected an ordinary telephone receiver, or better, a set of double head receivers similar to those used in wireless work.

It is important that both coils face each other, as they do not work when placed otherwise or at too great an angle.

It has been found that the ground G (shown in dotted lines) sometimes helps to increase the sound, this also is the case of condenser C, which improves the whole arrangement a good deal. This arrangement has little value and is only used for demonstration purposes, and works up to 50 feet.

The use of an aerial does not seem to improve it.

The outfit may be regulated by cutting in more or less battery at B till best results are obtained at the receiving end.

Closed Circuit Wireless Telephone

We must also mention another method of wireless telephony, namely, the closed circuit wireless telephone.

Like the induction telephone, the closed circuit systems are of little value, the greatest distance so far covered being 3½ miles. Fig. 19 shows the sender. Two copper plates 2 feet square are sunk in the ground about 4 feet deep. The plates must be sunk in edgewise at right angles with the surface of the earth as shown in illustration.

To telephone up to a distance of 500 yards, the two plates must be 25 yards apart from each other, for shorter or greater distances, the plates must be separated proportionately to above figures.

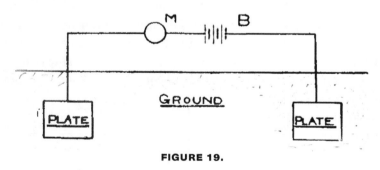

FIGURE 19.

M is the microphone, B the battery.

The receiver is shown in fig. 20. T is a telephone receiver of 50 or 75 Ohms. The lower the resistance the better the results. Two copper

3. While a condenser is similar to a battery in that it is used to generate voltage, the comparison ends there. Batteries create their voltage through chemical reactions between "wet" or "dry" elements until all of these reagents have been exhausted. A condenser, on the other hand (later known as a capacitor) temporarily stores a variable quantity of voltage that is directly applied to it.

plates are provided again, of the same size as those of the sender and spaced the same distance apart as those of the sending station.

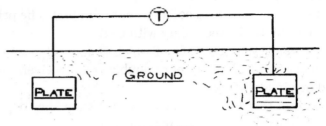

FIGURE 20.

It is very important that the two plates of the sender face the two plates of the receiving end, else no results will be had.

This system may be used successfully to telephone between two houses and will give the experimenters quite a little diversion.

Author's Experiments

The author, in 1903, was able to telephone over a distance of 3 miles by using the arrangement as shown in figs. 21 and 22.

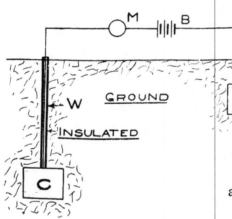

FIGURE 21.

Two plates, fig. 21, 3 feet square, one of copper and the other of zinc, were sunk in the ground in such a manner that C was buried 15 feet deep, Z only 3 feet below the surface. Microphone M and battery B were provided as shown. At the receiving side, two similar plates of similar size C' and Z', were sunk in the same fashion, the only difference being that the plate at the right was buried 15 feet, while that at the left only 3 feet below the surface.

The plates of the sender face those of the receiver.

The perpendicular distance of C and Z was 300 yards, that of Z' and C' the same.

The wires, W and W', leading to C and C' were carefully insulated and around all plates a saturated solution of chloride of zinc was poured. This had a double effect. Firstly, it set up an electromotive force between all the plates, and, secondly, chloride of zinc being very hygroscopic, it kept the earth moist and conductive around the plates.

The system proved very successful and was in use for over 8 months.

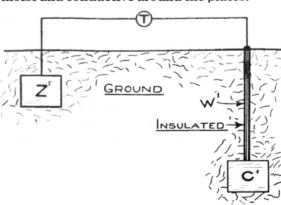

FIGURE 22.

Armstrong & Orling System

Another method to telephone wirelessly by the conductive method is described in U. S. Patent No. 744,001 by J. T. Armstrong and A. Orling.

This invention relates to means and apparatus for the transmission of speech and other articulate sounds to a distant receiver or receivers without the employment of wire or other like connection between the transmitter and the receiver, and has for its object the improvements hereinafter set forth.

In carrying out this invention the inventors provide the transmitting apparatus with two or more earth connections, through which are conducted a combination of high-potential discharges and low tension currents whose circuit or circuits are completed through numerous lines of current-flow which traverse the earth. The transmitted impulses enter the earth by one of the same earth connections and after traversing the same return to complete the other. This receiving apparatus (hereinafter described) is also provided with two or more earth connections which are adapted to cut the said lines of current-flow (in the neighborhood of the distant receiver) at points of different potential, causing some of the transmitted energy to pass through them and to actuate the receiver.

Fig. 22a shows one form of transmitter. Fig 22b shows a modified form thereof, and Figs. 22c and 22d are respectively different forms of receiving apparatus adapted for use with the transmitter.

According to the arrangement shown in Fig. 22a, they employ an electric circuit a, which is provided with a battery b, or other convenient source of energy, a self-inductance coil c, a microphonic or any other suitable telephone-transmitter d^x, and a switch e^x which is

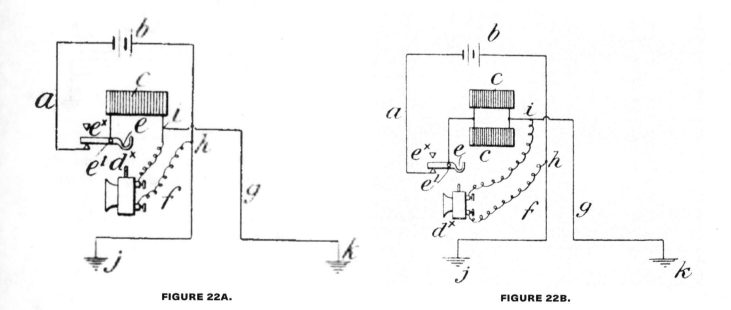

FIGURE 22A.   FIGURE 22B.

pivoted at e[1] and provided with a hook e[z], upon which the transmitter d[x] is adapted to hang when not in use, so that its weight may cause the said switch to open the circuit. Suitable connections f and g are made with the circuit a at h and at i on opposite sides of the transmitter d[x] and are respectively led to earth at j and k in such a manner that the earth is in shunt-circuit.

According to the construction shown in Fig. 22b, a plurality of self-inductance coils c is employed.

In operation the varying resistance of the transmitter d[x] causes a fluctuating current to pass through the coil or coils c, with the result that "extra" currents are induced, the circuit of which is completed through the transmitter d[x] and the earth, which is in shunt thereto. At each increase in the resistance of the transmitter d[x] a larger proportion of the low-tension-battery current will pass through the earth with a large proportion of the induced direct high-potential extra current. At each decrease in the resistance of the transmitter d[x] a larger proportion of the battery-current will pass through the coil or coils c and a large proportion of the induced inverse high-potential extra current will pass through the earth, owing to the resistance of the transmitter d[x] and the opposing electromotive force of the battery-current. The electrical effects that are thus led to earth at j and k (which connections are preferably as far apart as practicable when communication is carried on over great distances) set up lines of current-flow which extend to very considerable distances and are intercepted in the neighborhood of the receiver (see Figs. 22c and 22d) by its earth connections j[x] and k[x]. These connections, with the receiving part o of the circuit constitute a species of shunt through which some of the transmitted energy flows.

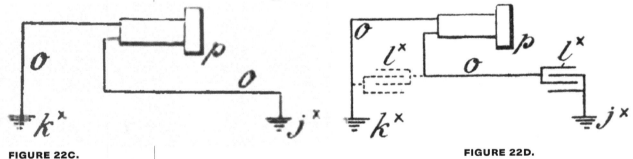

FIGURE 22C.   **FIGURE 22D.**

The receiving apparatus consists of a telephone p, by means of which the passage of the transmitted energy through the receiving shunt o may be detected.

In some cases they employ a condenser l[x], as shown either in full or dotted lines in Fig. 22d, to obviate the effect of earth-currents.[4]

4. The remainder of this chapter describes the construction of a wireless apparatus patented in 1903 by James T. Armstrong and Axel Orling, "System of Telephonic Communication," November 1903.

THE PERVERSITY OF THINGS

The Born and the Mechanical Inventor

Modern Electrics, vol. 3, no. 11, February 1911

We have spoken several times of inventions and the present editorial is one on cultivating inventions.

Generally speaking, there are two kinds of inventing. One is achieved by the man who can't help inventing new things; he is a born inventor, just like a musician to whom music comes naturally and who never needed a teacher. The second kind of inventing is the mechanical kind. By this is meant the kind of inventing that is done by persons to whom inventing does not come natural, but those who are suddenly confronted with a certain device that to their minds seems imperfect, whereupon they will bend their energies towards improving the existing device.

The born inventor differs from the mechanical one in that to him ideas come suddenly, without the slightest suggestion. The writer, for instance, who has invented a number of devices will suddenly have a certain idea, of which he perhaps never thought before. Within less than ten seconds after the first impression of the idea, he will see the entire device, down to the smallest detail clearly before his mental eye and from that mental picture a complicated sketch of the device can be drawn immediately without reflection or real thinking; just like copying an existing drawing.[1]

---

1. In his depiction of invention by those born to do so as a form of individual creativity and expression, Gernsback overlooks the importance of cultural forms, social groups, and institutions in deciding the legibility of the problem an invention is designed to solve, as well as the available means for solving it. Thanks to the scholarship in science, technology, and society (STS), we now emphasize the ways technology is socially constructed, rather than celebrating the creativity of the lone genius inventor. Any invention must take place within the context of what Wiebe Bijker calls a technological frame, one model that STS has produced to explain how "inventions" become possible. A technological frame consists of "the concepts and techniques employed by a community in its problem solving. Problem solving should be read as a broad concept, encompassing within it the recognition of what counts as a problem as well as the strategies available for solving the problems and the requirements a solution has to meet" (168). For instance, the technological frame of Bakelite plastic (Bijker's example in the above-quoted essay)—its "theories, tacit knowledge, engineering practice (such as design methods and criteria), specialized testing procedures, goals, and handling and using practice" (168)—didn't come into existence until the 1910s, even though, conceivably, it was a *material* possibility as early as the 1870s. Wiebe E. Bijker, "The Social Construction of Bakelite: Toward a Theory of Invention," in *The Social Construction of Technological Systems: New Directions in the Sociology and History of Technology*, 159–89 (1987; repr., Cambridge, Mass.: MIT Press, 2012).

Despite this, Gernsback's stated emphasis on "cultivating inventions" is very much in line with a renewed interest today in creativity as an academic discipline or even a measurable learning outcome. Take for example the College of Creative Studies at the University of California, Santa Barbara, or the International Center for Studies in Creativity at Buffalo State College. The tension between the exceptionality of the individual inventor struck by divine inspiration and the step-by-step cultivation of that mind-set in Gernsback's publications is still with us in books like Steven Johnson, *Where Good Ideas Come From: The Natural History of Innovation* (New York: Riverhead Books, 2010), and Jonathan Lehrer, *Imagine: How Creativity Works* (New York: Houghton Mifflin, 2012). For a stinging critique of creativity experts, see Thomas Frank, "TED Talks Are Lying to You," *Salon*, (October 13, 2013), http://www.salon.com/2013/10/13/ted_talks_are_lying_to_you/.

2. In cultural studies of media and technology today, we might think of these two styles of invention as Gernsback formulates them—the whole-cloth innovation of shiny new tools versus incremental acts of modification and repair—as reflections of a hemispheric divide. Steven J. Jackson argues that "a Western and productivist imagination" biases much of media studies today, obscuring the vast majority of technological practices in the developing world: "breakdown, maintenance, and repair constitute crucial but vastly understudied sites or moments within the worlds of new media and technology today." For more on "the distinctive repair ecologies of the developing world," see Steven J. Jackson, "Rethinking Repair," in *Media Technologies: Essays on Communication, Materiality, and Society*, 221–39 (Cambridge, Mass.: MIT Press, 2014).

3. Susan J. Douglas suggests that many companies were perceived to have failed in their pursuit of wireless telephone systems—a pre-broadcast radio, point-to-point model of wireless voice communication—despite their many technical achievements, partially because "public expectations as shaped by the press had outdistanced actual achievement" (145). In addition, several wireless companies were indicted for stock promotion and manipulation schemes. Douglas quotes a December 1907 *World's Work* article: "Wireless stocks, at large, are to be regarded by the public as little better than racetrack gambling. Most of these wireless telegraph stocks have been put through a long period of juggling, washing, manipulation, fraud, and malfeasance that should effectively remove them, for good and all, from the field of investment." For more on this history, see Douglas, "Inventors as Entrepreneurs: Success and Failure in the Wireless Business," chapter 5 in her *Inventing American Broadcasting, 1899–1922*, Johns Hopkins Studies in the History of Technology (Baltimore: The Johns Hopkins University Press, 1987).

4. This is perhaps a reference to Lee De Forest, who incorporated the oscillating arc transmitter first designed by Valdemar Poulsen into his early system for transmitting and receiving the voice with his Audion tube. After a falling out with Abraham White (infamous founder of the fraudulent United Wireless Telegraph Company), De Forest formed the Radio Telephone Company in 1907 around this new system. De Forest would later become a frequent contributor to the Gernsback publications. Douglas,

Considered all in all, the two kinds of inventors, as far as their inventions go are nearly equally matched. The born inventor will usually invent a great many things, three fourths of which are useless: he becomes guilty of over-production. The mechanical inventor invents very few things as a rule and most of them are usually useful. Thus nature tries to maintain the universal equilibrium.[2]

A few words of advice to the two kinds of inventors does not seem out of place.

The fundamental test of any invention should always be whether it is better or cheaper than existing devices and whether it will be profitable to market the invention.

Most inventors, on account of being far too enthusiastic and optimistic, fool themselves by not considering in cold blood all the defects of their devices. There was never an invention that had not its bad points and weak spots. These should be considered most critically by the inventor, because if he does not do it, the world will soon enough do it with surprising thoroughness, usually to the dismay of the misguided inventor.

Never market an invention before it is completely worked out and "fool-proof." If it has weak spots, try and improve on same, if you don't do it, your competitors will do it for you at your expense. It is fallacious to think that as long as the device works after a fashion it should be put on the market and the improving done afterward. Nothing is more preposterous. Witness the sad fiasco of the Wireless Telephone, exploited by several American companies, who are now defunct.[3] One of them erected costly steel towers from 100–200 feet high in dozens of cities in this country, and the great defects of the "arc" wireless telephone must have been well known to the technical staff as well as to the promoters. Nevertheless, they plunged along, trusting to good luck that the improvements were only child's play and would find themselves. However, the improvements did not materialize in time to avert the final crash and the tall steel towers to-day are sad monuments of inventors' folly and shout their warning to inventors who would market inventions before they are ripe.[4]

Radio Telephone Company aerial on top of the Terminal building near 42nd street and Park Avenue, New York City. From *Modern Electrics*, October 1908.

Inventing American Broadcasting, 171–72. Brian Regal, *Radio: The Life Story of a Technology* (Westport, Conn.: Greenwood Press, 2005), 35.

Ralph 124C 41+, PART 3

Modern Electrics, vol. 4, no. 3, June 1911

Synopsis of Preceding Installments

Ralph 124C 41, living in New York in the year 2060 while in conversation with a friend at his Telephot, an instrument enabling one to see at a distance, is cut off from his friend and by mistake is connected with a young lady in Switzerland, thus making her acquaintance by Telephot.

The weather engineers in Switzerland who control the weather decided to strike against the Government and turned on the high depression of their Meteoro-Towers, thereby snowing in a large district. An avalanche threatens to sweep away the house in which the young Swiss lady, Miss 212B 42, lives and she appeals to the great American inventor, Ralph 124C 41+, to save her, which he promptly does by melting the avalanche by directed wireless energy from his New York laboratory.[1]

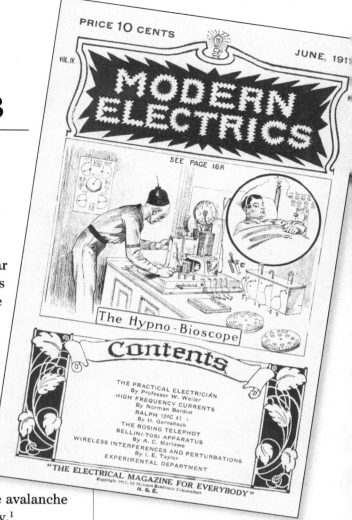

June 1911 cover of *Modern Electrics*.

RALPH 124C 41, after all the excitement of the last hour, felt the need of fresh air.

He walked up the few steps separating his laboratory from the roof and sat down on a chair beneath the revolving aerial. From down below a faint hum of the bustle of a great city rose up to him. Aero-flyers clotted the sky wherever one happened to glance. From time to time, trans-oceanic or trans-continental air liners would pass the horizon with a maddening swish.

Sometimes some great air-craft would come close up to him—within 500 yards perhaps—and he could observe how all the passengers craned their necks to get a good view of his "house," if such it could be called.

Indeed, his "house," which was a round tower, six hundred and fifty feet high, and thirty feet in diameter, built entirely of crystal glass-bricks and steelonium, was one of the sights of wonderful New York. A thankful city, recognizing his genius and his benefits to humanity, had erected the queer tower for him on a plot where, centuries ago, Union Square had been.

1. Gernsback prefaced the first installment (April 1911) of *Ralph* with the following note: "This story, which plays in the year 2660, will run serially during the coming year in *Modern Electrics*. It is intended to give the reader as accurate a prophesy of the future as is consistent with the present marvelous growth of science, and the author wishes to call especial attention to the fact that while there may be extremely strange and improbable devices and scenes in this narrative, they are not at all impossible, or outside of the reach of science." For a textual history of *Ralph*'s many editions, see Gary Westfahl, *Hugo Gernsback and the Century of Science Fiction* (Jefferson, N.C.: McFarland & Company, 2007), 97–148. Westfahl argues that the many different narrative modes explored by Gernsback in the original 1911 serial, as well as its gradual refinement in the 1925 mass-market edition and 1950 rerelease, reflect in miniature the history of science fiction itself.

The top of the tower was twice as great in circumference as the main building and in this upper part was located 124C 41's wonderful research laboratory, the talk of all the world. An electromagnetic tube elevator ran through the entire tower on one side of the building, and all the rooms were circular in shape, except for the space taken up by the elevator.

124C 41, sitting on top of his tower, mused about things in general. He really had no complaint to make; he had no hard feelings against anyone—only he was paying the penalty of fame, the penalty of greatness. He had everything; he had but to ask and the government would give it to him. His wishes were law.

But,—and this grieved him most—he was but a tool, a tool to advance science. To benefit humanity. He did not belong to himself, he belonged to the government, which fed and clothed him. He was not a free man. He was not allowed personally to make dangerous tests which would in any way endanger his priceless life. The government would supply him with some criminal under sentence of death who would be compelled to make the test for him. If the criminal was killed during the test, nothing was lost; if he was not killed, he would be imprisoned for life.

Being a true scientist to the core, this treatment took the spice out of 124C 41's life. He must submit to anything. His doctors must watch over him day and night, for he must not be sick. He must not indulge in any of the little vices that make life endurable; he must not smoke, he must not drink, he must have no undue excitement—the government would not have it.

He was a prisoner, sentenced for life to invent, to benefit humanity— a bird in a golden cage. Sometimes it became maddening; he could not endure it any longer, it seemed. He would remonstrate. He would call up the Planet Governor, the ruler of 90 billion human beings, and would ask him to be relieved of his work. The Governor would then call in person, and that powerful but kind man would reason with the great inventor until he would see that it was his duty to sacrifice himself for humanity. Twice already the Governor had called on 124C 41, and he knew it was vain to expect a release from him.

After all, he knew he was working for a great cause, and that it was his duty to keep his good-humor and master his weakness.

For some time he sat engrossed in his thoughts, while he watched the air-craft about him. He was awakened from his reverie by the voice of his faithful butler.

"Sir," he said, "your presence in the transmission-room would be appreciated."

"Why, what is the matter now?" 124C 41 exploded.

"It seems the people have heard all the details about your Switzerland exploit of an hour ago and desire to show their appreciation."

"Well, I suppose I must submit," the tired inventor dejectedly responded, and both stepped over into the round steel car of the electromagnetic elevator. The butler pressed one of the 28 ivory buttons and the car shot downward, without noise nor friction. There were no cables nor guides, the car being held and propelled by magnetism only. At the 22nd floor the car stopped, and 124C 41 stepped into the transmission-room.

No sooner had he made his appearance in that room than a deafening applause of hundreds of thousands of voices greeted him, and he had to hold his hands to his ears to muffle the sound.

The transmission room was entirely empty. There was nothing in it except a chair in the center.

Every inch of the wall, however, was lined with large-size telephots and loudspeaking telephone devices.

Centuries ago, when people tendered an ovation to some one, they would all assemble in some great square or some large hall. The some one to be honored would have to appear in person, else there would not be any ovation—truly a barbaric means. Besides, in those years, people stationed far away from the to-be-cheered-one could neither see nor hear well what was going on.—

It seems that, that afternoon on which our story plays, some enterprising news "paper" had issued extras about 124C 41's latest exploit, and urged its readers to be connected with him at 5 p. m.

Of course everyone who could spare the time called up the Teleservice Co. and asked to be connected to the inventor's trunk-line and the result was the ovation by distance.

Ralph 124C 41 stepped into the middle of the room and bowed in all the four directions of the compass, in order that everybody could get a good look of him. The noise was terrific; it seemed everyone was trying to out-hurrah and out-scream everybody else, and he beseechingly held up his hands. In a few seconds the vocal applauding stopped and some one yelled—"Speech!"

124C 41 spoke a few words, protesting that he did what everyone else would have done, and that he really was not entitled to this ovation. He also added that he did not consider himself a hero for saving the young Swiss lady, as he had done so without in the least endangering himself, which, by the way, was forbidden him by law, anyhow.

Nobody, however, seemed to share his opinion, for everyone began the applause anew and shouted himself hoarse.

Ralph 124C 41 could of course not see all his admirers on the telephot face-plates. There were so many thousand faces on each plate that nearly each face was blurred, due to the constant moving and shifting of the people at the other ends. They of course saw the great inventor plainly, because each one had the "reverser" switched on, which made it possible to see only the object at the end of the trunk line, as otherwise everybody would have seen everybody else, resulting in a blur.

In this case the blurs were in the inventor's transmission-room.

Ralph 124C 41 was obliged to make another speech and then retired to the elevator, the deafening applause still following him.

He then went down in the library and asked for the afternoon news.

His butler handed him a tray on which lay a piece of material *as large as a postage stamp*, transparent and flexible like celluloid.

"What edition is this?" he asked.

"The 4 o'clock *New York News,* sir."

Ralph 124C 41 took the "news" and placed it in a small metal holder which was part of the hinged door of a small box. He closed the door and turned on a switch on the side of the box. Immediately there appeared at the opposite side of the box, on the white wall of the room, a twelve-column page of the *New York News* and 124C 41, leaning back in his chair, proceeded to read as one would read a letter projected on a screen in a moving-picture show.

The *New York News* was simply a microscopic reduction of a page of the latest news, and, when enlarged by a powerful lens, became plainly visible.

Moreover, each paper had 8 "pages," not eight separate sheets, as was the fashion centuries ago, but the pages were literally on top of each other. The printing process was electrolytic, no ink whatsoever being used in the manufacture of the "newspaper." This process was invented in 1910 by an Englishman, and as improved upon by the American, 64L 52 in 2031, who made it possible to "print" *in one operation* eight different subjects, *one on top of another,* as it were.

These eight impressions could be made visible only by subjecting the "paper" to different colors, the color rays bringing out the different prints. The seven colors of the rainbow were used, while white light was employed to show reproduced photographs, etc., in their natural colors. With this method it was possible to "print" a "newspaper" fully 10 times as large in volume as any newspaper of the 21st century on a piece of film, the size of a postage-stamp.

Each paper published an edition every 30 minutes, and if one did not possess a projector, one could read the "paper" by inserting the *News* in a holder beneath a powerful lens which one carried in one's pocket, folded when not in use. To read the eight different pages, a revolving color screen was placed directly underneath the lens, to bring out the different colors necessary to read the "paper."

124C 41, glancing over the headlines of his *News,* noticed that considerable space was given up to his last exploit, the paper showing actual photographs of the Swiss Alpine scene, which a correspondent had taken while the now famous avalanche rushed down the mountain. The photographs had been sent by *Tele-radiograph* immediately after the occurrence in Switzerland, and the *News* had printed them in all the *natural* colors twenty minutes after Ralph 124C 41 had turned off the Ultra-power in New York.

These photographs seemed to be the only thing that interested 124C 41, as they showed the young lady's house and the surrounding Alps. These, with the monstrous avalanche in progression photographed and reproduced in the natural colors, made quite an impressive view.

124C 41 soon grew tired of contemplating this and revolved the color screen of his projector to green—the technical page of the News—his favorite reading.

He soon had read all that interested him, and as there was only one hour till supper time he began to "write" his lecture: "On the prolongation of animal life by π-Rays."

He attached a double leather head-band to his head; at each end of the band was attached a round metal disc which pressed closely on the temples. From each metal disc an insulated wire led to a small square box, the *Menograph,* or mind-writer.

He then pressed a button and a slight hum was heard; simultaneously two small bulbs began to glow in a soft green fluorescent light. 124C 41 then grasped a button connected with a flexible cord to the Menograph and leaned back in his chair.

After a few minutes' reflection he pressed the button, and immediately a wave line, traced in ink, appeared on a narrow white fabric band, the latter resembling a telegraph recorder tape.

The band which moved rapidly, was unrolled from one reel and rolled up on another. Everytime [sic] 124C 41 wished to "write" down his thoughts, he would press the button, which started the mechanism as well as the recording tracer.

Below is shown the record of a Menograph, the piece of tape being actual size.

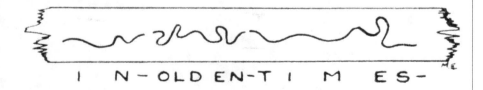

FIGURE 1.

Where the waveline breaks, a new word or sentence commences; the three words shown are the result of the thought which expresses itself in the words, *"In olden times."* . . .

The Menograph was one of 124C 41's earliest inventions, and entirely superseded the pen and pencil. Anyone can use the apparatus; all that is necessary to be done is to press the button when an idea is to be recorded and to release the button when one reflects and does not wish the thought-words recorded.

Instead of writing a letter, one sends the recorded *Menotape*, and inasmuch as the Menolphabet is universal and can be read by anyone—children being taught it early—it stands to reason that this invention was one of the greatest boons to humanity: Twenty times as much work can be done by means of the Menograph as could be done by the old-fashioned writing, which required a considerable physical effort. Typewriters soon disappeared after its invention, as there was no more use for them, nor was there any use for stenographers, as the thoughts were written down direct on the tape, which was sent out as a letter was sent centuries ago.[2]

124C 41 had soon recorded his lecture on the Menograph, after which he had supper with his family.

In the evening he worked for some hours in the laboratory, and retired at midnight. He soon went to bed, but before he fell asleep he attached to his head a double leather head-band with metal temple plates, similar to the one used in connection with the Menograph.

He then called for his faithful butler and told him to "put on" Homer's *Odyssey* for the night.

Peter, the butler, then went down to the library on the 15th floor, and took down from a shelf a narrow box, labelled *Odyssey, Homer*. From this he extracted a large but thin reel on which was wound a long

2. In another example of the porous boundaries between fiction and the technical editorials, the May 1919 issue of *Electrical Experimenter* featured a discussion of a "thought recorder" very similar to the menograph, including "three scientists' views on thought transmission."

While in *Ralph 124C 41+* everyone has been taught to read the menotape, meaning thought can be directly transcribed to paper for reading, later versions of this speculative apparatus include women stenographers as part of the thought recording workflow, as shown in this illustration. In Gernsback's later *Baron Münchhausen* serial novel, Martian women assist in the "Tos rod" correspondence technology, a "sensitive recording mechanism" that is "in charge of a secretary." One would think the narrator means in *the* charge of a secretary, but the passive voice describing the entire process makes things unclear. The thought "is always recorded," "is carefully labeled and stored away." Hugo Gernsback, *The Scientific Adventures of Baron Münchausen* (Burlington, Ont.: Apogee Books Science Fiction, 2010), 137.

The Thought Recorder is an Instrument Recording Thoughts Directly by Electrical Means, On a Moving Paper Tape. Our Illustration Shows What a Future Business Office Will Look Like When the Invention, Which as Yet Only Exists in the Imagination, Has Been Perfected. By Pushing the Button A, the Tape is Started and Stopt Automatically So That Only Thoughts That Are Wanted Are Recorded.

From *Electrical Experimenter*, May 1919.

narrow film. This film was entirely black but for a white transparent wave-line running in the center of it.

Peter placed the reel containing the film in a rack and introduced the end of the film into the Hypnobioscope. This wonderful instrument, invented by Ralph 124C 41, transmitted the impulses of the wave-line direct to the brain of the sleeping inventor, who thus was made to "dream" the "Odyssey."

It had been known for centuries that the brain could be affected during sleep by certain processes. Thus one could be forced to dream that a heavy object was lying on one's chest, if such an object was placed on the sleeper's chest. Or one could be forced to dream that one's hand was being burnt or frozen, simply by heating or cooling the sleeper's hand.

It remained to 124C 41, however, to invent the Hypnobioscope, which transmits words direct to the sleeping brain, in such a manner that everything can be remembered in detail the next morning.

This was made possible by having the impulses *act directly and steadily on the brain.* In other words, it was the Menograph reversed, with certain additions.

Thus, while in a passive state, the mind absorbs the impressions quite readily and mechanically and it has been proven that a story "read" by means of the Hypnobioscope leaves a much stronger impression than if the same story had been read while conscious.

For thousands of years humanity wasted half of its life during sleep—the negative life. Since 124C 41's inestimable invention, all this was changed. Not one night is lost by anyone if anywhere possible, conditions permitting. All books are read while one sleeps. *Most of the studying is done while one sleeps.* Some people have mastered 10 languages, which they have learned during their sleep-life. Children who can not be successfully taught in school during their hours of consciousness, become good scholars if the lessons are repeated during their sleep-life.

The morning "newspapers" are transmitted to the sleeping subscribers by wire at about 5 a. m. The great newspaper offices have hundreds of hypnobioscopes in operation, the subscriber's wire leading to them. The newspaper office has been notified by each subscriber what kind of news is desirable, and only such news is furnished. Consequently, when the subscriber wakes up for breakfast he already knows the latest news, and can discuss it with his family, the members of which of course also know the same news, being also connected with the news-paper hypnobioscope.

(To be continued.)

The Alexander Wireless Bill

Modern Electrics, vol. 4, no. 11, February 1912

Elsewhere we print a copy of the new Alexander Wireless Bill.[1]

Before we go any further we might as well tell all those interested in wireless that it is not necessary to feel any anxiety over this or any other wireless bill. We have noted in the past that every time we have printed any of the various wireless bills there was almost a panic among the amateurs and other wireless interests.

Readers of this magazine well know the policy that *Modern Electrics* has always pursued, and as far as the interests of the wireless amateurs is concerned, it is not necessary to point out that the proprietors of this magazine have always stood up for the wireless amateurs.

This holds true of the Alexander bill, and as soon as the bill was made public, *Modern Electrics* at once took steps to safeguard these interests. A great number of protesting letters were written at once to Washington and the results accomplished so far are distinctly encouraging, in so far as the amateurs and other wireless interests are concerned. As far as the Alexander bill itself is concerned, it is no worse than previous ones; in fact if it should be passed, which we doubt, it would not hurt the wireless amateurs in the least, inasmuch as this bill covers only interstate business. Wireless amateurs living in the same state could communicate with each other the same as before.

The Government realizes fully the importance of the American amateurs to-day, and if any bill should be passed, it will be one that regulates wireless, but in no way suppresses it.

This, *Modern Electrics* has always advocated. In fact *Modern Electrics* is heartily in accord with any bill that should regulate wireless in such a way that the amateurs do not interfere with commercial or Government business, which is only right and fair.

As stated elsewhere in this issue, Government officials have already been busy in New York of late, to find out what equipment the various amateurs use, what their wave lengths are, what power they use, etc.

This is distinctly encouraging, because it shows that the Government is getting accurate information before trying to pass any bill.

There should be a bill passed restraining the amateur from using too much power, say, anything above 1 K. W.

The wave length of the amateur wireless station should also be regulated in order that only wave lengths from a few metres up to 200

1. The Alexander Wireless Bill was introduced for debate in congress on December 11, 1911, by Joshua W. Alexander, Democratic representative from Missouri. Alexander later served as the U.S. Secretary of Commerce and was succeeded by Herbert Hoover, who would play a decisive role in the creation of the Federal Communications Commission (see **Who Will Save the Radio Amateur?** in this book). The Alexander Bill served as the kernel that eventually became the Radio Act of 1912. While the bill included no mention of amateur stations, after months of debate among and agitation by wireless communities, the Radio Act of 1912 explicitly defined the acceptable range of activities available to amateurs.

could be used. Wave lengths of from 200 to 1,000 metres, the amateurs should not be allowed to use, but they could use any wave length above 1,000. If this is done, all interference with Government, as well as commercial stations, will be done away with and the wireless situation will then be the same as to-day. The amateurs will have the same liberty and perhaps greater liberty than today, and complaints against them from Government or Commercial stations will cease automatically.

Modern Electrics has prepared a very comprehensive article which will be published in the next issue showing just what equipment the amateur can use to keep from interfering with Government and Commercial stations, and there is no doubt that this article will, in a great measure, prevent a lot of future mischief.

It is significant that none of the other periodicals, who always shout that they have the interests of the amateur at heart, appeared to know anything about the Alexander bill, nor did any of them take any steps to serve their readers.

Modern Electrics has been the first in the wireless field and will continue to serve the wireless amateur and the independent wireless interests in the future, as in the past.

Wireless and the Amateur: A Retrospect

Modern Electrics, vol. 5, no. 11, February 1913

On December 13, 1912, the new wireless law went into effect.[1] The average wireless "fiend" who has not followed the topic from the start will be interested in the following facts:

The very first talk about Wireless Legislation in the country started in 1908. The writer in his Editorial in the November, 1908, issue of *Modern Electrics* pointed out that a wireless law was sure to be passed in a very short while.[2] In order to guard against unfair legislation as far as the wireless amateur was concerned the writer, in January, 1909, organized the "Wireless Association of America."[3] This was done to bring all wireless amateurs together and to protest against unfair laws. Previous to this time there was no wireless club or association in the country. In January, 1913, there were over 230 clubs in existence, all of which owe their origin to the "Wireless Association of America."

The association had no sooner become a national body than the first wireless bill made its appearance. It was the famous Roberts Bill, put up by the since defunct wireless "trust." The writer single handedly, fought this bill, tooth and nail. He had representatives in Washington, and was the direct cause of having some 8,000 wireless amateurs send protesting letters and telegrams to their congressmen in Washington. The writer's Editorial which inspired the thousands of amateurs, appeared in the January, 1910, issue of *Modern Electrics*.[4] *It was the only Editorial during this time that fought the Roberts Bill.* No other electrical periodical seemed to care a whoop whether the amateur should be muzzled or not. If the Roberts Bill had become a law there would be no wireless amateurs to-day.

That editorial quickly found its way into the press and hundreds of newspapers endorsed the writer's stand. During January, 1910, the *New York American*, the *New York Independent*, the *New York World*, the *New York Times*, the *Boston Transcript*, etc., all lauded and commended the writer's views.[5] Public sentiment quickly turned against the Roberts Bill and it was dropped.

The first wireless bill not antagonistic to the amateur, the Burke Bill, appeared on March 8, 1910. It had some defects, however, and was dropped also. The Depew Wireless Bill appeared May 6, 1910, but did not meet with general approval; as the writer pointed out in

1. Refers to the Radio Act of 1912, which placed the allocation of radio spectrum and the licensing of public and private stations under the authority of the U.S. Secretary of Commerce and Labor, then Charles Nagel, the last person to hold the position before the formation of the Federal Trade Commission in 1914.

2. See **The Wireless Joker** in this book.

3. See **The Wireless Association of America** in this book.

4. See **The Roberts Wireless Bill** in this book.

5. Gernsback: "See Editorial article February, 1910, *Modern Electrics*."

his Editorial in the June, 1910, issue of *Modern Electrics*, it had several undesirable features, and the bill was never seriously considered, although it actually passed the Senate.[6]

At last the Alexander Bill made its appearance on December 11, 1911. This bill as far as the amateur was concerned was not quite acceptable to the writer, who had the amateurs' rights at heart and steps were immediately taken to bring about an amendment as the writer, perhaps more than anyone else, realised that this bill, in some term or other, would become a law sooner or later. This is clearly stated in his Editorial in the February, 1912, issue of *Modern Electrics*.[7] In that Editorial is to be found also *the first and now historical recommendation* that if a wireless law was to be framed it should restrict the amateur from using a higher power than 1 kw, and his wave length should be kept below 200 metres. No one else had thought of it before, and it is to be noted that when Congress finally passed the present wireless law, *it accepted the writer's recommendation in full,* thus paying him the greatest compliment, while at the same time acknowledging the fact that he acted as the then *sole* spokesman for and in behalf of the wireless amateur.

In March, 1912 the writer, in a letter to the *New York Times* (See page 24, April, 1912, issue *Modern Electrics*), pointed out the shortcomings of the Alexander Bill, and protested against unfair legislation.[8]

The Times, as well as a host of other newspapers, took up the cry and published broadcast the shortcomings of the Alexander Bill.

All this agitation had the desired effect and Mr. Alexander for the first time realized that the amateur could not be muzzled, especially when there was such a periodical as *Modern Electrics* to champion his cause. Promptly in April the Alexander Wireless Bill, *amended,* appeared and here for the first time in history the amateur and his rights are introduced in any wireless bill.

Mr. Alexander and his advisers accepted the writer's recommendation as set forth in his Editorial in the February, 1912, issue of *Modern Electrics* and the new paragraph (15) in the amended bill reads thus:

Fifteenth. No private or commercial station not engaged in the transaction of bonafide commercial business by radio communication or in experimentation in connection with the development and manufacture of radio apparatus for commercial purposes at the date of passage of this Act, *shall use a transmitting wave length exceeding two hundred meters, or a transformer input exceeding one kilowatt,* except by special authority of the Secretary of Commerce and Labor contained in the license of that station.

It will be noted that it copied the writer's recommendations word for word.

6. Gernsback: "See Editorial, August, 1910, *Modern Electrics*."

7. See **The Alexander Wireless Bill** in this book.

8. The letter to the *Times* editor reads: I note your timely article on "The Wireless Control Bill." While you have taken your stand very well, as far as the commercial interests go, I would like to give the views of the wireless experimenters and amateurs whom I have represented for the last six years.

Although it is not generally known, there are to-day close to 400,000 wireless experimenters and amateurs in the United States alone. The Wireless Association of America to-day numbers 16,189 members, and there are now 122 wireless clubs and subsidiary wireless associations scattered from coast to coast.

Much has been written and said against the wireless amateur, and while it is true that some mischief has been done, report has never been made of any case where serious damage was done, except in a single instance a few years ago, when a Massachusetts amateur transmitted a false "S. O. S." call, which sent a United States Government boat out on a fool's errand. On the other hand, the amateur has done a great deal of good by taking up distress calls, and it should be understood that good work has frequently been done in the past by amateurs receiving and transmitting such calls where they would probably not have been received by the few commercial and Government stations. Furthermore, a great many amateurs, while not necessarily schoolboys, have done much to further the art in general, and many patents have been taken out during the last three years by such students interested in wireless.

The interests which would like to see the Alexander Bill passed have evidently not looked very much into the future. On account of the vast distances in this country, radio-telegraphic and radio-telephonic intercommunication will positively find a distinct usefulness, greater without doubt than that of the present telephone. An immense usefulness alone will be found in radio-telephonic stations between moving vehicles and fixed posts.

There is a feature in wireless which perhaps is not appreciated. It has been recognized that, on account of the great interest which a young man

The amateur had at last come into his own. This is all the more remarkable as this is the only country that recognizes the wireless amateur.

On May 7, 1912, the Alexander Bill, amended, now known as S-6412, passed the United States Senate and on May 8th was sent to the House of Representatives and referred to the Committee on the Merchant Marine and Fisheries.[9] The bill was signed on August 13th by President Taft, thus making it a law.

In the March, 1912, issue, *Modern Electrics* long before the passage of the wireless law and ahead of any other periodical published an article on "Limited Wave Lengths" preparing the amateur for the new law and paving the way towards standardizing amateur stations.

Finally in the November, 1912, issue, page 829, the full text of the new wireless law was published, and it was announced that the law would go into effect December 14, 1912.

In the December, 1912, issue (page 922), the new law was fully discussed and all phases explained.

Again *Modern Electrics* was the only periodical to publish the license blanks and to show the amateur how to fill them out. No other periodical had enough interest in the amateur to render this important service to him.

And last but not least in this issue we are printing a facsimile copy of an original license, which up to the present minute closes amateur wireless history in the United States.

This terminates the fight which the writer has waged single-handedly for almost five years in behalf of the American amateur. It must be apparent even to the layman, who has not followed the evolution of the present law, that unquestionably the entire credit for obtaining the amateur's rights belongs to *Modern Electrics*. This is freely admitted today by all. The indisputable facts enumerated in this article make this clear.

Now that it is all over, and that Uncle Sam has set his seal of approval upon the amateur's wireless, the writer cannot but extend his heartiest congratulations to the 400,000 American amateurs; and he furthermore wishes to extend his thanks to all the amateurs who have supported him in his fight to bring about a new wireless era in America.

Long live the Wireless! Long live the Amateur!!

finds in the study of wireless, this new art does much toward keeping him at home, where other diversions usually, sooner or later, lead him to questionable reports; and for this reason well informed parents are only too willing to allow their sons to become interested in wireless. If it were only for this reason, it would be worth while not to curb enthusiasms for the new art. The public fully shares this view, which has been proved time and again.

H. GERNSBACK
Business Manager, Wireless Association of America,
Editor *Modern Electrics*
New York, March 27, 1912

Hugo Gernsback, "400,000 Wireless Amateurs: To Discourage Their Work Would Check Progress in the Art," *New York Times*, March 1912: 12, http://search.proquest.com/docview/97272129/abstract/DEC375C239DE4859PQ/4?accountid=10226.

9. Gernsback: "See June, 1912, issue *Modern Electrics*, page 245."

Our Cover

The Electrical Experimenter, vol. 1, no. 1, May 1913

April 1915 cover of *The Electrical Experimenter.*

The idea of our cover was conceived by the writer with the intention of inspiring the electrical experimenter at large.[1]

There is nothing fantastic about this cover, nothing impossible. It will all be very real in a comparatively short time. It is up to our experimenters to make it an accomplished fact.

The scene is laid near the coast, in almost any part of the globe. The time, let us say is in the year 2013. It is night. The large aerial system in the foreground radiates, not feeble telegraph impulses, but a tremendous power. This power is furnished by the large "Powerhouse," beneath the aerial system, some 30,000 Kilowatts being radiated into the ether constantly. Naturally, such a tremendous power going out into the air gives rise to peculiar phenomenae. The air becomes luminous for several miles around and above the aerial. An

1. A reproduction of the black-and-white image on this issue's cover would form the first full-color cover on a Gernsback magazine in April 1915, accompanied by a similar explanation of its feasibility.

inverted bowl-shape light dome, with the aerial system at its center, is produced, and this light illuminates the landscape for miles around. The lower antenna acts partly as a reflecting aerial which prevents the energy from being absorbed by the earth. It has been found that by using a curious vibratory pulsating wave of tremendous amplitude, almost no energy is lost in transmission through the ether, and for that reason the etheric power station as illustrated can supply energy within a radius of several hundred miles. The power is derived solely from the tides of the ocean—a tremendous force, which lay unharnessed through the aeons.

On top of the "powerhouse" we see two towers with curious light balls.

These are the "*radiofers*." You must understand that the "powerhouse" which shoots forth such a colossal force, can not be frequented by humans. As a matter of fact no human being could come near the house, or within 500 yards. For that reason the power is entirely controlled from a distance, by wireless of course. The control is exercised through the "radiofers."

In the left foreground we see a curious wheelless railroad. The cars float actually in the air, some feet above the broad, single iron track. The power is obtained from the distant power aerial by wireless, of course. One will notice the aerial wires on top of the cars, which receive the energy. The train is suspended by electromagnetism and glides smoothly along at the rate of some 200 miles an hour.[2]

In the left foreground also we see an immense 1,000-foot "*optophor*" tower. This tower shoots a dazzling colored light shaft of some ten million candle-power straight into the sky. Such "optophor" towers are stationed exactly 50 miles apart along the coasts, and every tower has a different colored light shaft. This light beam can be seen some 500 miles out at sea and by its light, transatlantic aerial as well as aquatic craft, can steer with unfailing accuracy towards their point of destination.

2. Gernsback: "In 1912 patent No. 1,020,942 was issued to [Emile] Bachelet on such a suspended train system."

From *A Treatise on Wireless Telegraphy*

The Electrical Experimenter, vol. 1, no. 1–2, May–June 1913

Part Two: Wireless and the Layman

Receiving Wireless Messages

The question we hear from most beginners is:—"What outfit do you advise me to use? I know nothing about wireless."[1] We advise the use of ANY honestly built receiving outfits. Which one to choose depends upon yourself, your taste and your pocketbook. This is where YOU must decide. Of course the lower priced outfits have naturally a short range—they won't catch messages hundreds of miles away, and those without tuning coils cannot be used to "cut out" one of the messages when two of them are in the "air" at the same time.[2] It is self-evident though, that you can start with the very cheapest outfit,—say an E. I. Co. detector and a pony telephone receiver.[3] With such an outfit messages can be picked

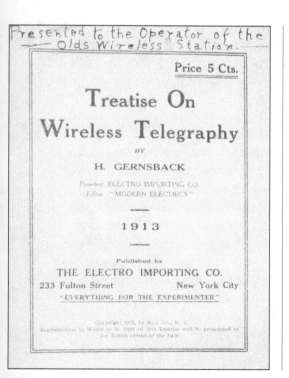

Pamphlet version of Gernsback's "Treatise on Wireless Telegraphy," published in 1913. Courtesy the Collection of Jim and Felicia Kreuzer, Grand Island, New York.

1. Published in the wake of the 1912 Radio Act and the growing popularity of wireless homebrew, this two-part editorial served as a primer for newcomers to the art, covering everything from the barest necessary equipment, code, and the secrecy of messages. It was printed the following year as a stand-alone pamphlet.

"A Treatise on Wireless Telegraphy" appeared in the first issue of *Electrical Experimenter,* Gernsback's new venture after the sale of *Modern Electrics* to Orland Ridenour, the business manager of Modern Publishing Company. Ridenour merged *Modern Electrics* with another publication to become *Modern Electrics and Mechanics,* which lasted for two years until it was acquired by *Popular Science Monthly* in April 1915, a magazine that still runs today. See Mike Ashley, *The Gernsback Days:*

A Study of the Evolution of Modern Science Fiction from 1911 to 1936 (Holicong, Penn.: Wildside Press, 2004), 31.

The Electrical Experimenter incorporated more photographs, broadened the scope of its appeal to include nonspecialist readers, and expanded its foray into fiction, even including a story contest among its readers. "The first prize of $5 went to Thomas W. Benson for a piece about how to set up a range of electrical equipment to play a trick on his sister's boyfriend." That story was "Mysterious Night" in the June 1914 issue. Michael Ashley, *The Time Machines: The Story of the Science-Fiction Pulp Magazines from the Beginning to 1950,* The History of the Science-Fiction Magazine 1 (Liverpool, UK: Liverpool University Press, 2000), 94.

The new publication included monthly sections like "Among the Amateurs," which reported on the activities of community wireless organizations around the country, a "How-to-Make-It Department," a "Question Box," and a "Patent Advice" column, which printed answers to readers' calls for input on their patent application drafts, for "a nominal charge of $1.00."

This issue's inaugural "Wrinkles—Receipts—Formulas—Hints" section included a primer on "How to Metallize and Electroplate Insects, Flowers, Small Household Goods, Etc.": "A nice Rose bud, an uncommon Insect, the first shoes of baby, and hundreds of other things can be conserved indefinitely by Metallizing them. The methods given below, enables any one to do this work very skillfully at a very small outlay."

2. The tuning coil provides a simple way to change the inductance and thus the frequency of a wireless set's circuit, thereby allowing the listener to tune in to different stations. By varying the length of the coil, it becomes resonant with different frequencies.

3. The "pony receiver" is the component of the candlestick-style telephones popular throughout the early twentieth century that was held to the ear. In a pinch, it could be used as a cheap and simple speaker for wireless telegraph sets.

up astonishingly well indeed. Many of our enthusiastic young friends started with such an outfit and kept on adding instruments till they finally had up-to-date stations.

The next question hurled at us is:—"How can I receive messages if I don't know the codes?"

A wireless telegram, no matter if it is in Chinese or English, "comes in" in dots and dashes. When you have the telephone receivers to your ear and a message is coming in, you hear a series of long and short, clear, distinct buzzes. A long buzz is a dash, a short buzz is a dot. The E. I. Co. sell a 10¢ code chart by means of which the dots and dashes are translated into letters. Thus (in the Morse code), dash, dash, dot, stands for the letter G; dash, dash means M, dash, dot, dash, dot means J and so forth. Any person with a few weeks' practice "listening to the wireless" can master the code, and read the messages with ease.

Remember that there are over two thousand high powered wireless stations in this country alone, each being able to transmit messages of over a thousand miles distance!

There are almost at any minute, during night and day, messages in the air, no matter where you are,—sending YOU messages, only waiting to be picked up by you. It is truly wonderful; it is the cheapest as well as the most elevating diversion known to modern man, the most inspiring example of the triumph of mind over matter.

"How about the Wireless Law?", you want to know next.

The law does not apply for stations used for *receiving* only. *There is no law which forbids you to receive all the messages you wish.* You can receive as many and as long as you please,—Uncle Sam doesn't mind. But you MUST preserve the secrecy of the message. You must not make use of any information you receive by wireless, if this information is of such a nature that makes it private property. Your own conscience will tell you which message to keep secret and which one you can make use of.[4] Here is the text of the Law:

Secrecy of Messages

"Nineteenth. No person or persons engaged in or having knowledge of the operation of any station or stations shall divulge or publish the contents of any messages transmitted or received by such station, except to the person or persons to whom the same may be directed, or their authorized agent, or to another station employed to forward such message to its destination, unless legally required to do so by the court of competent jurisdiction or other competent authority. Any person guilty of divulging or publishing any message, except as herein provided, shall, on conviction thereof, be punishable by a fine of not more than two hundred and fifty dollars or imprisonment for a period of not exceeding three months, or both fine and imprisonment in the discretion of the court."

4. Similar conversations emerged among early Internet communities around privacy and its maintenance. Eric Hughes, for instance, in his 1993 "A Cypherpunk's Manifesto," writes: "Privacy is not secrecy. A private matter is something one doesn't want the whole world to know, but a secret matter is something one doesn't want anybody to know. Privacy is the power to selectively reveal oneself to the world." The problem is how online communities should go about protecting privacy without restricting free speech, especially given that the very nature of computation depends on the copying of data. "If two parties have some sort of dealings, then each has a memory of the interaction. Each party can speak about its own memory of the encounter. How could anyone prevent this? One could pass laws against it, but the freedom of speech, even more than privacy, is fundamental to an open society. We seek not to restrict any speech at all. If many parties speak together in the same forum, each can speak to all the others and aggregate together knowledge about individuals and other parties. The power of electronic communications has enabled such group speech, and it will not go away merely because we might want it to." Eric Hughes, "A Cypherpunk's Manifesto," in *Crypto Anarchy, Cyberstates, and Pirate Utopias*, ed. Peter Ludlow (Cambridge, Mass.: MIT Press, 2001), 81–83.

Of late a great many stations are beginning to use the wireless telephone. This art is rapidly being perfected and is the coming thing in "Wireless." There is hardly a week that you do not read about some new wireless telephone and some new distance record established.

It is of course understood that any receiving apparatus that can receive wireless telegraph messages, 90 times out of 100, can receive wireless telephone messages. Of course, in that case no code is required as the voice comes through the receiver the same as through the regular telephone. (For further details on Wireless Telephony, see "The Wireless Telephone," by H. Gernsback, at 25¢; also Lesson No. 18 of the E. I. Co. Wireless Course.)

A primitive loose coupler, pictured in "Modern Wireless Instruments," *Modern Electrics* 1, no. 10 (January 1909): 359.

Distance

The question asked mostly by the layman is: "How far can I receive with such and such an outfit, my aerial being so high and so long?"

Nobody can correctly answer such a question. You can reason it out as well as we can. For example: Would you ask us: "How far away can I hear the steam whistle of the X & Y Cotton Mill?" No, you wouldn't, for it all depends. First, how hard the whistle blows, second, how good your hearing is, third, how the wind blows, and fourth, how many and how great are the intervening objects between the whistle and your ear. Some days you may hear the whistle two miles off with the wind blowing your way. Or if you are way down in the cellar you may only hear it faintly, although you are but two blocks away from it. It all depends. The one thing you are sure of is that the whistle blows with about the same strength each day. The same reasoning holds true for wireless to a very great extent.

As a rule, the higher up and the bigger your aerial, the better the wireless reception will be. Naturally if you are a thousand miles off from a station that can but send 500 miles, you won't hear it, no matter how good your instruments are. It's like trying to hear the sound of a whistle 10 miles away from you, that can at the very best be heard only within a radius of 5 miles.

Just use a little horse sense and you can do your own deducting: no wireless expert is required. It is also evident that the messages cannot come in with the maximum loudness unless the instruments are well in tune, and unless well designed instruments are used. Thus a loose coupler will give louder signals than a small tuning coil. It also depends a lot on the detector and its adjustment.[5]

5. The next stage in the development of tuning coils began with sets in the 1910s that featured a component known as the loose coupler. Loose couplers consisted of two interlocking coils: a primary coil remained stationary with a slider for varying its inductance, while a secondary, smaller coil slid in and out of the primary cylinder for further selectivity and sensitivity in tuning. The primary was wired to an antenna and the ground, and the secondary was hooked up to a detector. So, tuning this type of set consisted of a two-stage process (later sets would add more stages): first adjusting the slider on the primary and then calibrating the position of the secondary.

This is the way the detectors range according to their sensitiveness:

1st. The Audion (the most sensitive detector to date); 2nd, The Electrolytic; 3rd, The Peroxide of Lead; 4th, The Perikon; 5th, Zincite and Bornite; 6th, Silicon and Galena; 7th, Iron Pyrites (Ferron); 8th, Carborundum; 9th, Molybdenite. (See Lesson No. 10, of the E. I. Co. Wireless Course, on Detectors.)[6]

If you are entirely surrounded by high mountains or steel buildings, you naturally will not expect to receive messages as well as if you were on the top of a mountain. Also remember that wireless waves travel *twice as far over water as over land,* and that you can reach *twice as far after sundown than during the daytime.*[7]

Wave Lengths and Tuning

This seems to be the greatest stumbling block for most beginners. Again let us make a comparison. Take two pianos and place them in the same room. Or two violins will do as well. Tune two strings, one on each instrument, so both will give exactly the same note. Pick one of the strings in order to sound it, and the other "tuned" string, although 10 feet away will sound in unison, although you did not touch it. *Both are now in tune.* Both give out the same (sound) wave length. No mystery here. The secret lies in the fact that both strings ARE OF THE SAME LENGTH, and have the same tension, roughly speaking. Make one string longer than the other and both are "out of tune."

The same in wireless. Nearly all commercial stations operate on a wave length of from 300 to 600 meters. (A meter measures 39.37 inches.) Now in order that you can hear such a station, you must be able to tune up to 600 meters; roughly speaking your aerial should be 600 meters long electrically. That, however, would be a pretty expensive and cumbersome aerial. Besides it isn't required. We simply wind,

6. Detectors employing minerals such as silicon, lead, zincite, etc., were all referred to as "crystal detectors." Most crystal detectors were made of naturally occurring minerals. But in 1906 Henry Dunwoody, working for the De Forest Wireless Telegraph Company, first tried using carborundum (silicon carbide), a synthetically produced compound used today in bulletproof vests and LEDs. At that time, it was a mainstay of the electrical experimenter, commonly found around the workshop and "used for grinding and polishing components of electrolytic detectors or coherers." In effect, the chemical compound used to clean older forms of detectors was put into the construction of a new one. Although silicon (first used by Greenleaf Whittier Pickard, another employee of De Forest) outperformed all other minerals, it was incredibly difficult to procure at the time. Thus carborundum became the material of choice. For more on this story, see D. P. C. Thackeray, "When Tubes Beat Crystals: Early Radio Detectors," *IEEE Spectrum* 20, no. 3 (March 1983): 64–69, doi:10.1109/MSPEC.1983.6369844. For Pickard's attempts to procure silicon from chemical supply houses around the world (fused, not powdered, which is all he could find), see Alan Douglas, "The Crystal Detector," *IEEE Spectrum* 18, no. 4 (April 1981): 64–69, doi:10.1109/MSPEC.1981.6369482. Meanwhile, amateurs developed ingenious ways of constructing and modifying their own crystal detectors. In the August 1910 issue of *Modern Electrics,* for instance, an amateur named Edward Duvall suggests a method of stripping out an old telegraph sounder (the key pressed to send signals) in order to make a "double detector," recommending molybdenite as the best mineral to use.

7. Lingering belief in the existence of a luminiferous ether complicated understandings of weather's effect on wireless. It was clear that messages were received at greater distances and with less interference during the evening and during cooler weather. Whether the influencing factor was heat or humidity or an effect of the mysterious nature of light itself was a matter of debate.

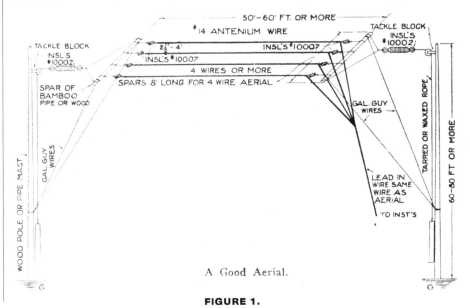

A Good Aerial.

FIGURE 1.

roughly speaking, 600 meters of wire on a coil or drum and our aerial can now be quite small, within certain limits of course, and we can for this reason "catch" the station having a 600 meter wave length, providing our other instruments are sensitive enough. Thus, for instance, it will be seen that the E. I. Co. No. 8486 tuning coil, as well as their No. 12002 loose coupler have sufficient wave length capacity to catch 700 meter waves. As they are both provided with adjusting sliders, more or less wire can be put into the circuit and therefore both these instruments can be used to catch wave lengths from 100 up to 700 meters, but not over this amount.

Therefore, if we should want to hear a station having 1400 meters wave length, we would connect two No. 8486 tuning coils in series, which would give us 700 + 700 = 1400 meters wave length. Or we would connect one No. 8486 tuner in series with the primary of the No. 12002 loose coupler and we would get the same effective wave length. As a rule only stations doing long distance work use excessive wave lengths, thus the Marconi Transatlantic station at Glace Bay has a wave length of about 7100 meters, while the new Government station at Washington, which sends messages over 3,000 miles, has a wave length of about 4,000 meters.[8] By consulting the *"Wireless Blue Book"* the wave length of all important wireless stations can be found, as each station normally uses a certain prescribed wave length. (See Lessons No. 4, 5, 6, 7, 8, 9, of The E. I. Co. Wireless Course.)

The best all around aerial is about 75 feet long, composed of four strands E. I. Co. "Antenium" wire No. 14, or stranded "Antenium" cable. One of the best forms is shown herewith. We recommend E. I. Co. No. 10007 insulators, although others can be used. For a 75 foot aerial, the strands should be about two to three feet apart. For a 150 foot aerial from three to four feet apart and so on. The strands should never be less than 1½ feet apart even for a very small aerial. All connections should be soldered if possible. Use as many insulators as feasible, remember you have but little energy when receiving; few and poor insulators waste 50 per cent. of the little incoming energy. If you have a good spacious roof it is not necessary to use poles to hold up the aerial. It may be stretched between two chimneys, etc. The spreaders to hold the wire strands apart may be of bamboo, wood, metal pipe, etc. If metal is used, the wire strands should be insulated from the former. (See Lesson No. 11, of the E. I. Co. Wireless Course, on Aerials.)

The ground is quite important. The best wire to use is a No. 12 copper wire run from the instruments to the water or gas pipe using one of the E. I. Co. No. 10003 ground clamps to make an efficient connection. If no water or gas pipe is to be had, bury a metal

8. Today, radio spectrum is an endangered resource with the rapid expansion and global adoption of mobile data. The FCC's National Broadband Plan of 2010, for instance, in the face of a "looming spectrum crisis," proposed a massive transfer of radio spectrum to the mobile broadband industry, which, they argued, already provides all the functionality of television and radio. The National Broadband Plan seeks to set up at least a more fluid system of spectrum management and at most a full-fledged marketplace in which industries can trade their allotted airspace.

plate, copper preferred, not less than three feet square, in a good moist ground; a number of these plates connected to the ground wire would be preferable. It should be buried at least six feet deep. Another good ground is a six to ten feet long iron pipe rammed into moist earth, the ground wire being connected to it, either soldered, screwed, etc. The ground wire running from ground to instruments should never be less than No. 16 B. & S. copper, and can, of course, be bare. Insulation on a ground wire is just that much waste.

Connections and Hook-Ups

The diagrams given in the E. I. Co. catalogue No. 11 show how to connect most of these instruments. Their $.25 book "Wireless Hook-Ups" and their Wireless Course (Lessons 12 and 13), gives hundreds more of them, while their Engineering Department, on receipt of 10¢ to cover postage, will be only too glad to furnish any hook-up to be used in connection with their instruments. Connections should be made with nothing finer than No. 18 B. & S. copper wire (Annunciator wire). All connections must be as short and straight as possible. Avoid all wire crossings as far as practicable, if you can't avoid crosses, the wires should cross each other at right angles; and NEVER wind the connecting wire in coils ("curls") which may *look* pretty, but kills all wireless messages. Make all connections as tight as possible, a loose connection is worse than no connection at all.

Reception of Messages

We presume you have a complete receiving set. You proceed thus:

First, you must know if your detector is adjusted to its best sensitiveness. If no message comes in you don't know if your detector is in its best receptive condition. For this reason, the up-to-date wireless man uses the "Buzzer test." Aside from giving imitation wireless buzzes, the buzzer may be used *to practice telegraphy*. It consists of three things: 1st—Any buzzers, such as the E. I. Co. No. 954, or better No. 965 or No. 950; 2nd—E. I. Co. No. 1118 key; 3rd—A dry cell. Connections MUST be made as shown. Now every time you press the key you will get a perfect imitation of a wireless signal and it becomes child's play to adjust the detector to its greatest efficiency. The buzzer test can of course be used with ANY detector. It saves lots of time and bother and is quite necessary. Sometimes a detector may have a "dead spot" and you will be "listening in" for hours, without being able to catch as much as one dot. The buzzer test makes such an occurrence impossible.

In order that the buzzer itself is not heard directly, it is usually wrapped in absorbent cotton or wadding which muffles its sound effectively, or it may also be placed in another room.

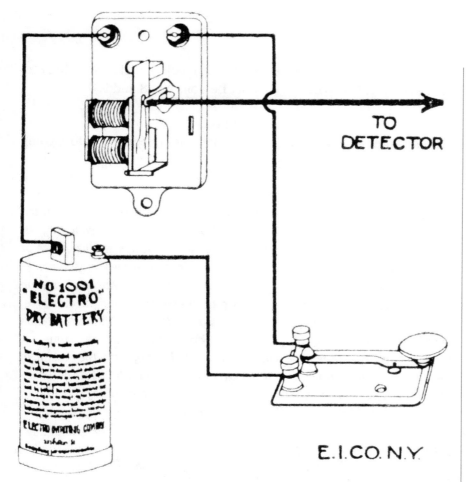

TO DETECTOR

NO 1001 "ELECTRO" DRY BATTERY

ELECTRO IMPORTING COMPANY

E. I. CO. N.Y.

FIGURE 2.

9. Variable condensers were additional components that heightened the selectivity of a simple tuning coil or loose coupler. According to a 1912 handbook, the variable condenser "consists of a number of fixed semi-circular metal plates between which swings a set of smaller movable semi-circular plates. The fixed plates form one half of the condenser and the movable plates the other. In this way the capacity of the condenser is very closely adjustable. The movable plates are provided with a pointer moving over a graduated scale so that the comparative amount of capacity in the circuit is indicated." Alfred Powell Morgan, *Lessons in Wireless Telegraphy* (Newark, N.J.: Cole & Morgan, 1912), 56–57.

Rotary variable condenser manufactured by the Electro Importing Company. The inner spherical drum is made of ebonite. The bottom has several stickers, reading "Property of Harold S. Greenwood, W6MEA" and "Assembler No. 14 Finished this Instrument. Any Defect should be Promptly Reported to Electro Importing Co. 233 Fulton St., N. Y." and a Foreman's Inspection Stamp. Courtesy of the Collection of Jim and Felicia Kreuzer, Grand Island, New York.

When the detector is adjusted the tuning coil (or loose coupler) is regulated by moving the slider or sliders back and forth till the sliders are heard the loudest. If the loose coupler is used the secondary is moved back and forth in addition, till the best position is reached. Now the variable condenser (or condensers) are adjusted if required.[9]

The variable condenser is of the greatest use during excessive "static," which sometimes interferes seriously, during summer weather, especially when "taking" a long distance message. It is also of invaluable help to "cut out" unwanted messages when two or more are "coming in" simultaneously. Thus by adjusting the tuner (or loose coupler) in conjunction with the variable condenser it is often possible to cut out all interference from unwanted stations.

It is an excellent idea to have several detectors in a station, arranged in such a manner that by means of a multi-point switch any one of them can be thrown into the circuit. It will thus be found that some stations, especially during interference, can be heard better on a certain detector than on another. Some will be found to work best for long distance work, others work best for medium distances, etc., etc. (See also E. I. Co. Wireless Course, Lessons No. 8 and 9.)

Sending Messages

Let us quote the law, as far as the amateur is concerned, before going any further:

The Wireless Act

"*Be it enacted by the Senate and House of Representatives of the United States of America, in Congress assembled; That a person, company, or corporation within the jurisdiction of the United States shall not use or operate any apparatus for radio communication[10] as a means of commercial intercourse among the several States, or with foreign nations, or upon any vessel of the United States engaged in interstate or foreign commerce, or for the transmission of radiograms or signals the effect of which extends beyond the jurisdiction of the State or Territory in which the same are made, or where interference would be caused thereby, with the receipt of messages or signals from beyond the jurisdiction of the said State or Territory, except under and in accordance with a license, revocable for cause, in that behalf granted by the Secretary of Commerce and Labor upon application therefor; but nothing in this Act shall be construed to apply to the transmission and exchange of radiograms or signals between points situated in the same State; Provided, That the effect thereof shall not extend beyond the jurisdiction of the said State or interfere with the reception of radiograms or signals from beyond said jurisdiction.*"

General Restrictions on Private Station

"*Fifteenth. No private or commercial station not engaged in the transaction of bona fide commercial business by radio communication or in experimentation in connection with the development and manufacture of radio apparatus for commercial purposes shall use a transmitting wave length exceeding two hundred meters or a transformer input exceeding one kilowatt except by special authority of the Secretary of Commerce and Labor contained in the license of the station; Provided, That the owner or operator of a station of the character mentioned in this regulation shall not be liable for a violation of the requirements of the third or fourth regulations to the penalties of one hundred dollars or twenty-five dollars, respectively, provided in this section unless the person maintaining or operating such station shall have been notified in writing that the said transmitter has been found upon tests conducted by the Government, to be adjusted as to violate the said third and fourth regulations, and opportunity has been given to said owner or operator to adjust said transmitter in conformity with said regulations.*

10. Gernsback: "Wireless Telegraph or Telephone sending stations included."

Special Restrictions in the Vicinities of Government Stations

"Sixteenth. No station of the character mentioned in regulation fifteenth situated within five nautical miles of a naval or military station shall use a transforming wave length exceeding two hundred meters or a transformer input exceeding one-half kilowatt."

Let us explain in plain English just what this means: As you notice from the first paragraph, the part which we underlined, it is pointed out to you that the law does not concern you unless you send messages <u>from one state</u> into another. You therefore do not require a license as long as your messages do not reach over the border of your state and if you do not interfere with a station's business (in your state) which receives messages from another state. Of course, you want to know how you can tell what your transmitting range is. We will explain.

It has been proved by experience with spark coils, that in almost all cases a one-inch spark cannot possibly reach over eight miles. From this information the following table has resulted:

Spark transmitting range.

SPARK COIL SIZE	MAXIMUM TRANS. DIST.
¼ inch	2 Miles
½ in	4 Miles
1 in	12 Miles
2 in	16 Miles
3 in	24 Miles
4 in	32 Miles

Suppose you live in the city of Columbus, Ohio. The nearest state line is Kentucky, about 86 miles in a direct line from Columbus. If you do not wish to be licensed you can use any spark coil up to 10-inch spark, or a ½ K.W. close core transformer.

Suppose your home is in Austin, Texas. The nearest state line is Louisiana, a distance of 230 miles. Thus you could with perfect safety use, for instance, an E. I. Co. No. SO 200 outfit, which does not reach more than 200 miles.

It is also pointed out that if you live within five miles of a Government wireless station you cannot use more than ½ K.W. power, though the next state border might be 100 miles or more distant.

Of course if you live close to another state, as for instance, in New York City, you are required to take out a license for any size transmitter.

What the License Is

The license has not been created to muzzle you; it is the other way around. Uncle Sam gives you a written order telling you that you can send messages to your heart's content, *and no one can tell you to stop sending,* as long as you do not create mischief.

The license is free, it costs not a penny. All that is required of you is that you are familiar with the law and that you can transmit messages at a fair degree of speed.

The law does not require that you take an examination in person if you are located too far from the next radio inspector. All you have to do is to take an oath before a notary public that you are conversant with the law and that you can transmit a wireless message. If you wish to be licensed—and we urge all amateurs to do so, as it is a great honor to own a license—write your nearest Radio Inspector (See below), and he will forward the necessary papers to you to be signed.

Radio inspectors are located at the following points: (Address him at the Customs House):

Boston, Mass., New York, N. Y., Baltimore, Md., Savannah, Ga., New Orleans, La., San Francisco, Cal., Seattle, Wash., Cleveland, Ohio, and Chicago, Ill. Also the Commissioner of Navigation, Department of Commerce and Labor, Washington, D. C.

In an interview with the *New York Times,* W. D. Terrell, United States Radio Inspector for the port of New York, said in discussing the new law:

> The new law regulating wireless messages will work no hardship to the amateur operator. It is the intention first, to classify the various operators and place each operator in his proper class. They will then be permitted to work or play as much as they please, but under an intelligent, general supervision. Only those stations are affected which are near enough to the coastal stations to offer interference, or which work across the state lines which brings them under the supervision of the inter-State laws. I would like to make it very clear that the license costs the amateur nothing, and that the Government is willing to facilitate the wireless operators in every way possible to secure their license.

So much for the law. Everybody will now understand that the law is just and fair and that it gives the amateur a distinct standing in America, a standing which he does not enjoy in any other country. He knows what he can do and what he can't do, and no one can come to him and boss or abuse him, as Government or Commercial wireless operators were wont to do before the enactment of the law.

With sending outfits the reasoning is almost the same as with the receiving outfits.

In order to select an outfit you must, of course, know where and how far you wish to send. Upon all this depends.

As a rule two or more friends get the "Wireless bug" and order two or more complete transmitting sets. Of course, the outfits selected must necessarily be powerful enough to cover the intervening distance between the houses of the friends, and this only you know.

Therefore if you and your friend decide to converse by wireless and if the distance between your two houses is 10 miles you will buy either the E. I. Co. outfits No. SO-10 or SO-15 (SO stands for Sending Outfit, the number indicates the mileage that the outfit will cover). Of course, a more powerful set may be used, although there is no particular advantage in doing so, except, perhaps, that the incoming signals of necessity will be louder with the more powerful sets. It goes without saying that almost ANY receiving outfit which the E. I. Co. list can be used with ANY of the sending outfits. Bear in mind that the selections which they give with their sending outfits do not have to be used if not wanted. Thus their "Interstate" outfit or even their "Transcontinental" receiving outfit can be used with their SO-¼ set. For if you and your friend live one-quarter of a mile apart and both of you have SO-¼ Sending outfits, you probably want to have a receiving outfit with which both of you can pick up messages 2,000 miles distant. In that case you would order two SO-¼ sending outfits only, and two RO-2000 outfits, or else two "Transcontinental" receiving outfits. If either you or your friend feel that you cannot afford such a set, why then get the set that you can afford best and that suits you best. As you see there is no hard and fast rule about the relation of sending and receiving outfits. On the other hand we don't have to tell you that if you wish to obtain two SO-200 outfits you require of necessity a good receiving outfit, else you couldn't hear the station 200 miles off. A little common sense will help everyone decide just what combination to order.

Like receiving sets, the transmitting sets are divided into two groups. The untuned (open circuit) and the tuned (closed circuit) ones.

The untuned ones have, 1st—a spark coil, 2nd—source of power, usually dry cells or a storage battery, 3rd—the spark gap, 4th—the key.

Such outfits can be used only for very short distances and should never be used above three miles. When connections are made by following the blue prints, which is supplied with all sets, the pressing of the key gives a strong spark in the spark gap. The spark gap (the open space between the zinc plugs) from the smallest to the largest sets, must never be more than one-eighth to three-sixteenths of an inch. A bigger gap does not work. Pressing the key long gives a dash, pressing it but for a fraction of a second gives a dot. Combinations of these represent the telegraphic characters; the code can be learned in a few weeks, practicing twice a day from one-half to one hour. (See Lesson No. 15, of the E. I. Co. Wireless Course.)

In the tuned outfits, we have in addition to the above enumerated apparatus: 5th—The Leyden jars, or condenser; 6th—The Helix, or oscillation transformer. The Leyden jars change the red spark obtained from a spark coil, into an intense blue-white crashing spark. The Leyden jars also create a train of fast oscillations and go to make the outfit far more powerful although no more battery power is required. The Leyden jars also give better "carrying power," as the signals can be heard more distinctly and not "mushy" as if no Leyden jars were used. For each outfit the best jars or condensers have been selected by the company and no changes should be made here.

The helix as well as the oscillations transformer, are, to the sending outfit, what the tuner and the loose coupler respectively, are to the receiving outfit.[11] The helix or the oscillation transformer is the tuning coil pure and simple for the transmitting station. Like the tuning coil the helix and the oscillation transformer have sliders or else clips by means of which more or less wire convolutions can be put in the circuit of the aerial. Therefore more or less wire, and consequently more or less wave length is added to your aerial. Again there is not much of a mystery here. Anyone understands it. (See E. I. Co. Wireless Course Lesson No. 14.)

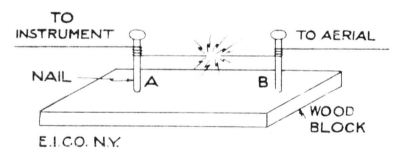

TO INSTRUMENT · TO AERIAL · NAIL · A · B · WOOD BLOCK · E.I.CO. N.Y.

FIGURE 3.

In the larger sets where the battery power is insufficient as well as un-economical we have two methods open to fill the gap. One is the Gernsback electrolytic interrupter working on 110 volts Direct or Alternating current, which supplies the spark coil (transformer coil) with the power;[12] the other method requires the use of a CLOSED core transformer operating without any kind of interrupter direct from the alternating current supply. This kind of transformer, however, does not work on the direct current, not even in connection with the electrolytic interrupter. The choice, for this reason lies entirely with you.

11. A sending station's helix, or oscillation transformer, performs the same function as the coil in a receiving set: it determines the frequency on which the signal transmits.

12. The design of this component is first detailed by Gernsback in **A New Interrupter** in this book.

The aerial switch is an absolute necessity where both a sending and receiving set is used in one station. If you are through receiving a message from your friend, you, of course, wish to answer him. You therefore, must switch the receiving set off from your aerial and switch the sending set onto the aerial. The aerial switch does all this in one operation.

For sets using nothing higher than a 2½-inch spark coil an ordinary double pole, a double throw switch may be used. For heavier sets using more power the E. I. Co. Antenna switch No. 8100 must be used, as the smaller switch cannot carry the necessary power.

Sending a Message

In order to send messages it goes without saying that you must know how to "tap the key." The easiest way to learn and the cheapest way at the same time, is to get a buzzer set as explained under "Reception of Messages." With this set, which represents a first class learner's outfit, you can send yourself dots and dashes to your heart's content until your wrist has limbered up sufficiently to do rapid sending. After a few weeks' practice it will be as easy to send a telegraphic message as to write on a typewriter.[13]

If your friend has a wireless and starts learning the code with you, it becomes very simple for both of you to soon become proficient in the art. Each will send to the other, the Morse or Continental alphabet, which is sent back and forth till the right speed is obtained. After this certain words are exchanged between the stations; later on short sentences are sent and so forth, till it becomes possible to converse freely by wireless.

There is but little adjusting to do when sending. As a rule amateurs converse with only one, seldom two, and rarely three stations. For this reason much adjusting is unnecessary. When using a small set comprising spark coil, Leyden jars, and helix it becomes necessary to adjust the Leyden jars. Either more or less jars (which adds more or less capacity to the circuit) are used till the spark sounds loudest in the spark gap and appears most powerful. A little experimenting will quickly tell when the right capacity is used. *It is important to understand that the capacity should be adjusted only when the spark gap is connected to aerial and ground.* (See Lesson No. 14, E. I. Co. Wireless Course.)

The next important adjustment is in the helix (or oscillation transformer if this is used in place of a helix). To change the clips around on the helix (or on the oscillation transformer) it is necessary that a small gap is first made in the aerial circuit. This is done best by driving two nails in a piece of very dry wood, and connecting the aerial

13. The cadence or rhythm characteristic of an individual telegraph operator's sending touch was known as their "fist." Individual operators were identifiable by their fist, which would become useful during wartime to track messages and their points of origin. The later term "ham radio" is a derivation, referring to the awkward keystrokes of ham-fisted amateurs whose wrists had not yet "limbered up sufficiently." David Kahn, *The Codebreakers: The Comprehensive History of Secret Communication from Ancient Times to the Internet* (New York: Scribner, 1996), 270. Friendships and even romantic relationships were also maintained via characteristic touches and abbreviations ("Hw r u ts mng?"). See "Friends They Never Met," *New York Times,* November 30, 1890. For more on telegraph-mediated love stories of the late nineteenth century, see Mark Goble, *Beautiful Circuits: Modernism and the Mediated Life* (New York: Columbia University Press, 2010). One romance novel, *Wired Love* (1880) by Ella Cheever Thayer, reads: "Nattie's breath came fast, and her hand trembled so she could not hold the scissors. With a crash they dropped on the table, making one loud, long dash." Mark Goble writes, "'One loud, long dash' would translate literally, in Morse, as 'AAAAAAA.' Her body speaks in code even when excited past the point of language" (56).

wires to each nail as shown in sketch. The two wires, A and B are brought close together now and when the key is pressed down a small spark will jump from A to B showing that you are changing the aerial and that energy is radiated from same. Now change the adjustment on the helix (or oscillation transformer) till the longest and fattest sparks jump between A and B. To do this A and B are separated until a point is reached where the spark cannot jump any further. You know now that you are radiating the maximum of energy and the point on the helix (or adjustment on the oscillation transformer) should be carefully marked so you will know at any time just where the maximum is. It goes without saying that you should also note how many Leyden jars (or how many condensers) you are using when making the test and you should write this information down, for if you were to use more or less Leyden jars (or condensers) you would have to change the adjustment on the helix (or oscillation transformer) as explained above. Now after the maximum "radiation" has been ascertained, the test block with the nails is discarded and the break in the aerial wire connected again. You know now that your station is radiating the maximum energy and adjustments of the sending set will not be required for some time to come. Indeed they may be left undisturbed indefinitely.

We believe that we have made everything as plain as possible and that by reading this treatise the elementary points of "Wireless" must become plain to even the layman. If, however, you desire additional information, we will be only too glad to answer your questions promptly and explicitly. Now it's up to you to get busy and "start something"!!

The Radioson Detector

The Electrical Experimenter, vol. 1, no. 10, February 1914

It is a well-known fact that the Electrolytic Detector has always been one of the most sensitive detectors invented since detectors first came into general use.[1] The reason why it has not been adopted as the universal detector is partly due to the fact that the ordinary Electrolytic Detector, as it has been known in the past, was not a really commercial article, for it cannot be denied that even the best Electrolytic Detectors, as manufactured heretofore, had some serious defects. One of the reasons, and perhaps the main reason why it was not used universally, is that in all such detectors manufactured heretofore it was always necessary that a certain amount of acid was handled; this naturally is a serious objection, as not everybody likes to have acid around the instrument table, and for the reason, also, that the acid in the Electrolyte (or rather the water in it) evaporates quite readily, and therefore makes continuous adjustment necessary.[2]

The Bare-point detector, while excellent in many respects, is subject to every draft of air, as the exceedingly fine platinum wire, which can hardly be seen by the naked eye, is usually subject to drafts, and, as a matter of fact, even the operator's breathing against the detector will readily throw it out of adjustment. Of course, this is not the case if the detector should be encased by a glass bell or other cover. However, it cannot be denied that the Electrolytic Detector as a whole is the most sensitive detector if it is put together in its correct fashion.

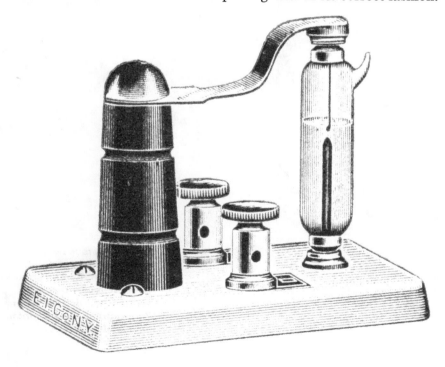

1. Gernsback: "From the Greek Radio—Radius, and Sonus = Sound."

2. The electrolytic detector, or "bare-point detector," was first developed by the Canadian inventor Reginald Fessenden. Fessenden began his career as chief chemist for Thomas Edison and in 1900 was hired by the United States Weather Bureau to develop a system of wireless transmission stations along the nation's coast to relay weather information and, perhaps, better predict hurricanes and floods. The electrolytic detector emerged out of this research as a replacement for the conductive minerals of the crystal detector (see **A Treatise on Wireless Telegraphy** in this book) with an electrically conductive liquid. The electrolytic detector was Fessenden's answer to the problem of early wireless telegraph systems that couldn't tune to particular frequencies. These early systems transmitted on one frequency only, creating a mess of interference for any two sets working in proximity with one another. The electrolytic detector was sensitive enough to "tune." Soon thereafter, De Forest appropriated (some would say "stole") this technology in 1903. See Tom Lewis, *Empire of the Air: The Men Who Made Radio* (New York: HarperCollins, 1991), 38, 49.

For Wolfgang Hagen, Fessenden was a member of a post-Edisonian wave of inventors (including Tesla and Marconi) who trafficked just as much in the fantastical and the occult as they did in capitalist entrepreneurialism and self-promotion. A specifically American form of radio emerges from what Hagen calls the "law of alternating current," developed by the electrical engineer and socialist Charles Proteus Steinmetz in 1893. Drawing on Lacan's tripartite division of the psyche into the imaginary, the real, and the symbolic, Hagen argues that the law of alternating current coalesces in a period of rapid development around the real of functioning machinery; the symbolic language of current, voltage, and resistance; and the imaginary of electrical phantasms. Thus, in the same breath, Steinmetz could praise Tesla as a genius and dismiss him as a crackpot. "From this point forward, one can say that electricians were able to fulfill their dreams of radio, which they perhaps once found in [Edward] Bellamy," Hagen writes, referring to the utopian novel *Looking Backward: 2000–1887* (1888) that included a prefiguration of something like broadcast radio, piped to individual homes via cable: "They could do it if they followed a law—the law of alternating current—and

Many inventors have busied themselves in constructing an Electrolytic Detector that would have only the good features of same and none of its bad ones, but not since the advent of the Radioson has it been possible to produce a really satisfactory article.[3] Even the Bare-point detector, which heretofore has always been considered as the most sensitive detector of this class, is only really sensitive in the hands of an operator who is very familiar with its working and knows exactly all its functions. The writer might state that there are mighty few operators who are fully conversant with the theoretical as well as the practical side of such a detector, and that is the reason why the Electrolytic Detector, as it has been known heretofore, was not as successful as it deserved to be.

The Radioson Detector has been the outcome of years of experimenting and it is interesting to note that only a platinum wire of a certain size, which has been found by experiment, will produce the best results. A few hundred thousandths of an inch variation in thickness will make an enormous difference in the sensitiveness of the Radioson Detector. It might be stated that only one in about four manufactured will come out fit to pass inspection, and the other three must be discarded as useless; this, perhaps, is the reason that this detector costs more to manufacture, and therefore is more expensive than the regular detector.

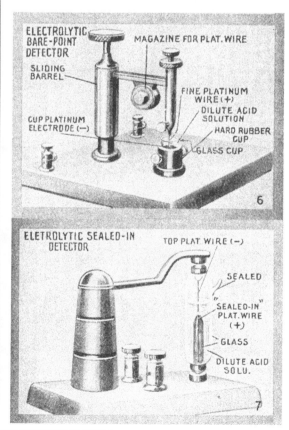

Radioson versus Fessenden's original bare-point electrolytic detector. From "Radio Detector Development," by H. Winfield Secor, *The Electrical Experimenter*, January 1917, 652–53.

from this law their new machines could be built. [. . .] All dreams now have a reference; not merely symbolic, but one that is actually constructed." [*Man kann also sagen: Ab jetzt können die >>Electricians<< ihre Träume vom Radio erfüllen, die sie vielleicht bei Bellamy fanden. Sie Können es, wenn sie einem Gesetz, nämlich dem Gesetz des Wechselstroms folgen und nach diesem Gesetz ihre neuen Maschinen bauen (lassen).*] Wolfgang Hagen, *Das Radio: Zur Geschichte und Theorie des Hörfunks—Detuschland/USA* (Munich: Wilhelm Fink Verlag, 2005), 176.

For information on Steinmetz's political activism, see Sender Garlin, *Three American Radicals: John Swinton, Crusading Editor; Charles P. Steinmetz, Scientist and Socialist; William Dean Howells and the Haymarket Era* (Boulder, Colo.: Westview Press, 1976).

3. Gernsback's "Radioson" was an improvement on Fessenden's bare-point electrolytic detector, the main benefit being that the necessary acid solution was sealed in a protective glass vial, or a "sealed-point." H. Winfield Secor describes the design in a later issue: "Another form of electrolytic detector which will stand considerable rough usage is that known as the Sealed-Point Electrolytic Detector. The commercial form of this instrument, as here illustrated, is known as the Radioson. The operation is the same as in the bare-point electrolytic type of detector and a battery of two dry cells is used with it, together with a pair of high resistance telephone receivers and having the battery potential preferably regulated by means of a high resistance potentiometer. The advantage of this type of electrolytic detector is that the acid is sealed in, consequently does not spill or evaporate." Harry Winfield Secor, "Radio Detector Development," *Electrical Experimenter* 4, no. 9 (January 1917): 652–53, 680, 689–90.

THE PERVERSITY OF THINGS

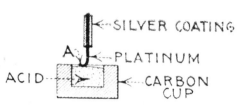

FIGURE 1.

Why is the Radioson more sensitive than the ordinary Electrolytic Detector? Consider the following:

Fig. 1, greatly exaggerated, shows the elements of the ordinary bare-point "Electrolytic," using the finest wire. By observing the extremely fine (0.0001 inch) Wollaston wire under the lens, it will be seen that the contact between the fine wire, "A", and the surface of the acid is never a mere point-contact, but as the fine wire is so very light it curves around and a considerable portion—about ⅛ inch—usually floats or lays on top of the acid, see sketch.[4] This gives a contact of 0.0001" x 0.125" = 0.00003927 sq. inches, which is far too much for high sensitivity. For

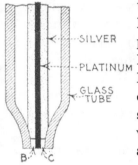

FIGURE 2.

this reason some makers tried to seal in the Wollaston wire into a glass tube and then grinding the point so that only a point of the wire is exposed. However, this was not an improvement. Consider Fig. 2. If the Wollaston wire is sealed in, the silver coating, as well as the platinum wire, comes to the surface. What happens? The acid eats away the silver, and a space, "B," "C" remains between the glass and the sides of the fine platinum wire. The acid by capillary action fills up this space and consequently the contact on such a detector is as large as the one obtained with the bare-point detector. This "sealed-in" detector, therefore, shows no improvement whatever. Now, consider Fig. 3—the Radioson way. By an absolutely new process we succeeded in melting a 0.0002" platinum wire (without silver coating) into a tube made of a specially prepared glass. The acid does not attack platinum, as is well known. Consequently the contact of the Radioson can under no circumstances ever be more than the area of 0.0002" diameter, or 0.0000000314 square inches. Consider this figure with the former one! The Radioson is, therefore, 1246 times smaller than the contact of the best bare-point Electrolytic.

It is, therefore, not surprising that the Radioson Detector is so marvelously sensitive.[5]

The writer has found, and his opinion has been shared by several Radio experts, that the Radioson to-day is unquestionably the most sensitive detector, even far surpassing the Audion, which heretofore was considered the most sensitive

4. Named for its inventor—William Hyde Wollaston, the nineteenth-century British chemist known for the discovery of the elements palladium and rhodium—Wollaston wire is an incredibly thin strand of platinum used in electrical devices: "Platinum wire is drawn through successively smaller dies until it is about .003 inches (0.076 mm, 40 AWG) in diameter. It is then embedded in the middle of a silver wire having a diameter of about 0.1 inches (2.5 mm, 10 AWG). This composite wire is then drawn until the silver wire has a diameter of about .002 inches (0.051 mm, 44 AWG), causing the embedded platinum wire to be reduced by the same 50:1 ratio to a final diameter of .00006 inches (1.5 µm, 74 AWG). Removal of the silver coating with an acid bath leaves the fine platinum wire as a product of the process." See Thomas H. Lee, "A Nonlinear History of Radio," in *The Design of CMOS Radio-Frequency Integrated Circuits* (Cambridge: Cambridge University Press, 2004); "Wollaston Wire," Wikipedia, http://en.wikipedia.org/w/index.php?title=Wollaston_wire.

5. Robert P. Murray, editor of the *Antique Wireless Association Review*, reports on his attempts to reconstruct this Radioson detector in 2005: "I fashioned one after that shown in the 1914 E. I. Co. catalog. This had a similar arrangement of electrodes in a glass cup, but in this case the cathode was a coil of about 3 inches of the same platinum wire. The anode was about ¼ inch of platinum wire soldered into the advancing screw. This detector worked, but very faintly. I found that it worked best when the wire was just about drawn up out of the acid, and possibly pulling on the surface tension. I do not know why the device worked this way. I mention this only as a speculation. I can hardly see 0.001 inch platinum wire without a magnifying glass. I can certainly not see when it dips into the acid, but can hear the result in the headphones. When I hold the wire between my fingers I can not feel it." Robert P. Murray, *The Early Development of Radio in Canada, 1901–1930: An Illustrated History of Canada's Radio Pioneers, Broadcast Receiver Manufacturers, and Their Products* (Chandler, Ariz.: Sonoran Publishing, 2005), 3.

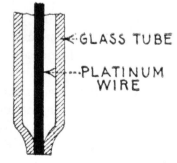

FIGURE 3.

detector manufactured.[6] It is a matter of record that by connecting a double-pole, double-throw switch on one side of the Radioson and connecting on the other side of the switch to an Audion, it will be found that the Radioson is far more sensitive than the Audion. In some cases signals that can not be heard at all with the Audion come in fairly loud with the Radioson.

The Radioson is, to-day, the only detector known that needs no adjusting whatsoever. An important point is that messages come in clearly and distinct even while the detector is shaken, and for this reason it is, of course, never subject to shocks and it is, therefore, indispensable for portable sets, in automobiles, railroad trains, ships, aeroplanes, etc. The acid as well as other sensitive parts are sealed into the detector cartridge. For this reason there is never any spilling of the acid nor any danger of the acid coming into contact with the hands of the operator. The Radioson is adjusted to its highest sensitiveness at the factory, and for that reason it is quite impossible to put it out of adjustment except if the cartridge is broken or unless a high tension discharge is put through the detector.

The Radioson practically requires no attention, it is always ready for use and the operator never loses part of a message on account of bothersome as well as annoying adjustments common to EVERY OTHER detector.

The Radioson is clean as well as very compact. It works on a shaky table as well as on a steady foundation. An interesting fact is that the Radioson does not require the use of a Potentiometer, but it is necessary to use two dry cells (three volts) in connection with the detector. These cells may be of very small size, such as a flashlight battery.

In order to get the best results with the Radioson it is necessary to use it in connection with at least a 200 ohm head set, or a higher resistance set up to 8000 ohms: either set may be used, but nothing less than 2000 ohm must be used, as too much current would flow, which, in time, would destroy the very fine platinum wire; this naturally would make the detector useless.

The writer, who designed this detector, found that by placing the anode, that is, the member carrying the fine platinum wire (contrary to other sealed-in electrolytic detectors), upside down, better results obtained. This is done for the reason that it allows the microscopic gas bubbles to disengage themselves more readily from the anode point than if the sealed-in anode was placed in the usual position, namely, point down. In the latter case, the gas bubbles sometimes adhere to the point, which, of course, decreases the sensitiveness of the detector, as has been often found by many experimenters.

A very interesting fact about the Radioson is, that when it has been

6. This is a huge claim. De Forest's Audion would soon replace all other forms of detector and revolutionize the way we listened to radio. But could this idea that the electrolytic detector might outperform the vacuum tube, a media historical path not taken, actually hold weight? While Gernsback would admit by 1919 that the Audion was a technically superior form of detector, he held out on the importance of other designs that more easily invited amateur experimentation. The Audion tube was a complex and expensive component (at least initially), one that heralded the coming of fully-assembled radio sets built for passive listening in living rooms across the country. Thus the divide between vacuum tubes and crystal or electrolytic detectors at the end of the 1910s was less about technical superiority and more about technical literacy. Gernsback writes in a retrospective 1919 editorial on the coming of the "radio telephone": "Using the Audion as a generator for undampt waves and as a RT transmitter is of course a great accomplishment in itself. And the device works well, better than anything else so far. But it is not the ultimate goal. Vacuum tubes of the Audion type are tricky as yet, and not too practical. Unless you use special tubes, and you can't just now, due to a complicated patent situation, the speech is not always clear, and far from satisfactory. At the critical period, the tubes often go bluey and refuse to talk. Amateurs therefore should look for substitutes of vacuum tubes or devise other tubes employing entirely different principles." Hugo Gernsback, "Developing the Radiophone," *Radio Amateur News* 1, no. 6 (December 1919): 269. Readers continued, however, to be interested in crystal set designs well into the 1920s. See, for instance, **The Detectorium** in this book.

used for several months, it is sometimes found that it is not quite as sensitive as it was originally. All that is necessary to do then is to take out the cartridge and shaking it violently by holding it between two fingers and shaking it in the direction of its axis. This immediately restores its full former sensitiveness for the following reasons: Although the acid, as well as the other ingredients used in making the electrolyte are chemically pure, there is always a chance that some microscopic particle of material might partly cover the anode, but by shaking the electrolyte, this particle will readily come off, and, besides, the shaking has the effect of also cleaning the glass as well as the anode point in a very efficient manner. For this reason the Radioson has a very long life, and if it is handled carefully it will last for years; furthermore, the electrolyte used does not affect the platinum wire in any manner whatsoever, even if the detector is used continuously.[7]

Persons familiar with the Electrolytic Detector might be of the opinion that as the acid as well as the anode is sealed in airtight, sooner or later the working of the Radioson might be affected, on account of accumulation of gas. However, this is not the case, as the gas bubbles on account of the extraordinary small dimension of the anode are microscopically small. By looking at the figures above, giving the amount of anode area exposed, this will be readily understood, and, while it is not to be denied that there must be a certain amount of gassing, the same is so very slight that, for practical use, it does not come into consideration at all.

7. By the time of his 1922 book, *Radio for All,* Gernsback admits the inherent technical limitations of the Radioson detector: "Unfortunately the Radioson, once subjected to strong signals or even too strong static currents, would burn out the exceedingly fine Wollaston wire, after which the instrument became inoperative. Although the Radioson was perhaps one of the best electrolytic detectors ever designed, no means could be found to keep it from burning out and the manufacture of it was given up by the makers. Soon after the invention of the electrolytic detector, crystal detectors came into vogue." Hugo Gernsback, *Radio for All* (Philadelphia: J. B. Lippincott Company, 1922), 58.

The Radioson Detector as advertised in the November 1914 issue of *Popular Electricity and Modern Mechanics.*

Another interesting point in connection with this detector is, that, by placing several Radiosons in parallel, this will increase the volume of the sound, and, although the increase is not more than 10 or 15 per cent., it is quite noticeable. Placing the detectors in series cuts down the efficiency.

Another very important fact is that heating the Radioson cartridge increases its sensitiveness enormously. Placing it very near to a steam radiator or letting the sun shine upon it, will bring in the signals sometimes fully 200 per cent. louder. This phenomenon was discovered by Dr. Branley [*sic*] of Paris some years ago.[8]

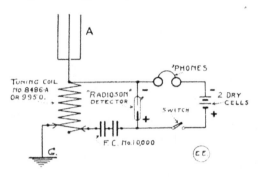

All in all it may be said that without exaggeration the Radioson Detector is, to-day, the most sensitive detector that has been devised as yet. The Electro Importing Co., the manufacturers of this detector, guarantees each and every detector in all respects, and the Company furthermore guarantees that every Radioson is absolutely uniform, and it will be observed that all of them, when compared, will be equally sensitive. This is a very important feature, especially if comparative tests in the intensity of received signals are required.

The author will be glad to answer any questions concerning the Radioson, and he shall be glad to furnish such information as is consistent to give in connection with this detector.

8. Not only did Édouard Branly discover the effect of heat on a coherer in the 1890s, he was the first to describe and publish on its operation. See **The Dynamophone** in this book for more about coherers.

August 1915 cover of *The Electrical Experimenter.*

Sayville

The Electrical Experimenter, vol. 3, no. 4, August 1915

Sayville and Tuckerton are at present the only links connecting the American Continent directly with the German Empire. Both stations can now receive and transmit wireless messages to and from Germany, without any relaying means, at any given hour of the day.[1]

First Tuckerton and now Sayville have been taken over by our watchful Government with the evident intention of preventing the two stations from committing an unneutral act, in conformity with international law practice.[2]

The layman will immediately ask, that if the United States Government takes over the wireless plants, why does it not also take over the various transoceanic cable stations? He furthermore cannot see what harm the wireless stations can do, because they have been required for some time to use plain English or plain German, no code messages having been allowed since the outbreak of the war.

1. Sayville, New York, is located on Long Island. Tuckerton, New Jersey, is on the southeastern Jersey shore, just north of Atlantic City. Both towns were the sites of massive wireless telegraph stations owned by the German company Telefunken.

2. The following month's issue explains exactly why the Sayville station was closed. Charles A. Apgar, an amateur inventor and contributor to Gernsback's magazines, had devised a way to record wireless telegraph signals on phonograph cylinders, the first permanent record of a wireless message ever produced. Using this method, Apgar produced recordings of covert German transmissions that were sent via Sayville to U-boats operating in the Atlantic Ocean. His presentation of these findings to the U.S. Navy precipitated the closing of Sayville and a law banning all amateur wireless activities. Both a model of the value of the amateur wireless tinkerer and the cause of the entire amateur community's work being put in jeopardy, Apgar was a divisive figure: "Meanwhile, the part played by the magazine [*Electrical Experimenter*] aroused sharp resentment from the old Sayville officials. Dr. K[arl] G. Frank, head of the station, wrote a bitter letter to the editor, the point of which was a little blunted by the fact that by the time it was printed, the Government had already closed Sayville. Dr. Frank, incidentally, was later convicted as a German Intelligence agent." T. R. Kennedy Jr., "From Coherer to Spacistor," *Radio-Electronics* 29, no. 4 (April 1958): 45–59. For the story of German cryptography channeled through American wireless plants during World War I, see Grant Wythoff, "The Invention of Wireless Cryptography," *The Appendix* 2, no. 3 (July 2014): 8–15, http://theappendix.net/issues/2014/7/the-invention-of-wireless-cryptography.

The answer is simple enough. A cable message during the time of its dispatch stays on the cable. It has only one destination; no one can "tap" the message without serious difficulties. Not so with the "wireless." Its waves being propagated in every direction, a thousand stations, or more, if properly equipped, can catch the message anywhere within the receiving radius of the sending station.[3]

No doubt the far-seeing shrewd German Government long ago foresaw the possibility of a war with England and the inevitable isolation of the empire which would result in consequence thereof.[4] It clearly foresaw the cutting of its cables by the enemy and it acted accordingly. Sayville and Tuckerton was the answer. Thus when England actually did cut the German cables early in August, 1914, Germany was by no means isolated telegraphically. Thanks to the wireless, which a wise government had long before planted on foreign soil, telegraphic traffic between America and Germany goes on the same as before, with the difference that the messages now travel over the very heads of Germany's enemies. With the inauguration of the German submarine warfare, the German Admiralty doubtless found a powerful weapon in the shape of the Sayville and Tuckerton wireless stations. The two stations being controlled almost entirely by German capital, it was reasonable to expect that they would support the German navy and its submarines to the best of their ability. Before the United States Government took over the Sayville station, early in July of this year, we had been reading a lot of nonsense as to some new devices being employed at that station which were supposed to be used in sending out messages by means of a special time-spacing system between the dots and dashes, as well as by varying the length of the dots and dashes themselves, which latter make up the telegraphic alphabet. Such a thing is, of course, not impossible, but why should it be done if much simpler means are at our disposal?

Let us imagine the following: A German spy is located on the ocean liner *Adriatic* headed for Liverpool. When the ship is two days out the spy learns that the ship, on account of submarine danger, will not dock at Liverpool but at Greenock (Scotland) instead. This is important information. Within ten minutes he has sent a wireless to a stockbroker in New York as follows:

H.P. Frye & Co.
235 Wall Street, New York
Sell at once 2,000 shares U.S. Steel at 58.
　—John Miller.

3. This issue includes the latest installment of Gernsback's serial novel, *Baron Münchhausen's New Scientific Adventures,* in which the narrator establishes a wireless connection with the Baron as he lands on the surface of the moon. The next pages switch back to hard technical description with a feature that details precisely how submarines send wireless messages from the depths.

4. In a 1918 letter published in the *Chicago American* and later the *New York Times,* Gernsback stresses to the American public the importance of his home country Luxembourg in stemming the aggression of Germany. Four years after the German invasion of Luxembourg, Gernsback writes: "For a generation back Germany has cast covetous glances at the little country—no doubt on the theory that 'he who does not honor the pfennig is unworthy of the thaler.' Due to its insidious and clumsy methods, however, little headway was made in Germanizing the Luxembourgeois. . . . Of course the Luxembourgeois greatly resent the German occupation, just as much as do the Belgians. Since the invasion the inhabitants have been in a more or less ugly mood, as testified on good authority by the information which filters through them from time to time." Hugo Gernsback, "The Case of Luxembourg," *Chicago American* (March 1918).

Paul Lesch argues that Gernsback actively cultivated a Luxembourgish identity throughout his life: "It is particularly interesting that Gernsback, born of German parents, not only insists on the anti-German sentiments existing in Luxembourg at the time, but seems to share them as well." Luc Henzig, Paul Lesch, and Ralph Letsch, *Hugo Gernsback: An Amazing Story* (Mersch: Centre National de Littératur, 2010), 18.

When Frye & Co. receive the message they consult their code book and find that it reads thus:

Adriatic will dock at Greenock.

Within twenty-five minutes after John Miller handed his message to the operator on the *Adriatic* Sayville has received and dispatched the following message to its Nauen (Germany) station:[5]

F.S. Schneider & Co.,
Friedrichstrasse, Berlin.
Cannot dispose 2,000 shares U.S. Steel at 58. Are bid 55½. Advise.

This message is promptly received by the German commander of the submarine U-69, bobbing up and down not far from the south coast of Ireland.

He also consults his code book and deciphers the harmless message thus:

Adriatic will dock at Greenock next Tuesday.

With this intelligence the German submarine commander is enabled to change his course in order to successfully hunt his quarry.

This is only one of the ways how the wireless stations at Sayville and Tuckerton can be used successfully to violate our neutrality; there are undoubtedly scores more.

In view of the above it is not quite clear how the United States Government, even by exercising the greatest care in censoring messages, can hope to prevent at all times the dispatch of unneutral wireless messages.

5. The Nauen station's power was renowned throughout the world and was the subject of much interwar speculation over secret research conducted by the Germans into "rays." See **Predicting Future Inventions** in this book.

Baron Münchhausen's New Scientific Adventures, PART 5

"Münchhausen Departs for the Planet Mars"

The Electrical Experimenter, vol. 3, no. 6, October 1915

"THERE are more things in heaven and earth, Horatio, than are dreamt of in your philosophy."

So sings Shakespeare. One of these "things in heaven" is the Planet Mars, the most fascinating, the most astounding revelation to the feeble human intelligence. Shakespeare, the master of the drama, never conceived anything like a drama of an entire world—millions of intelligent beings—fighting a heroic battle, a battle for existence. Yet this drama was going on right before his very eyes, but 35 million miles away; for the Martians have been fighting for water ages ago, and the available supply becomes smaller each year.

There is nothing more inspiring, nothing more gripping to the imagination, than this wonderful battle between organized intelligence on one side and unrelenting nature on the other.

Mr. Münchhausen's scientific lecture gives you the latest facts—now almost universally believed—about Mars. You can spend no better half hour than turning your mind from your humdrum existence towards a subject which is as absorbing as it is lofty in its grandeur.[1]

Once upon a time, a grouchy old gentleman with a grievance for fiction writers, presumably because the latter received more emoluments for their stuff than the former for his poetry, thus vented his resentment in immortal song:

"'Tis strange, but true; for truth is always strange—stranger than fiction."

From this, some coarse soul, totally oblivious of any poetic infection whatsoever, took it upon himself to mutilate the above passage of one of Lord Byron's poems and taught us unsuspecting mortals to hawk, parrot-wise, ever after until the end of fiction, thusly: "Truth is stranger than fiction!"

With all due regard to the memory and genius of Byron, I, I. M. Alier, a citizen of a free country, take it upon myself to correct his Lordship at this late and quarrelsome date, to wit:

"There is no fiction."

1. The character of Baron Munchausen was inspired by the real-life Hieronymus Karl Friedrich, Freiherr von Münchhausen (1720–1797). Rudolf Erich Raspe created a series of fictionalized accounts of Baron Munchausen's life in the 1780s, tall tales about "Singular Travels, Campaigns, Voyages, and Sporting Adventures." After Raspe, the character was picked up by a number of other artists in various media, including Terry Gilliam's 1988 film *The Adventures of Baron Munchausen.* Gernsback uses the original spelling of the real Baron's name.

If, as often—no, always—has been proved that the most violent fiction at some time or other invariably comes true, then by all proceeds of modern logic, there cannot be such thing as fiction. It simply does not exist. This brings us face to face with the startling result that if fiction always comes true some time or other, why then, bless their dear souls, all fiction writers must be prophets!! Hurrah for the F. W.!! But hold on, boys; don't let our enthusiasm run away with us on a Ford. The spark plug has run afoul somewhere. While it's nice to be a prophet, don't forget that a prophet is never, never recognized in his own country. Thus the New Testament teaches; so I think it will be safer for all F. W. to remain F. W., rather than to be honorless prophets.

May 1915 cover of *The Electrical Experimenter.*

However, that is not what I had in mind when I started—it's so hard for me to say what I mean, and a good deal harder for me to keep my thoughts running on the track. They ramble from one nothingness into another. My mind in that respect is a good deal like a one-eyed, religious old cow on a pasture. She eats up whatever she sees alongside of her, but when she finally turns around she perceives with astonishment that there is still a lot to graze on the other side; so she steers around to starboard and returns to her original starting point.

But I am rambling again. So let's return to the original starting point.

Seriously speaking, and by way of emphasizing how much stranger truth is than fiction, I have but to point to Jules Verne's famous stories. When 45 years ago he wrote *"Twenty Thousand Leagues Under the Sea,"* no one took him serious. It is doubtful whether he himself believed that the submarine which he invented in that story would ever become practical. It was just fiction. Yet 45 years later we see how a submarine, almost exactly as his vivid as well as prophetic mind conceived it, down to the most minute detail, emerges from a German harbor and travels under its own power over a distance of 4,000 miles, through the North Sea, the English Channel, down the Atlantic, through the entire length of the Mediterranean and up through the Dardanelles to Constantinople! And by way of diversion it manages to sink several battleships of the enemy by means of its torpedoes.

Now, bold as he was, Jules Verne never conceived such an "impossible thing," and while his famous *Nautilus* was equipped with almost every other modern submarine necessity, the infernal automobile-torpedo was missing. Truth is indeed very much stranger than fiction. Hundreds of similar instances could be cited, but lack of space prohibits it; besides, I mustn't ramble.[2]

Münchhausen, as will be remembered, had explained the mysteries of the moon to me, and he had also mentioned the great danger of falling meteors, which had been increasing alarmingly in number for some time. ⁺The moon's attenuated atmosphere offers no protection from meteors, as did the earth's thick air. But few meteors ever reach the surface of the earth; the colossal friction between the meteor and the air ignites the former, and most of it falls down on the earth as a fine dust. The burning of the meteors represents the shooting stars we see. On the moon, however, the meteors crash down bodily, causing tremendous havoc, and this terrible bombardment goes on forever, without let-up.⁺ Consequently, when Baron Münchhausen stopped short that evening in the midst of a sentence, I naturally was alarmed not a little. Great, therefore, was my joy when, sitting before my radio set the next evening, 'phones clapped tight over my ears, my eyes glued on the clock, the familiar high, whining spark suddenly reverberated in my ears at the stroke of 11 o'clock.

It was Münchhausen. But his usual sonorous voice to-night had an unfamiliar metallic timbre that puzzled me greatly; in a short time, however, the mystery was cleared, and this is what poured in my astonished ears:

"My dear Alier. No doubt you thought I had been killed by a meteor last night. Well, as you Americans put it, I certainly had 'a close shave.' A meteor crashed down on my aerial 50 feet from where I was sitting; of course it went up in smoke—metallic vapor, to be correct—due to the tremendous heat generated by the impact of the meteor on the granite rocks. The whole meteor itself went up in a fiery cloud of red vapor and I was blown headlong a distance of over 50 feet, right down into the mouth of a giant crater, by the colossal resulting blast of the concussion.

"Now, this long-extinct crater is a very deep one; how deep, I was soon to learn! I went down head first and kept on falling at a terrible rate of speed. I must have been falling down that awful abyss what seemed to me like hours. As I kept on plunging down, I was gloomily reflecting what an inglorious death it was to die at the bottom of an unromantic crater on a dead and dried-up moon. I thought of many things, when I suddenly became conscious of a terrific cold. Call it instinct or presence of mind, as soon as I had started on my downward

2. Gernsback: "In order to distinguish facts from fiction in this installment, all statements containing actual scientific facts will be enclosed between two ⁺ marks.—AUTHOR"

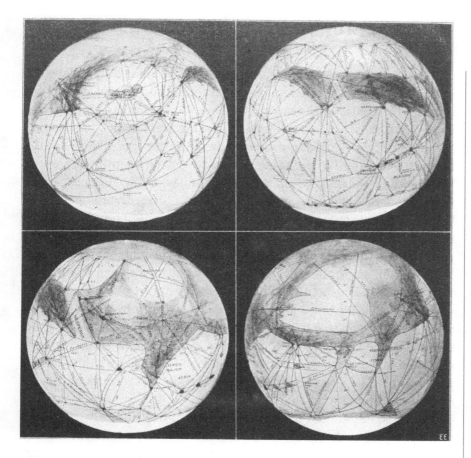

FIGURE 1. "Four different views of the planet Mars. As Mars turns on its axis once every 24 hours the same as the earth, we are enabled to see the entire surface of the Martian globe during that time. The four views, as shown, are therefore taken six hours apart from each other. These pictures were made during the last "opposition" in 1911, when Mars was some 47 million miles distant from the earth. It never comes nearer than 35 million miles to the earth. In 1924 the two planets will be but 35 million miles apart. In the views shown, *the top is south, the bottom north*, for through the telescope all objects are turned upside down. The white patch at the bottom is the north Polar snow cap, the southern cap is not in evidence for it has melted already. The melted water has been conducted equatorward by the "canals." Note that the canals run through the dark areas, which are not oceans therefore, but land with vegetation. The light areas are deserts. Nearly all canals are perfectly straight, the ones near the edges of the pictures appear curved only because we are looking on a globe and not on a plane surface. Photos courtesy of Prof. Percival Lowell, Flagstaff Observatory, Flagstaff, Ariz."

journey I had jerked my body in such a manner as to righten it; in other words, after a few attempts, I succeeded in falling feet down. It was indeed a fortunate circumstance that the sun was almost directly over the crater, for it saved me the anguish of plunging into a pitch-black abyss. While it was not nearly as light as it was at the top, still I could see where I was falling, and that was some consolation. Thus, when I glanced down toward my feet after a while, I am sure that my heart, which had stopped beating, stood still entirely for some seconds. It took me a few seconds to collect my bewildered senses, for this is what I had seen:

"*The crater had no bottom at all,* but went right through to the center of the moon, where it connected with another crater, *which went to the opposite side of the moon.* I knew this must be so because when I had looked down I had seen several stars shining through brilliantly from the night side of the moon. Then the awful truth flashed through me and I almost swooned. *I was falling through the whole length of the moon!* I had been in many tight quarters before during my somewhat exciting career, but this experience indeed bade well to be the inglorious end of my adventurous life. However, my far-famed presence of mind and my cool head soon asserted themselves, as was naturally to be expected of me.

"I knew the diameter of the moon to be 2,164 miles. A quick mental calculation proved that it would take my falling body about 24 minutes to reach the center of the moon. As there was nothing to stop my fall. I must naturally continue to fall, due to the tremendous momentum acquired, till my body would *almost* emerge at the opposite side of the moon at the mouth of the other crater. At this point my speed would be zero and I would have fallen for 48 minutes. If I could not manage then to grasp a projecting rock, I would commence to fall back again toward the center of the moon. I reasoned that once more my momentum would carry me past the center and I would then be almost carried to the mouth of the opposite crater—my original starting point.

"I say almost, for the friction of my body against the air would tend to retard my fall. If at this point, where my speed was again zero, I could not succeed in taking hold of a projecting rock of the crater's side I would begin to fall down once more, the same as before. I would then continue falling back and forward exactly like a bouncing ball, each time, however—just like a rubber ball—a little less than on the previous plunge. Thus my drops would become of shorter and shorter duration, and finally I would fall no more.

"As I had mentioned before, the sun was almost overhead, shining down into the crater. I also remembered that it was almost exactly 12 o'clock midnight, terrestrial time, when the meteor smashed my aerial; this, then, was the time I started on my remarkable journey into the bowels of the moon. With a tremendous effort I pulled out my chronometer and noted that it was 12.23 a. m. In another minute I would fly past the center of the moon. Looking about, I saw in the uncertain light that I went whizzing through an immense hollow, proving to me that the center of the moon was far from solid, due no doubt to the centrifugal force of the moon at the time when it had not solidified, some millions of years ago. I estimated later that the moon was an immense hollow sphere with a solid crust about 500 miles thick. By way of a homely comparison, the moon therefore must be a hollow globe like a rubber ball. Like the latter, it is filled inside with air, while its crust can be compared to the rubber of the ball.

"In another minute I had passed the center and was now dropping toward the other side of the moon. If I continued falling in my present position, I must naturally emerge at the opposite side with my feet toward the sky, as a little reflection will reveal to you. So once more I jerked my body about, and I was falling 'up,' with my head at the top, my feet pointing to the sun. At the end of another 24 minutes I could feel my body slowing up from the terrific speed. As the crater at this side of the moon was fortunately rather narrow, I found little difficulty in reaching for a projecting rock as soon as my plunge had come to a dead stop. I held on for dear life and clambered up a narrow ledge, where I fell down exhausted and panting from my dreadful experience.

"My sensations in falling through 2,164 miles of space, going over 16 miles per second at the center of the moon, you would, of course, like to know. Well, the first minute it is rather unpleasant. Highly so. The place where your stomach should be by right is one vast area of nausea. But once you become accustomed to it, it becomes bearable, for there is nothing else to do. You might think that the rush of air would kill you in a few seconds, or else draw all the air out of your lungs, and thus asphyxiate you. But neither is the case, for the air is so thin on the moon that the rush is not as terrific as it would be on earth. Also, by keeping the mouth tightly shut and breathing—with difficulty, it is true—through the nose, one does not die in 48 minutes. The friction of the air against my body did not ignite either, for, as I told you some time ago, the temperature inside the moon is near the absolute zero, the awful cold of the stellar world. But neither did I freeze to death, for the simple reason that the friction of my body through the attenuated air was just sufficient to keep me comfortable. Thus you see that if it had not been so cold, I would have burned up; and vice versa, if the friction of the air against my body had not heated it, I would have frozen to death long before I reached the center of the moon. Then, too, another important point to consider is that on the moon, as explained previously, my body weighed 27 pounds, against 170 pounds on earth.[3] This is, of course, a rather small weight, and for that reason my fall was not as terrible as it would have been if my body had weighed 170 pounds, as on earth. For that reason, too, I was not attracted so much to the sides of the crater as I would have been if my weight had been greater. Also it was fortunate that the two craters widened out considerably the further down they went into the moon's interior. As a matter of fact, the 'hole' of each crater at no point was less than three miles in diameter. This was indeed very lucky for me, for the following reason:

†"If we drop a stone into a very deep and narrow shaft, as has been shown experimentally on earth, this stone will never reach the bottom. Instead, *it will bury itself in the eastern wall* of the shaft long before reaching bottom, providing the shaft is deep enough. The explanation is that the earth rotates on its axis from west to east at a speed of 1,524 feet per second at the equator. Thus, it is apparent that the earth revolves quicker than the stone can fall in a few seconds.[4] It therefore intercepts the stone's flight, with the result that the stone must of necessity strike the eastern wall of the shaft. This phenomenon is termed 'the falling of a body toward east.'†

"Now, precisely the same condition exists on the moon, of course.

3. Gernsback: "†An object weighing 1 lb. on Earth weighs 0.167 lb. on the Moon.†"

4. Gernsback: "†The speed of a falling body at the surface of the earth after the first second is 16½2 feet. In 6 seconds a stone would have traveled but 579.†"

Fortunately, I started falling at the western side of the crater, but as the latter was so wide I never came near enough to its eastern wall to hit it. Likewise the other crater, at the opposite side of the moon, measured some four miles in diameter and, while I finally did reach the eastern wall, my flight had come to an end as explained already. Indeed, nature favored me all through, for the moon rotates with a velocity of but 15⅓ feet per second at its equator, against a speed of 1,524 feet of the earth. For this reason there was no danger that my body would collide with the sides of the crater somewhere in the interior of the moon, for my flight was far more rapid than the speed of the moon's rotation on its axis.

"But in the meanwhile my troubles were far from being terminated. No sooner had I regained my breath than I became conscious of the terrible cold; for I was now but a few feet from the surface of the moon, but on that side which was turned away from the sun, where nothing but icy cold, darkness and desolation reign. Aside from this, I was some 2,160 miles away from Flitternix, my companion, and our 'Interstellar.' Walking around half of the moon was out of the question; neither could I stay where I was without freezing to death. So I climbed up to the surface of the moon with considerable effort. Then by the aid of the starlight I ran rapidly around to the western side of the crater, for I had to run in order to keep warm. After having obtained my bearings by the aid of the stellar constellations, to make

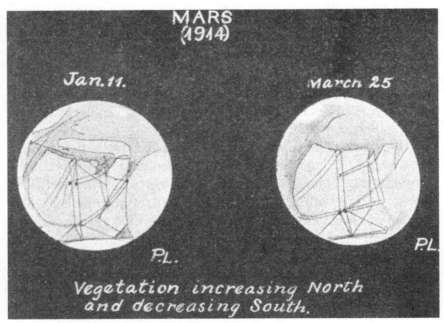

FIGURE 2.

sure that I was at the western side of the crater, I took a deep breath, looked down in the abyss through which the sun was shining from the other side, and dived head down into space once more.

"You see, I had reasoned that it was far better to attempt the flying journey through the moon once more than to perish with the cold on the dark side of the moon. Besides, I had experience now and having been successful once, it was natural for me to expect success again. I had nothing whatever to lose, and everything to gain.

"My first experience was repeated without any incident; furthermore, I calculated that I should land at the eastern wall of the far crater within 48 minutes if everything ran smoothly. But I had left our good old sun out of my calculations. You see, the gravitational attraction of the sun controlled the fall of my body in the same proportion as it controls the rotation of the moon and the earth. I mentioned how in my former flight I had risen to the top of the moon; as a matter of fact, somewhat higher, for the opening of this crater was higher than the surface of the moon. But now I was falling *toward* the sun, and the sun was aiding and accelerating my flight; for I moved constantly nearer to it.

"For this reason at the end of 48 minutes I did not strike the eastern end of the crater. Instead, I whizzed right past the eastern wall, almost brushing it, and continued to rise up into the air about 100 feet before my speed was spent. I promptly prepared myself to plunge down into the crater again. Indeed, before I realized it, I had begun to fall down once more when the unexpected happened.

"I suddenly felt a rope encircling my body, and before I had time to think, I was jerked sideways, and in another second I had fallen on a heap of sand and looked with astonishment into Professor Flitternix's eyes, who stood over me grinning sheepishly!

"This is what had happened: Flitternix had, of course, seen me fall into the crater, and as he had rushed to the edge, he had seen how I dropped down with lightning speed. Looking closer, he also noticed what I saw, namely, that the crater went right through the entire mass of the moon, for he could see the stars shining through from the other end. He was loath to believe that the fall would kill me, and, as a scientist of note, he calculated exactly in advance what was likely to happen to me. He reflected that it would take me some two hours to make the round trip, as he knew that I could not possibly stay at the other side of the moon. He reasoned, correctly, that in case I was not killed I would come swinging through the crater in due time. Unperturbed as he is by such mere details, he went to the 'Interstellar' and had his lunch. Within two hours he returned to the crater, armed with a telescope and a long rope. It did not take him long to locate me down in the abyss by means of his glass, for I was rapidly coming to the surface then. Attaching one end of the rope to a near-by rock, he fashioned a sliding noose on the other side and waited.

"Now it must be said to the credit of Flitternix that in his younger days he had lived in the West on a ranch, and there had become an expert in the science of lassoing. He boasted that once he lassoed a common sparrow by its left hind leg, but this I believe to be somewhat exaggerated. Be that as it may, when I finally emerged to the surface, a living piece of lava ejected out of an extinct crater, Flitternix had little trouble in lassoing me as I came whizzing up. Whereupon I thanked him and asked him if lunch was ready, for the trip had given me quite an appetite, as you may well imagine. Luncheon over, we decided right then and there to quit the moon, for Flitternix, as well as myself, were of the opinion that there was little further to be explored on this dead world. Besides, the meteors had become so alarmingly frequent that it would be only a matter of time when one of us would be killed.

"Flitternix wanted to return to earth at once, for he itched to give a lecture before the American Astronomical Society, whose honorary president he was. I, however, had more ambitious plans. I once had looked through the great telescope of the Lowell Observatory at Flagstaff, Ariz. If I live to be a thousand years old I will never forget the glorious sight which then presented itself to my eyes.

†"I saw a ball, lighted up dazzlingly at both extremities. I saw great patches of an ochre red scattered over the surface of the sphere and I had seen dark blue areas among the vast ochre patches. Over the latter runs a mass of fine lines, nearly all of them connecting with the white caps at each extremity. Moreover, these fine lines cause one to gasp involuntarily, for they are as straight and true as if laid out with a rule and pencil. More astonishing yet, some of these lines run absolutely parallel with other ones for the whole length of their extent. And more wonderful yet, whenever two or more lines meet in a junction there is invariably a round black point.

"The ball I had been looking at transfixed for a long time was Mars, the nearest planet to earth, then 37,000,000 miles distant from the latter. The late Prof. Percival Lowell, the great authority on Martian research work, had convinced the scientific world that the dazzling white caps at the poles of this planet are the polar snow fields. The great ochre patches are desert land, while the dark blue areas represent large tracts of fertile land and its resulting vegetation.

"Now, according to well-known physical laws, proved beyond discussion, the smaller a body, the quicker it will cool off. All planets and their moons once were white-hot like our sun. The smaller ones cooled off first and the larger ones are not cold yet. Thus the earth, which measures 7,912 miles in diameter, is still red-hot in its interior, as is proved by its active volcanoes. The moon, which is but 2,164 miles across, cooled off ages ago. The oceans once filling their beds then filtered down into its bowels, there to freeze solid, for there was no heat to keep the water fluid. Its atmosphere, which was formerly as dense as that of our earth, was gradually thrown off into space, til to-day practically no atmosphere remains. Thus the moon to-day rolls on through space a dead world.

"The planet Mars, measuring 4,363 miles in diameter, as will be seen, is only twice the diameter of our moon and much smaller than the earth. Consequently it must be rapidly nearing its extinction, the same as the moon. Its oceans are already dry, while most of its land is desert. The atmosphere has nearly all gone too, proved by the fact that we seldom observe clouds on Mars through the telescope. But there must still be water on the planet as yet, this being irrefutably proved by its polar snow caps. This view is further strengthened by the fact that these caps undergo seasonal changes. As the sun beats down upon them, we see first the one, then the other, grow smaller in size, till at the end of the Martian mid-summer, the northern one has disappeared almost entirely. During the next hot season, the same happens to the southern one. Where has this water—*the only remaining water on Mars*—gone? It cannot have filtered into the interior, for if it had, we could not possibly witness the reappearance of the polar snow fields every Martian year, as we actually do. Where, then, does the water go?

"Dr. Lowell solved the problem in a brilliant as well as ingenious manner.

"His view—and it is shared by most of our scientists to-day—was that Mars is inhabited by a thinking people, fighting a heroic battle for their existence. Without water, life, as we know it, cannot exist. Now, ages ago, the shortage of water had made itself felt on Mars. Long before the first cave man appeared on earth Mars had been an old world, where civilized peoples had reigned for centuries. While our ancestors were still jumping from limb to limb among the trees in primordial forests and jungles, the water problem on Mars had become acute. The fertile lands were fast turning into deserts, for rains had become more and more infrequent, until they had stopped almost entirely. Furthermore, as Mars is flat without mountains or elevations of any sort, there could not be any natural rivers to convey the water to the plains and valleys as is the case on our world. The Martians, seeing utter extermination staring them in the face, proceeded to save their race. They did precisely the same thing that we are doing in Western America and the Egyptians are doing in Egypt, namely effecting the irrigation of deserts or semi-deserts on a large scale. Our recent Roosevelt dam in Arizona offers a good example of this. Our engineers on earth have to bring the water to the deserts, precisely as the Martian engineers must have been doing for centuries past.

"On earth, however, this is a comparatively simple matter, for here we have rivers and lakes in abundance which can be tapped with ease. Not so on Mars. The only remaining water there is found around the poles; by sheer necessity, therefore, the Martians had to go to the poles for their water supply, and this is exactly what our telescopes reveal that they did. For the long unswerving straight lines which we see are part of the canals bringing the water down from the poles to the desert land there to irrigate it. So far the Lowell observatory has discovered almost 600 canals, but there are doubtless many more. They criss-cross the entire surface of the planet in every conceivable direction, most of them, however, running due north and south in the direction of the poles. Not only do the canals cross the desert lands, but we see them carried bodily across the dark blue areas which we know to be irrigated vegetation tracts. The fact that the canals run across these areas is another proof that they are not oceans, as had been thought at one time.

"Now the lines which we see running over the planet are really not the canals themselves, but are simply wide strips of vegetation fertilized and kept alive by the water from the canals. The average width of the canals proper Dr. Lowell estimates to be about six miles. There are some of them, however, which are thought to be much wider than this. The length of these canals, however, is stupendous. There are some canals which actually measure 3,400 miles in length. A great many are over 2,000 miles long. Dozens of them run for 1,000 miles, and nearly all of the canals run in absolutely straight lines.

"The circular black points, mentioned above, which we see almost invariably at the juncture of one or more canals, are termed oases. They also represent vast tracts of vegetation and probably contain large cities, farms and so forth.

"It must convince the strongest opponent of Dr. Lowell's theory, when viewing Mars and its canals through a first class telescope, that these wonderfully straight lines cannot by any possible chance be the work of Nature. Its counterpart is found nowhere on earth, nor in the heavens. And if by any chance, for argument's sake, these lines should be of a natural origin, it is inconceivable that so many of them could join and meet as they do and form these exact circular areas. Their artificial origin is too apparent and cannot be otherwise considered to-day. Dr. Lowell's theory has so far withstood the onslaughts of nearly all opponents, as a matter of fact, his explanation is to-day accepted almost universally.

"But how do the Martians move the tremendous masses of water through their canals? Mars is entirely level, and water does not flow on a level surface without a 'head.' Moreover, during one season it must needs flow from the north towards the equator, when the northern polar snow cap melts under the influence of the sun's heat. *During the next season, however, this flow must be reversed* for now the south polar snow cap melts with a resulting flow of the water from the south to the north.

"But how do the Martians succeed in moving the water? We don't know. Even Professor Lowell could tell us nothing on this point. Terrestrial science simply has as yet not advanced enough to offer an explanation.†

"Well, to make a long story short, Flitternix and I decided to voyage to Planet Mars. My little astronomical lecture was given solely for the purpose of refreshing your mind as to Mars in order that future reports which I shall make to you from the planet will be better understood by you and your friends.

"As long as our 'Interstellar' was able to succeed in reaching the moon without mishap, I felt sure that the trip to Mars would not be an unduly difficult undertaking. Flitternix was of the same opinion. We calculated that the intervening 50 million miles separating the moon from Mars should be negotiated by our space flyer within 30 days, barring accidents. While this may seem like a short time to cover such an immense distance, our speed of 1,600,000 miles a day, or 66,666 miles an hour, is only a trifle greater than †the speed of the earth (65,533 miles an hour) as it travels in its orbit around the sun.†

"We immediately made our preparations and within six hours after I had emerged from the crater, the 'Interstellar' had left the moon.

"And now for a little surprise! No doubt you noted that my voice does not sound the same as usual. You will have observed, furthermore, that I did not stop talking since I started. To break the news gently to you, I am not talking at all! While you are listening to my voice at this minute, I will be some 1,100,000 miles distant from the moon, heading towards Mars!

"The explanation? Simple as usual!

"Before leaving in our 'Interstellar,' we stretched an immense aerial inside of the canyon, the one of which I spoke to you several days ago. As you will remember, I told you then that it was open but a few feet across its opening at the top. It thus formed a long, narrow slit at the top, into which there was little likelihood of any meteor dropping, which could destroy the aerial. We stretched four wires in all along the inside of the canyon, spacing the strands six feet apart. Each strand is 6,000 feet long in order to give the required long wave length in transmitting as well as receiving impulses between Mars and the Moon, as well as between the latter and the Earth.

"To this aerial I connected my latest invention, my *Interplanetarian Radiotomatic*. It is nothing but an ingenious adaptation of modern tele-mechanics and works as follows:

"When the aerial receives a certain number of equally spaced dashes, an ultra-sensitive detector is actuated upon which in turn operates a gas-valve relay. This relay then closes its contacts, which sets in motion the well-known telegraphone, invented by [Valdemar] Poulsen. A second ultra-sensitive detector, also connected to the aerial, is in series with the registering electro-magnets of the telegraphone; in front of these magnets runs the moving steel wire, on which are then recorded the impulses coming in over the aerial. You will observe that no message can thus be recorded unless the original *key dashes* unlock the telegraphone mechanism. At the end of the message the same number of equally spaced dashes will lock the telegraphone mechanism. The recorded message is now ready for re-transmission at any time desired. This is accomplished in a simple manner too.

"I took our 300-day clock and fastened upon it a contact which would be closed at exactly 11 p. m. every night and would be opened again at 12 o'clock midnight. This contact closes a circuit in which is included the telegraphone mechanism. As soon as it starts, the steel wire with its recorded message begins to reel off in front of the two *reproducing* electromagnets, which in turn are connected with a special telephone receiver. Thus the telephone receiver will begin to talk its message (if one was sent during the day) every evening at 11 o'clock.

"But connected to the telephone receiver are several amplifiers, arranged in cascade. The last amplifier is attached to the mouthpiece of the transmitter of my wireless telephone. Thus the weakest recorded talk on the telegraphone wire will cause the telephone of the last amplifier to talk into the wireless transmitter louder than myself.

"Now my 300-day clock every night at 11 o'clock also closes the contacts of a powerful relay, which in turn operates the generating plant of my wireless telephone, disconnecting it at midnight. Therefore when the amplifier with its telephone begins to talk into the transmitter of the wireless telephone, there will always be enough power to transmit it to you on Earth.

"As soon as we arrive on Mars we will in all probability find all the necessary materials to erect a giant Radio telephone plant, and if we succeed we will send daily messages to the Moon, and my radiotomatic relaying plant will transmit the messages to you every night.

I might also mention that my ultra-sensitive detector contains two radio-active substances, making the detector such a marvelously sensitive instrument that it will work a set of amplifiers in cascade *when an electric pocket buzzer is operated one hundred and fifty miles away from it, connected to the ground only and using no aerial!*

"You might say: 'Why use the relaying plant on the Moon at all? Why not transmit from Mars to the Earth directly?'

"The reason is that when the weak impulses arrive from Mars, after having traveled from 50 to 60 million miles, they cannot be sufficiently strong to pass through the Earth's thick atmosphere. It is far better that the weak impulses should operate the relaying plant first and send out from it very strong impulses which have to travel only some 238,000 miles to Earth.

"We tested the plant thoroughly and after we had satisfied ourselves that it would work for at least 300 days I opened the telegraphone circuit and began to register this message to you. It will be the last one which you will receive for 30 days or more. As it must needs take us from five to 10 days to build a transmitting plant on Mars, you need not expect to hear from us for from 35 to 40 days. You might, therefore, commence to 'listen in' beginning with the 35th day from to-night. No message can ever be repeated, for the *'wiping'* electro-magnets of the telegraphone wipe out the magnetic impulses from the steel wire as quickly as they pass the transmitting magnets. Neither can you transmit a message to me, for no provisions were made to relay your messages to us while on Mars.

"I will now bid you adieu, my boy. Think of us during the next 30 days! Good-bye—good-bye . . . . .!'"

There was a silence for some seconds, and as I was still listening awestruck, I was suddenly startled by another voice breaking in:

"Hallo, there Alier, this is Professor Flitternix. How's Yankton? Beastly old town! Was once forced to sleep on a billiard table in the Palace Hotel, as all the rooms were full. The robbers charged me $2.50 for the 'room' plus the regulation rate of 50 cents an hour for the use of the billiard table! Mean town, that Yankton! Well, good-bye . . . . ."

There was a snapping noise and the rhythmic, low, sizzling sound stopped abruptly. All was quiet once more.

(To be continued.)

Phoney Patent Offizz

The Electrical Experimenter, vol. 3, no. 7, November 1915

Bookworm's Nurse

U. R. WRIGHT OF WHEREATIN VA.

No. Umsteen hundred and forty 'leven   Pat. applied for 10 minutes 'fore lunch.

Specification of Phoney Patent— Application Sandpapered September the Tooth

To those who—consarn it all, here goes:

1, U. R. Wright, of the Burg of Whereatin, Va., do hereby swear dreadfully, and affirm firmly that I have invented means whereby and by which a confirmed Bookworm may be relieved of all anxiety about reading too late at night, getting wet in the rain, getting overheated or run over by vehicles.

Full description of this wonderful apparatus follows:

A small but powerful dynamo is strapped to the small of the back. On each end of the shaft is an aluminum flywheel covered with fly-paper and having teeth around the inner edge of the rim. Pawls fastened to the legs of the wearer engage these teeth and spin the dynamo when the victim walks along either fro or to, hither or thither. The current thus generated is led by small wires to storage batteries concealed in the high stove-pipe hat which goes with the outfit. So a man, thus fitted out, really has *"bats in his belfry."* The top of this hat is the most ingenious part of the whole mechanism. It consists first of a shallow hard rubber pan, shaped much like a friction tight molasses bucket lid. In this lid are laid narrow strips alternately of copper and zinc, these being connected in parallel. On top of this is an image of Theodore Roosevelt, rampant, carved from a lump of copper sulphate or bluestone. More about this later.

Between the shoulders of the unfortunate is an umbrella which normally hangs down, closed, behind him. On the handle end of the umbrella is a segment of gears which engage with a small motor. Now the action is thus: When a shower starts, and the Bookworm is ambling along, face buried in a volume of "Deadwood Dick," the rain drops trickle over the Bluestone image of "Teddy," partly dissolving him and covering the zinc and copper strips with bluestone solution,

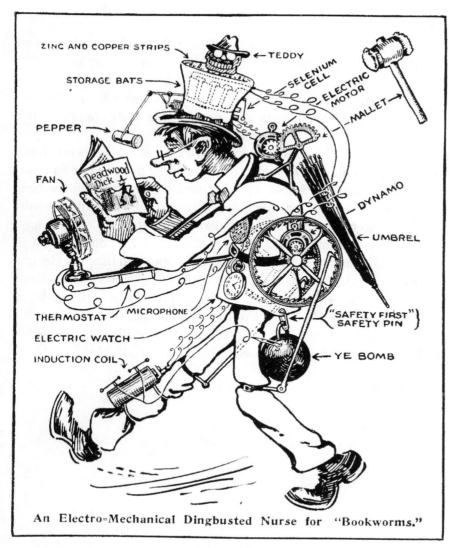

An Electro=Mechanical Dingbusted Nurse for "Bookworms."

FIGURE 1.

thus forming an electric cell. (The bluestone image, after months of constant use, will acquire a likeness to William Jennings Bryan.)[1] The electric current thus set up actuates a switch which cuts in the storage battery to the motor which hoists the umbrella. There is a tiny hole in the bottom of the hard rubber lid or pan which contains the zinc and copper strips so that the bluestone solution trickles very slowly out, and down the back of the wearer's neck. For this reason the patient must wear blue acid proof overalls. When all of the solution has run out, the switch is released and the umbrella is automatically let down to its normal position.

Strapped to the chest (if he has one) of the victim is a small electric fan in circuit with a thermostat so when it is warm the thermostat starts up a delightful mountain breeze.

1. That a sculpture of Theodore Roosevelt out of electrically conductive sandstone might melt into the likeness of William Jennings Bryan when exposed to the rain is probably a dig at the latter, who had recently resigned from his position as United States Secretary of State due to his opposition to American entry into World War I after the sinking of the *Lusitania*.

Reading so much a man is likely to forget to wind his watch, so an electric watch is provided.[2] Injury to the eyes by reading in too dim a light is prevented by a selenium cell, which in dim light releases a switch operating an electro-magnet which pulls a lever to and fro rapidly, on the end of which is a pepper shaker. This shakes pepper in the eyes of the reader and thus calls a halt.

While crossing the street, interested in a book, one pays no attention to approaching autos and cars. A sensitive microphone worn on the person actuates a spark, which explodes a bomb under the coat tails of the bookworm, hefting him gently into the air, while the said vehicle saunters nonchalantly on beneath him upon its wonted way.

To prevent reading in bed too late at night, a 10-pound mallet is fastened in the umbrella socket, and at the appointed time the electric watch connects the circuit and the mallet descends upon the noodle of the victim, giving 40 swift swats, knocking him insensible till morning, when he is awakened by a shock of 10,000 volts from an induction coil.

In testimony whereof, I have hereunto appended my nom-de-feather this day, O Lord, preserve us from further attacks.

U. R. Wright.
By his attorney,
Stanley H. Covington, Lynchburg, Va.
Witnesses: I. B. Darn, Whooda Thoughtit, Ischga Bibble.

2. The addicted reader shown wearing the bookworm apparatus is holding a volume of the popular nineteenth-century dime novel series Deadwood Dick (1877–97), authored primarily by Edward L. Wheeler (1854–86). This nod by Gernsback and his staff to the popular print tradition upon which they built is fitting, as Wheeler was unique among dime novel authors in his direct appeal to the reader, according to Christine Bold. This would make his stories especially seductive for hungry bookworms, "In his work, the characters take over all the authorial functions: they invent their own stories, they acknowledge the conventions of their existence and they recognize, finally, that these conventions derive directly from the commercial contract." Christine Bold, "The Voice of the Fiction Factory in Dime and Pulp Westerns," *Journal of American Studies* 17, no. 1 (April 1983): 29–46, http://www.jstor.org/stable/27554256. Bold cites *Deadwood Dick as Detective* (1879) in particular: "Fear not that man, for in your hour of need Deadwood Dick is on deck. When you least dream of it, he is lurking near, watching for your welfare, with a brother's care, and removing such obstacles as will be apt to trip you and throw you into the power of enemies."

THE PERVERSITY OF THINGS

Hearing through Your Teeth

The Electrical Experimenter, vol. 3, no. 9, January 1916

Using the Head as a Reproducer on a Phonograph.

FIGURE 1.

The following interesting experiment can be performed by anyone who has an ordinary disc phonograph. It is interesting, in so far as it shows the transmission of sound through the teeth, and through the bony substance of the human skull, which in turn reacts upon the auditory nerve.[1] It is not well known but it is a fact nevertheless that sounds do not necessarily have to enter through the aural opening in order that we can hear sounds. Physicians in testing for hearing sometimes use a tuning fork which, after struck, is pressed with its lower part against the back of the skull, right behind

"Every Martian is required, for reasons which you will understand presently, to wear a peculiar soft metallic cap. From the back of the latter a thin metallic cable runs down the Martian's back and is fastened there to his metallic coat. All Martian clothing, as well as footgear, is invariably of metallic weave. Now as all pavements and all flooring, carpets, as well as rugs, are metallic on Mars, for reasons which will be apparent to you later, a metallic connection with the earth or 'ground' is always effected.

"The metallic cable of which I just spoke does not make contact with the cap itself, but it is insulated therefrom. It connects, however, with a small reddish metallic plate about the size of an American silver dollar. This plate in turn, by means of a flat spring, presses against the temple of the wearer; the cap itself holds the plate in place. A similar plate presses against the other temple, but this plate, unlike the other, is connected metallicly [*sic*] to the cap itself. From this description you will gather that the metallic cap performs the function of a wireless aerial, while the metallic clothing forms the ground. The two reddish plates pressing against the bare temples are made of two metals unknown on earth, and the metals are distributed over the surface of the plate in honeycomb fashion without touching each other. Now if the two plates are pressed against the temples and when wireless waves are passing through them, the waves are translated into vibrations of a certain frequency. It has been found that if these vibrations reach the conscious sense of hearing which is located in the *Temporal Lobe* of the brain, sounds can be impressed upon the brain without requiring the ear and its auditory nerve. In other words, the sound is "heard" directly within the brain without the agency of the ear's mechanism.

"If this should be somewhat hazy to you a homely (though inaccurate) illustration will not be amiss here. At first blush one would think that the ear is absolutely essential for hearing, but this is not the case. Try the following simple experiment: Stop up both your ears as tightly as possible with cotton so that you will not hear a sound from outside. If you are partly deaf—and I trust you are not— all the better for the experiment. Place a darning needle between your teeth by biting on it hard and take care that your lips do not touch the needle. The needle itself should project about 1 inch from your mouth. Now operate an ordinary disc phonograph and with care press down

---

1. This issue's installment of *Baron Münchhausen's New Scientific Adventures* includes a distant extrapolation of the bone-conduction technique in the form of a technology that allows Martians to communicate as if telepathically. The story also contains a digression in the form of an instruction manual very similar to this one:

"Not far from the equator of the planet a central music plant is operated by a single Martian, who, of course, is a musical genius. He operates one of the organ-like instruments of which I spoke before. The 'plant' comprises besides the instrument, two Tos rods each 20 feet high. These rods work exactly as the ones just described, except that they are operated at an enormous frequency. I have stood in front of them while they were operating, so close, in fact, that I could have touched both rods with my hands. However, my ears detected not the slightest sound. But, incredible as it seems, millions of Martians were listening to the wonderful music at that minute, produced by these same rods, *but not with their ears*. They were listening with their brains! [. . .]

the ear. The sound is then heard inside of the head the same as if it had actually entered through the opening of the ear itself. This principle is made use of in the experiment described here, and while it is not electrical by any means, it probably will interest every experimenter who owns a phonograph.[2]

Stop up both of your ears with cotton as tightly as possible so that no sound will be heard from the outside. Now place an ordinary darning needle between your teeth by biting on it; hard, taking care at the same time that the lips or tongue do not touch the needle. The latter is important because if either lips or tongue touch the needle the sound will be decreased considerably. For best results the needle itself should project not more than 1 or 2 inches from the mouth. For that reason the darning needle should be broken off about one and one-half inches from its sharp point. It goes without saying that the sharp point should project out of the mouth while the broken off end should be inside of the mouth.

Now start an ordinary disc phonograph and carefully press down upon the record with the needle's point held at the same angle as the reproducer's needle is held ordinarily. With a little practise one will become proficient in moving the head at the same ratio of speed as the ordinary reproducer arm is moved from the outside of the record towards the inside. As soon as the needle touches the record with sufficient pressure, the inside of the head will be filled immediately with music exceedingly loud and clear.[3]

A curious result of the experiment is that a person standing near by can hear the music, the head acting as a reproducer in this case.

Of course, it will be understood that a totally deaf person will not be able to hear any sound if the auditory nerve is dead or inactive. It is, however, interesting to note that partly deaf people can hear the music quite well. This is particularly true of persons hard of hearing who cannot ordinarily hear the sounds of a phonograph.

The writer should like to hear from readers, particularly from those who are partly deaf, who have tried this experiment; the *Electrical Experimenter* will be glad to publish the results in subsequent issues.[4]

upon the record with the needle's point held at the same angle as the reproducer's needle is held ordinarily. Your whole brain will be filled immediately with music, exceedingly loud and clear. Of course, in this case you still hear with your ear's mechanism, the sound vibrations being carried to the eardrum through the bones of the head; but it is interesting to note that if a truly deaf person tries the experiment he will be enabled to "sense" the sounds, although not perfectly.

"Of course, the function of the Martian apparatus is based on an entirely different principle, and the above experiment is only cited by way of comparison."

2. Google Glass, the prototype head-mounted-display glasses released in 2013, used the same principle of bone conduction technology to send audio to the user via a small oval-shaped component positioned behind the ear. Nathan Ingraham, "Google Glass Headset with Bone-Conduction Speakers Revealed in FCC Filing," *The Verge*, January 2013, http://www.theverge.com/2013/1/31/3938182/google-glass-revealed-in-fcc-filing.

3. In 1923, Gernsback was awarded a patent for a device he called the Osophone and its "sound vibrations transmitted directly to the osseous tissue of the body." Hugo Gernsback, "Acoustic Apparatus," December 1924, http://www.google.com/patents/US1521287.

4. A profile of this experiment was published in the November 1923 issue of *Science and Invention* as the cover story. See **Are We Intelligent?** in this book for an editorial that evokes what might be possible with such alternative forms of communication.

ABOVE:
November 1923 cover of *Science and Invention.*

LEFT:
Cover story in the November 1923 *Science and Invention.*

The Future of Wireless

The Electrical Experimenter, vol. 3, no. 11, March 1916

t the present state of the art, Wireless can be sub-divided into three classes:

1st. Wireless Telegraphy or Radiotelegraphy.
2nd. Wireless Telephony or Radiotelephony.
3rd. Wireless transmission of Power. The latter we may term as *Radiokinetics*.

As is well known, the first two are already in everyday use, all over the world. The third is as yet undeveloped, but already it looms large above the horizon.

Like all great things, Wireless has had its share of trials and tribulations. It takes time to develop an entirely new art. Moreover, Wireless received a black eye in its earliest infancy in this country. As will be recalled, a number of unscrupulous individuals unloaded millions of dollars of worthless stock on a credulous public, before the art had sufficiently advanced to make possible a successful commercial exploitation.[1]

Practical commercial Wireless Telegraphy is not much older than ten years to-day. Commercial Radiotelephony has but made its appearance during the past one or two years, while Radiokinetics does not exist at all as yet.

But let us consider how the three classes of Wireless line up as far as their ultimate usefulness and commercial practicability are concerned. Let us look the problem square in the face and let us see what we shall find.

It is our opinion that a purely Wireless Telegraph Company can never reach such immense proportions as our wire telegraph companies. The reason is obvious. The wire telegraph companies are too well intrenched [*sic*] to be driven out of the field; it is quite certain that wireless telegraphy can no more hope to supersede wire telegraphy, than the telephone superseded the wire telegraph. Aside from this it seems hopeless for any one large central wireless plant to send out and receive within a single hour, 8,366 separate messages, as is the case for instance with one of the New York offices of the Western Union Telegraph Company. Wireless will probably never lend itself to such exploitation. Its greatest use will always be long distance

---

1. For more on these failed companies, see **The Born and the Mechanical Inventor** in this book.

transmission of intelligence, either over land or water or both, and between land and ship or vice versa, or between ships. This is its true field and here the wire companies cannot compete. This naturally limits its possibilities. Thus, while the future of Wireless telegraphy does not seem too rosy, we need not feel discouraged. The young man who embarks in radiotelegraphy to-day, will use it only as a stepping stone towards something infinitely greater. This was the exact case of T. N. Vail, the present head of the American Telephone and Telegraph Co., popularly known as the Telephone Trust. Vail was originally a telegraph man when he was called in by Bell and his associates; had he not known all about telegraphy he probably would not be the president of the huge corporation to-day.[2]

This brings us to Radiotelephony. To us there does not seem one field in the entire electrical industry that is destined to a greater and speedier development than this one. We venture to say that within the next fifteen years, Radiotelephony will become one of the greatest electrical industries, for it supplies one of the predominating wants of the times.

The radiotelephone can be used by anyone, just as easily as the wire phone. To operate the instrument it is only necessary to take down the receiver and talk. But three months ago it was demonstrated that it is eminently practical to catch the wireless voice—on the wing as it were—and connect it with an existing wire telephone line. Vice versa, President Vail talked into a wire telephone at New York, where his voice was transmitted to Arlington; here it "took wings" and was wafted without wires to Honolulu, some 4,000 miles distant. This accomplishment more than anything else has opened the public's eyes.

We prophesy that in less than 15 years every automobile, whether pleasure or commercial, will carry its small radiophone outfit. Its occupants will thus be in constant touch with their homes or offices and vice versa, a convenience much needed to-day. Imagine the immense usefulness of such a device. Nor is this an idle dream. There is at least one company in existence to-day capable of filling an order to equip autos with radiophones having a 20 mile range. Nor will there be much confusion of voices becoming mixed up in transit; our tuning apparatus is becoming more accurate each day and it will be an easy matter to tune out unwanted voices. It will take considerable capital and a host of trained men to turn out enough radiophones to equip several million automobiles, aeroplanes, motorboats, yachts, and large vessels, but it will be done nevertheless and soon at that. Every farmer will have his Wireless Telephone to talk with his neighbors. Every train will have its radiophone enabling passengers to talk to their homes or offices. The radiophone will link moving humanity

2. Theodore Newton Vail (1845–1920) presents a good example of the translatability of technical expertise across numerous fields during the period. Before becoming president of AT&T in 1885, he worked as a telegraph operator in New York and railway mail service clerk in Nebraska. Further, he came from a long line of machinists. Theodore was the cousin of Alfred Vail (1807–1859), an employee of Samuel Morse who was responsible for some of the most intricate aspects of the earliest telegraph receivers and transmitters. Alfred Vail received his technical education in his father Stephen's machine shop, the Speedwell Iron Works. For more on the gradual nineteenth-century evolution of machine-shop culture into telegraph manufacture and operation, the latter represented by trade publications like *The Telegrapher* and *The Operator*, see Paul Israel, *From Machine Shop to Industrial Laboratory: Telegraphy and the Changing Context of American Invention, 1830–1920* (Baltimore: The Johns Hopkins University Press, 1992), 24–86.

with the stationary one, the same as the wire telephone linked humanity together before. To us there is nowhere a brighter future than in the vast possibilities of the Radiophone.

As to Radiokinetics, this will surely follow the Radiotelephone in due time. Its future is probably even brighter than the latter. Already Tesla speaks of transmitting energy by the thousands of horsepower wirelessly. Who dares predict what this branch of wireless will bring during the next twenty years?[3]

3. Despite the fame of Nikola Tesla's plans for the World Wireless System, which would enable the global distribution of electricity without wires, the technology never materialized. Ghislain Thibault uses the example of Tesla's inventions, which circulated largely in the form of visions or promises, to explore the question of how the history of technology can incorporate discourses that explain, support, precede, and sometimes overshadow material machines: "debates on the ambiguous notion of invention in science and technology in the fin-de-siècle expert community and popular science narratives created a space of undecidability for the materiality of invention, a space that Tesla's controversial discursive inventions occupied." Ghislain Thibault, "The Automatization of Nikola Tesla: Thinking Invention in the Late Nineteenth Century," *Configurations* 21, no. 1 (2013): 27–52.

Imagination versus Facts

The Electrical Experimenter, vol. 3, no. 12, April 1916

We are asked not infrequently why *The Electrical Experimenter* lends so much space to the exploitation of the future, or in other words, why we make so much of things which as yet exist but in the imagination.

We admit that such is the case and furthermore we believe that more matter of this kind is printed in our pages than in any other kindred publication. The reason should be obvious. The electrical art and its allied branches are but a comparatively new and unexploited science. An immense number of discoveries and inventions remain to be made, each succeeding day broadening our vision and showing how little we really do know as yet. Every new discovery immediately leads to hundreds of other inventions and each one of them opens up new fields. There seems to be no end or let up; indeed, there cannot be such an end in a world where everything is infinite. We will never reach a period where everything will be known, where nothing is left to be discovered, nothing further to be invented! Progress in science is as infinite as time, it is inconceivable how either would stop.

Some well-meaning people with a shrunken horizon may disagree with us here, and to them we would like to quote the case of one of our government patent officials living in Washington one hundred years ago. This worthy individual left the patent service as he was certain that almost everything of importance had been invented. He felt equally certain that on account of this deplorable state of affairs the Patent Office would shortly be forced to close its doors forever! Since that time almost one million patents have been issued.[1] The telegraph, the telephone, the phonograph, the electric light, electric trains, the storage battery, the X-ray, wireless telegraphy, radium, and scores of other inventions of the first magnitude have been made. And each succeeding year brings new wonders. We are fully aware of the fact that some of the imaginary articles which we publish are wildly extravagant—now. But are we so sure that they will be extravagant fifty years from now? It is never safe in these days of rapid progress to call any one thing impossible or even improbable. The telephone would have been considered ridiculous fifty years ago, while the aeroplane, prophesied and predicted for generations, was declared a total impossibility by men of science as late as fifteen years ago. Some even published long scientific dissertations proving beyond doubt that

1. The idea that "everything that can be invented has been invented" is often misattributed to Charles H. Duell, Commissioner of the U.S. Patent and Trademark Office from 1898 to 1901. An 1843 report from then–Patent Office Commissioner Henry Ellsworth comes closer: "The advancement of the arts, from year to year, taxes our credulity and seems to presage the arrival of that period when human improvement must end." For a brief history of this mytheme, see Samuel Sass, "A Patently False Patent Myth," *The Skeptical Inquirer* 13 (Spring 1989): 310–13.

such a machine could not possibly be made to remain aloft in the air. Jules Verne who forty-six years ago in his "20,000 Leagues Under the Sea," imagined the submarine down to its very battery for propelling it under water, was ridiculed and called a dreamer in his day. Nevertheless, the aeroplane as well as the submarine are very much in evidence these days.

Of course, we could publish nothing but facts, nothing but experiments. This would be a very simple as well as easy matter and much less costly. Our self-imposed, infinitely harder task, however, has helped to make this journal what it is to-day: The most widely read and circulated electrical publication on earth. And all this within three short years.

It is no easy matter to think out new things of the future and illustrating them adequately by means of expensive washdrawings or three-color cover illustrations. Indeed, there is nothing more difficult connected with the publication. But if we succeed—and we think we sometimes do—in firing some experimenter's imagination to work in a new direction, due directly to our imaginary illustrations, then indeed we feel amply repaid for our toil.

A world without imagination is a poor place to live in. No real electrical experimenter, worthy of the name, will ever amount to much if he has no imagination. He must be visionary to a certain extent, he must be able to look into the future and if he wants fame he must anticipate the human wants. It was precisely this quality which made Edison—a master of imagination—famous.

Imagination more than anything else makes the world go round. If we succeed in speeding it up ever so little our mission has been fulfilled. There can be no progress where imagination is lacking.

What to Invent

The Electrical Experimenter, vol. 4, no. 1, May 1916

f late, we are receiving a great many inquiries from experimenters, would-be inventors, inventors, as well as others, asking us to publish a "list" of useful electrical devices which as yet require to be invented. Most of our correspondents state that they are of an inventive turn of mind and quite a few admit very frankly that in the past they have lost a good deal of money and time in trying to develop ideas which afterwards turned out to be of no earthly practical use. By boiling down the various inquiries, this is what our correspondents desire: "What electrical inventions are urgently required at present, and which ones are the most desirable from a financial viewpoint?"

As most everyone is familiar with the important problems as yet unsolved, such as: Electricity direct from Coal; Harnessing of the Sun's and the Ocean's Energy; Cold Light, etc., we do not for the present wish to dwell upon these.[1] For that reason the "list" which we suggest below will probably be more in keeping with our would-be inventors' desires. We make no claim that the suggestions are highly original, or that they could not easily be improved upon. We do, however, think that it would be quite profitable to invent and market any one of the ideas and devices cited. At least that is our humble opinion.

Wire Insulation. At present we use either cotton, silk, rubber, or enamel to cover wires. There is needed a covering, having all the good qualities of silk and cotton as well as enamel, but none of their bad ones, i.e., the insulation must take up a minimum of space, it must be tough and must not crack or break.

Storage Battery Casings. 98% of all portable storage batteries are now encased in wood. Wood is cheap and if well impregnated it is fairly acid proof for a limited time. As a whole the material, however, is not satisfactory. There must be something better. What is it?

Heavy Current Microphone. Wireless telephony is retarded at present because there is no practical transmitter that can handle from 5 to 10 amperes continuously. The microphone should be small and should not require water cooling, as this makes it highly undesirable. Preferably no carbon should enter into its construction.

Marble Substitute. There is an immense demand for instrument bases and parts, switch and switchboard bases, etc. At present very expensive marble, slate, wood, or composition is used. Porcelain is

1. Gernsback poses the question of cold light to Edison in their interview. See **Thomas A. Edison Speaks to You** in this book.

cheap, but never presents a good appearance, especially for instrument bases. Marble dust is cheap and can be readily had in large quantities. Who will be the first *to mold* a real cheap marble base, that takes a good polish? We are aware of the fact that artificial marble is in existence. It is, however, almost as expensive as the natural.[2]

Telephone Muffler. A device is needed whereby you can talk into your telephone transmitter in such a manner that a person sitting close by cannot hear what you say. Every business office, for obvious reasons, can use such an attachment. At the present time the business man must use a cumbersome, as well as expensive telephone booth. There have been telephone mufflers on the market in the past, but all died a quick death; there was just one trouble with them; they didn't muffle!

Tele-Music. An "industry" rivaling the moving picture business can be created when some genius perfects a means supplying telephone subscribers with all kinds of music from a brass band down to a violin concert. The requisites are that ten or 100,000 subscribers can listen in, all at the same time, without the sound weakening as more telephone lines are put in the circuit. The subscriber must be able to use his regulation instrument. No expensive attachments should be used; only, perhaps, let us say, a low priced horn, quickly attachable to the telephone receiver. The music should be heard loudly all over the room. No expensive nor complicated plant should be used at the point where the music originates. A two-wire line should connect the plant with "central."

These are only a very few suggestions. If required we will publish more from time to time.

2. Bakelite, one of the earliest synthetic plastics (patented in 1909), would eventually take on the function Gernsback calls for here. The material became the default for radio cabinet and knob construction by the 1930s, allowing for ever more unique designs in radio sets, especially portable sets. According to Michael Schiffer, the 1932 International Kadette Convertibles Midget was "the first small radio with a Bakelite (plastic) cabinet to be produced in large quantities." Michael B. Schiffer, *The Portable Radio in American Life* (Tucson: University of Arizona Press, 1991), 110.

The Electrical Experimenter's Patent Advice section served as a marketplace for amateur inventors, with editorial advice to readers on their designs, and advertisements from patent attorneys and solicitors looking to buy and sell new ideas. From the April 1916 issue.

The Perversity of Things

The Electrical Experimenter, vol. 4, no. 4, August 1916

This is by no means a new subject. Much has been said and written about the recalcitrant behavior of things in general toward us humans. Much remains to be said and written. Very much remains to be explained.

Of all individuals, the experimenter, perhaps, suffers most under the tyranny of the inanimate things and it happens not infrequently that he succumbs to their wiles and to their devilish snares.

Let us take an everyday case, familiar to every experimenter, worthy of their name. You have planned the construction of a certain instrument and you know just how to go about making it.

You have shaped the wooden base which is now ready to be drilled for its several holes. You have laid out the holes on the base and while doing so you have broken the point of your pencil twice. This has caused you some annoyance. After you have drilled every hole but one, you break the only two drills you have in the attempt to drill through a knot! Of course *that* hole just had to come on a knot! But you must have that hole. What is there to do? In the absence of the drill, a bright idea strikes you. The hole will be drilled without the drill! Therefore you take a nail, heat it red hot, and after much effort the hole is finally well under way. As the process is slow and weary, you decide not to burn the hole all the way through. So when the nail has penetrated three-quarters through the board you take recourse to the hammer, intending to drive the nail through the remaining portion of the wood. In this you are successful indeed—but, alas, with the result that the entire base is split in two! Another base is made now and everything seems to run smoothly, including the holes. After the base is finished you give it several good coats of shellac and it is then placed in the sun to dry, while you go to lunch. On your return you look at the base. Somehow it does not look the same. Sure enough—it has begun to warp badly already and you know that by tomorrow morning it will be bent into almost a semi-circle.

By this time you are angry clear through and you slam the base into the furnace, accompanying the action by much profanity. You mutter of "Hoodoo" and "Hard luck," but if you are a true dyed-in-the-wool experimenter, you will be working on a third base before long. You do not fare much better with the balance of the work. Everything you touch seems to be "hoodooed." Screws don't fit, nuts will not go on their respective screws, the brass casting cannot be fitted for hours at a time, you can't seem to drill or tap a hole without snapping off either the drill or the tap. To add insult to injury, just when the instrument is about to be completed, the screwdriver slips and makes a nasty gash in your left hand. This, of course, puts you out of action for the time being and you are, indeed, worthy of our admiration if you do not let loose a "blue streak" of English not found in Webster's unabridged dictionary!

Now let us analyze this seeming perversity of things. Why do inanimate things act thus? Why does the eternal wisdom of nature, seemingly always interpose obstacles in his way whenever man desires to invent or construct a certain new thing?

The answer is that it does not. It is not the things that are perverse, it is ourselves who make them seem perverse.[1]

In our example just mentioned, it was not the fault of the base which kept you from completing it. It was purely your lack of forethought, and mostly your impatience that caused all your trouble. For if you had used a carpenter's pencil, instead of a frail drafting pencil, the point would assuredly not have broken twice. Simply lack of attention here. Also, had you turned the base around the other way, your drill would, in all probability, not have struck the knot hole. Again lack of attention. Once you knew that you had not the proper tools to drill holes through a hard knot, you had no right to attempt the work in spite of it. Forethought would have told you the inevitable result. This holds true for the second base. Common sense should have told you not to use the "green" wood for an instrument base; painting shellac on one side of it, taught you a graphic lesson.

And so it does all the way through. So it has gone for aeons and centuries. Man always stands ready to blame inanimate things for his blunders; everything is blamed on the perversity of things, when it should be blamed on the perversity of man.

It took the human race several million years to construct an automobile. The material, the things, existed for millions of years long before the human race was heard of; it was not for lack of things the first automobile was not built sooner. It was for the lack of man's intelligence. To-day, the same man with a little acquired intelligence and a little acquired experience turns out several thousand automobiles each working day—no perversity of things here.

Summing it up the failure of most experimenters and workers who do not accomplish anything can usually be directly traced to their lack of knowledge of inanimate things. At best few people thoroughly understand materials. Few people can tell offhand what certain materials will do and what they cannot do, under conditions to which they

1. The anthropologist Tim Ingold distinguishes between the properties and the qualities of materials in his essay on the importance of attending to "the stuff things are made of" in studies of material culture. Similar to Gernsback's lesson here, Ingold argues that the properties of things are objective while the qualities are a matter of subjective perception: "If, as I have suggested, we are to redirect our attention from the materiality of objects to the properties of materials, then we are left with the question: what are these properties? How should we talk about them? One approach to answering this question has been proposed by the theorist of design, David Pye. His concern is to examine the idea that every material has inherent properties that can be either expressed or suppressed in use. This idea is frequently enunciated by sculptors and craftspeople who assert that good workmanship should be 'true to the material,' respecting its properties rather than riding roughshod over them. . . . Pye argues that it is not really the *properties* of materials that an artist or craftsperson seeks to express, but rather their *qualities*." Ingold quotes from Pye's book: "The properties of materials are objective and measurable. They are out there. The qualities on the other hand are subjective: they are in here: in our heads. They are ideas of ours. They are part of that private view of the world which artists each have within them. We each have our own view of what stoniness is." Tim Ingold, *Being Alive: Essays on Movement, Knowledge, and Description* (London: Routledge, 2011), 29, citing David Pye, *The Nature and Art of Workmanship* (London: Cambridge University Press, 1968), 47.

have never been subjected. Men like Faraday and Edison have this exceptional instinct developed in a high degree. They know intuitively what a certain metal will do in a vacuum when subjected to a high potential electrical discharge. With most other workers it is a case of long experience and intimate contact with things that gives them the true insight into their characteristics.[2] Furthermore, the man who accomplishes things is the man who doesn't lose his temper and who doesn't get impatient. The successful experimenter's motto should be: *Patience*.

If people would only stop to think how infinitely little we know about everything about us, and how thoughtless we are in our relations to all inanimate things, we would not be so apt to complain about the fabled Perversity of Things.

2. Commenting on his hands-on training in early nineteenth-century Edinburgh, the engineer James Nasmyth writes, "The truth is, the eyes and fingers—the bare fingers—are the two principal inlets to sound practical instruction. They are the chief source of trustworthy knowledge in all the materials and operations that the engineer has to deal with. No book knowledge can avail for that purpose. The nature and properties of materials must come in through the finger-ends; hence I have no faith in younger engineers who are addicted to wearing gloves. Gloves, especially kid-gloves, are non-conductors of technical knowledge." James Nasmyth, *James Nasmyth, Engineer: An Autobiography*, ed. Samuel Smiles (New York: Harper, 1883), 99–100. Quoted in *The Mindful Hand: Inquiry and Invention from the Late Renaissance to Early Industrialisation*, History of Science and Scholarship in the Netherlands 9, ed. Lissa Roberts, Simon Schaffer, and Peter Dear (Amsterdam: Koninklijke Nederlandse Akademie van Wetenschappen, 2007), 309.

War and the Radio Amateur

The Electrical Experimenter, vol. 5, no. 1, May 1917

The Radio Act of 1912, under section 2 states:
Every such license shall provide that the President of the United States in time of war on public peril may cause the closing of any station for radio communication and the removal therefrom of all radio apparatus, or may authorize the use or control of any such station or apparatus by any department of the Government, upon just compensation to the owner.

e now stand on the threshold of war; indeed, before this issue is in the hands of our readers war will have been declared, or what is equivalent, this country will be in a state of war.[1]

Let us then be perfectly frank with each other, and let us face the situation as it behooves upright, patriotic, law-abiding citizens. The European war has taught us that messages sent from secret radio plants by spies have been of priceless value to the enemy. Small wonder then that hysteric officials of all the warring nations have exterminated every possible as well as impossible private wireless plant in their respective countries. But to what good? True, every stationary outfit has been dismantled or confiscated by the warring Governments, but as always: where there's a will there's a way. When the German spies in England and in France found that it was not very healthy to operate their outfits in attics or in house chimneys—for a sending outfit is soon located—they simply put their radios in touring cars, cleverly concealing the aerial wires inside of the car bodies. The apparatus too were easily concealed, and the English and French were outwitted simply because you cannot locate a moving radio outfit except by pure chance.

Which brings us face to face with the question: Did it pay the warring nations to kill the few private Radio stations they had before the war? We are honestly inclined to believe that far from being an advantage, it proved an actual disadvantage. No one at all familiar with the technique of the radio art, doubts for one minute that if a spy has the courage as well as the funds—and spies always have both—he cannot be stopt from sending wireless messages if he elects to do so. Working under cover and by moving from one place to another, nothing will stop him.

1. With Germany's decision to resume unrestricted submarine warfare in January 1917 and the publication of their secret Zimmerman telegram (which offered a military alliance to Mexico), the United States declared war on Germany on April 6, and on the Austro-Hungarian Empire the following day.

If we recognize this truth we realize how absurd it is to close all privately owned radio stations during the war. It will do no earthly good and can do only actual harm. Now we do not wish to appear selfish, nor do we wish to be classed as unpatriotic. Very much the contrary. If the administration, after carefully considering all the facts, decides to close all privately owned radio stations in this country, we will not as much as raise a single word of protest. The administration knows what is best for the welfare of the country and in time of national peril we would be the last ones to annoy our officials.

But is it not true that our splendid body of over 300,000 patriotic American Radio Amateurs, scattered thickly all over the country, can be of inestimable value to the Government? Can not our red-blooded boys be trusted to assist our officials in running down spies, who probably would not be readily located otherwise? In our big cities thousands of ears listen every minute of the day to what is going on in the vast ether-ocean. Trust our very capable American youths to ferret out the senders of questionable signals or strangely worded messages. The very multitude of these amateurs is a priceless protection. Then again both our Army and Navy badly need Radio operators. What other country can furnish such a vast army of well trained and intelligent operators as ours, thanks to the amateurs?[2]

When in 1916 the writer organized the *Radio League of America,* he incorporated in its statutes that every member should pledge in writing his station to the Government. Up to this moment the League has forwarded to Washington thousands of such pledges, among them every important amateur station in the country. These stations can be used by the administration at a moment's notice. At least our amateurs are fully prepared.

Would it not be questionable wisdom to shut down all these stations that can and will do enormously more good than possible harm?

Let our officials ponder and let them consider fairly the facts in the case. That is all that we desire.

To All Radio Amateurs:

The Department of Commerce of Washington, by its Secretary, the Hon. Wm. C. Redfield, has kindly sent us the following information of particular interest to all amateurs in the United States at the present time.

Secretary Redfield has issued orders that for the present no new licenses to radio amateurs will be issued and the renewal of outstanding amateur licenses will be granted only by the Department upon special favorable reports by the radio inspectors. (This refers to sending outfits only.)

2. Part of what worried the government about amateur signaling and open airwaves was the possibility of malicious interference, surveillance, and covert communications. Strategic planning for these problems began during debates over the 1912 Radio Act. By the time the United States entered the war five years later, it had already "begun construction of a massive network of stations from the Philippines to Puerto Rico" in preparation to take charge of all public and private long-distance radio communications. "The pace of change had quickly exceeded the ability of the government to keep up. The most recent legislation on radio, the 1912 Radio Act, did not dictate the citizenship of station owners. Theoretically, any government could open a radio station in the United States or subsidize a trusted company to act on its behalf. From such a lodgment, a station could monitor the navy's operations or jam its signals, while to all outward appearances remaining a legitimate commercial firm. . . . As Admiral [Robert S.] Griffin and others believed at the time, 'radio is a natural monopoly.' There was chaos in the new field of radio, but service seemed to be most efficient if under a single authority. That single authority, in the view of the U.S. Navy, ought to be the U.S. government. The security of the country compelled it." Jonathan Reed Winkler, *Nexus: Strategic Communications and American Security in World War I* (Cambridge, Mass.: Harvard University Press, 2008), 97, 62–63.

The Department also informs our readers, reminding them of the fact that the operation of transmitting radio instruments without licenses is prohibited under severe penalties, which, under the conditions of the time, would be exacted in the case of those who showed no regard for the requirements of the law.

Up to the time that we go to press, the Department has not formulated final plans as to what steps will be taken in regard to radio amateurs as a whole, and whether they will be allowed to continue to operate the same as before. It is our personal impression, however, that no drastic steps are likely to be taken by the Government as long as the amateurs cooperate with the department.

In view of this we most urgently and earnestly request all amateurs at the present time to refrain from using their transmitting stations except for regular work. In other words, all unnecessary gossip and fooling should be rigidly suspended for the present, particularly the "Q.R.M." nuisance which at best, only serves to irritate our officials, and makes their work harder.[3] If amateurs do not voluntarily stop such annoyance the Government will certainly prohibit the use of all privately owned radio outfits.

These are no times to use the ether for a lot of nonsense; we all wish to help our country as much as we possibly can until normal conditions are restored again.

Always remember, that our Government has granted the radio amateurs more powers than any other country in the world, and in times of stress, it is up to the amateurs to show of what stuff they are made by cooperating with our officials to the fullest extent of their powers.

THE EDITOR.

3. QRM is the Q code (a set of standardized messages encoded in three Morse code letters) for "Are you being interfered with?" These codes are often used among wireless communities as nouns rather than complete sentences, as is "QRM" in this paragraph: a stand-in for "willful interference."

Silencing America's Wireless

The Electrical Experimenter, vol. 5, no. 2, June 1917

s all our readers are aware the United States Government, thru the Navy Department, has issued orders thruout the land to cause the immediate dismantling of all radio stations, whether large or small, commercial or amateur, sending or receiving. All aerials have been ordered dismantled and apparatus packed away.

This action came as a great surprise to all patriotic amateurs, who for years past had been encouraged by the Government and who were certain that in time of war they would be allowed to "do their bit" with their outfits for the country.[1]

That the Government should silence all *sending* outfits was eminently proper, and we have as yet to hear the first complaint on that score. But why the *receiving* outfits should he dismantled by the Navy Department is very puzzling indeed.

President Wilson's Executive Order is based upon the Radio Act of 1912, which act however, mentions nothing about closing receiving stations during the time of war. That purely receiving stations were considered harmless by the framers of the law, is best proved by the fact that such stations do not require to be licensed as do all sending stations. Moreover, in President Wilson's Executive Order of April 6, no mention is made of receiving stations. Indeed, the following passage strikes us as very significant:

> —and furthermore that all Radio Stations not necessary to the Government of the United States for *Naval Communications may* be closed for radio communication.

The italics are ours. Particularly the one word MAY. In the same paragraph the President uses the command SHALL, while the word *may* does not imply that every radio station should be taken over by the Navy Department. Indeed, the longer we study the third paragraph of the President's Executive order, the more we become convinced that the closing of every amateur station, or even commercial stations, was remote from President Wilson's mind when he issued his order.

1. President Woodrow Wilson's executive order not only shut down all sending and receiving stations in the country not yet under government control, it also assumed ownership of all wireless patents ostensibly so that private firms could cooperate on new devices. "A government-imposed patent moratorium instructed all suppliers to make use of the best components, no matter who owned the patent. The government guaranteed to protect all suppliers against infringement claims and encouraged the inventors not to be oversensitive to relatively free use of their apparatus during the national emergency. Under this arrangement, with the inventors and radio companies concentrating less on marketing strategies and litigation and more on research and development, significant advances in continuous wave technology were achieved. Civilian military cooperation produced apparatus more ideally suited to the navy's special needs." Susan J. Douglas, *Inventing American Broadcasting, 1899–1922*, Johns Hopkins Studies in the History of Technology (Baltimore: The Johns Hopkins University Press, 1987), 278.

EXECUTIVE ORDER

Whereas the Senate and House of Representatives of the States of America, in Congress assembled, have declared that a state of war exists between the United States and the Imperial German Government; and

Whereas it is necessary to operate certain radio stations for radio communication by the Government and to close other radio stations not so operated, to insure the proper conduct of the war against the Imperial German Government and the successful termination thereof

Now, therefore, it is ordered by virtue of authority vested in me by the Act to Regulate Radio Communication, approved August 13, 1912, that such radio stations within the jurisdiction of the United States as are required for Naval Communications shall be taken over by the Government of the United States and used and controlled by it, to the exclusion of any other control or use; and, furthermore, that all radio stations not necessary to the Government of the United States for Naval Communications may be closed for radio communication.

The enforcement of this order is hereby delegated to the Secretary of the Navy, who is authorized and directed to take such action in the premises as to him may appear necessary.

This order shall take effect from and after this date.

> The White House,
> 6 April, 1917.
> (Signed)
> Woodrow Wilson.

2. While the presence of local stations and a well-trained citizenry were important, Jonathan Reed Winkler argues that the Great War necessitated a holistic understanding of the United States' strategic interests within an increasingly global communications network: "It was only during World War I that the United States first came to comprehend how a strategic communications network—the collection of submarine telegraph cables and long-distance radio stations used by a nation for diplomatic, commercial, and military purposes—was vital to the global political and economic interests of a great power in the modern world." Jonathan Reed Winkler, *Nexus: Strategic Communications and American Security in World War I* (Cambridge, Mass.: Harvard University Press, 2008), 2. The airwaves were thus increasingly seen as another territory to be defended in war and peacetime, especially at a moment in which the available broadcast spectrum—thanks to primitive tuning circuits—was thought to be limited to no more than twelve long-distance stations in the world at a time. See ibid., 251–53.

Also relevant here is Markus Krajewski's media archaeological account of attempts to configure and conceptualize the globe as a single system around the turn of the twentieth century. Markus Krajewski, *World Projects: Global Information before World War I,* trans. Charles Marcrum II (Minneapolis: University of Minnesota Press, 2014).

In conformity to the Radio Act of 1912, the President in time of war, may authorize any department of the Government to close all radio stations. But the President's order of April 6, was not to the Department of Commerce, which in the past controlled the nation's radio affairs, *but to the Navy Department.* Why? Because the President, it seems to us, had only the radio communications of the Navy in mind. If, therefore, the Navy Department had caused the closing of all radio stations, particularly sending stations along our sea borders, such action would have seemed perfectly logical. But why the Navy Department should wish to close stations a thousand miles removed from the sea borders, seems to us very puzzling. Furthermore, why all college radio stations, and those belonging to radio apparatus manufacturers as well, should be dismantled seems far fetched. Then there are cases like the one of the Lackawanna Railroad, which is one of the pioneer railroads in the United States to use wireless for train dispatching. Is it wise to dismantle such stations on which the safety of passengers depends?[2]

We certainly have no quarrel with the Navy Department; quite the contrary. We wish to help, but we sincerely hope that its officials will soon find a way to modify its recent sweeping order.

There are, indeed, encouraging signs already. Certain commercial stations on the Pacific Coast have recently resumed operation, and it is to be hoped that amateurs will be allowed to operate their receiving stations, at a not too distant future.

The Magnetic Storm

Electrical Experimenter, vol. 6, no. 4, August 1918

Why" Sparks had stopped reading the New York *Evening World*: He contemplated his old meerschaum pipe meditatively while with his long and lanky index finger, stained by many acids, he carefully rubbed a long, thin, and quivering nose. This was always a sign of deep, concentrated thought of the nose's owner. It also, as a rule, induced the birth of a great idea.

Again, and very slowly he re-read the article, which millions that same day had read casually, without a quiver, let alone, a nose quiver. The newspaper item was simple enough:

> NEW YORK, Aug. 10, 1917.—An electromagnetic storm of great violence swept over the eastern section of the United States last night. Due to a brilliant Aurora Borealis,—the Northern Lights,—telegraph and long distance telephone, as well as cable communications were interrupted for hours. No telegraphic traffic was possible between New York and points West. It was impossible to work any of the transatlantic cables between 12:15 A. M. and 9:15 A. M., every one of them having "gone dead." The Aurora Borealis disturbance affected all telegraph and telephone lines extending between Chicago and the eastern cities. On telegraph wires of the Postal Telegraph Co. without regular battery being applied at terminal offices, grounded lines showed a potential of 425 volts positive, varying to 225 volts negative; the disturbance continuing between 12:15 A. M. and 9:15 A. M.
>
> At Newark, N. J., in the Broad Street office a Western Union operator was severely shocked, trying to operate the key, while long sparks played about his instruments.

Sparks rose excitedly and began pacing the cement floor of the vast Tesla laboratory, totally oblivious to the fact that he was sucking a cold pipe. The more he paced about, the more excited he became. Finally he flung himself into a chair and began feverishly to make sketches on big white sheets of drawing paper.

"Why" Sparks had been just an ordinary "Bug," an experimenter, when he entered Tesla's great research laboratory at the beginning of the great war in 1914. Tesla liked the keen, red-haired tousled boy, who always seemed to divine your thoughts before you had uttered five words. His clear blue eyes, lying deep in their sockets, sparkled with life and intelligence and what Sparks did not know about electricity was mighty little indeed. I believe there is no electrical book in

FIGURE 1. ". . . The President of the Glorious French Republic Shouts Dramatically: "Messieurs . . . le Jour de Gloire est Arrivé . . . VIVE-LA-FRANCE!!"—-and Throws in the Huge Switch With Its Long Ebonite Handle. . . ."

existence that Sparks had not devoured ravenously in his spare hours, while having lunch or else while in bed, in the small hours of the morning. His thirst for electrical knowledge was unbounded, and he soaked up every bit of information like a sponge. Yes, and he retained it, too. In short, the young prodigy was a living electrical cyclopedia and highly valued by his associates. No wonder Tesla in three short years had made him superintendent of the laboratory.

Of course, Sparks' first name was not really "Why." But some one had dubbed him with this sobriquet because of his eternal "But *why* is this,"— "*Why,* why should we not do it this way"— "*Why* do you try to do that?" In short his first word always seemed to be "Why,"—it had to be, in his unending quest of knowledge. And his "Why" was always very emphatic, explosive-like, imperative, from which there was no escape.

Ah, yes, his first name. To tell the honest truth, I don't know it. Last year in the spring when I went up to the laboratory, I thought I would find out. So when I finally located the young wonder, behind a bus bar, where he was drawing fat, blue sparks by means of a screwdriver. I told him that I intended to write something about him and his wonderful electrical knowledge. Would he be good enough to give me his real first name?

He was watching a big fuse critically, and in an absent-minded manner exploded: *"Why?"* That finished my mission. And for all I know his real name is "Why" Sparks.

But we left Sparks with his drawings, in the laboratory. That was on a certain evening in 1917. To be exact it was about 10 o'clock. At 10:05 Tesla accompanied by two high army officials strolled into the laboratory where Sparks was still feverishly engaged with sketches lying all about him.

Tesla who was working out a certain apparatus for the Government had dropt in late to show Major General McQuire the result of six weeks' labors. The apparatus had been completed that day and the General, a military electrical expert, had come over specially from Washington to see the "thing" work.

But before Tesla had a chance to throw in the switch of the large rotary converter, Sparks had leaped up, and was waving excitedly a large drawing in Tesla's face. He gushed forth a torrent of sentences, and for fully five minutes Tesla and the two Army officials were listening spell-bound to the young inventor. For a minute or two the three men were speechless, looking awe-struck at Sparks, who, having delivered himself of his latest outburst, now became normal again and lit up his still cold pipe.

It was Tesla who first found his voice. "Wonderful, wonderful. Absolutely wonderful, Sparks. In a month you will be the most talked of man on this planet. And his idea *is* sound." This to the General. "Absolutely without a flaw. And so simple. Why, oh why! did I not think of it before? Come, let me shake the hand of America's youngest and greatest genius!" Which he did.

There then followed an excited thirty-minute conversation with the two army men and an endless long distance talk with the War Department at Washington. Then there was a rush trip to Washington by Tesla and Sparks, conferences at the War Department, and finally a few days later Sparks went to the White House and was presented to the President, who was highly enthusiastic about the model which Sparks and Tesla demonstrated to the head of the Nation. Still later there were certain rush orders from the War Department to the General Electric and Westinghouse Companies for many big, queer machines, and these same machines were shortly . . . But here the Censor bids us an emphatic "Halt." One may not even now divulge certain military information. You appreciate that.

Baron von Unterrichter's flying "Circus" was getting ready to bomb a certain American depot behind the lines. The Americans of late had shot down entirely too many of the Baron's flyers. Only yesterday von der Halberstadt—a German ace himself—and one of von Unterrichter's closest friends had been downed, and killed inside of the German lines. So the Baron was out for blood this sunny morning. As he put it:

"*Verdammte Yankee Schwienehunde,*[1] we will show them who is master of the air hereabouts," shaking his fist at the American lines beyond.

"*Sie, Müller,*" this to an orderly.

"*Zu Befehl, Herr Leutnant,*" replied the young orderly as he came on the run and stood at attention, clicking his heels together, hand at his cap.

"*Versammlung, sofort,*" barked the chief, as he hastened Müller off to summon post haste every man of the aerial squadron for the usual conference before the attack.

In less than ten minutes the thirty flyers were standing drawn up at military attention before their chief, forming a half circle about him. Von Unterrichter's instructions were simple enough. This was a reprisal raid; von der Halberstadt's death must be avenged, fearfully avenged. No quarter was to be given.

"*Dieses Amerikanische Gesindel!*"—here his voice rose to a shrill pitch, "must be taught to respect us, as never before. The orders are to bomb every American base hospital within the sector. . . ."

At this several of the men recoiled involuntarily, which did not escape the keen eye of von Unterrichter, who now incensed to blind fury, by this show of "softheartedness," as he put it, exhorted his men in his harshest possible terms. "And as for their flyers, you must not give quarter. You must not be satisfied with disabling their machines. Kill them! *Schiesst die Lumpen zusammen!* Pump nickel into them, if you see that they may land unharmed"—this in direct violation of all flying etiquette—a thing abhorred by any decent flyer as a rule. It is bad enough to have your machine shot down, but "sitting on a disabled enemy's tail," and pouring machine gun fire into a helpless man, struggling in mid-air,—where was German prestige coming to with such methods. Plainly the men did not like such liberties with their honor, but orders were orders. They grumbled audibly and cast not very encouraging looks at their chief. Even his parting shout: "*Vorwärts—für Gott und Vaterland,*" failed to bring the usual cheers.

1. Gernsback: "For translation of foreign terms see end of this story."

Promptly on the minute of 10 the fifteen flyers of the "Circus" rose, like a flock of big white sea gulls heading in "V" formation towards the American lines. Von Unterrichter was leading his herd in a big *Fokker*. He was out for blood and he meant to have it. His face was set, his jaws clenched like a vice. Hate was written in large characters over his face. . . . Why didn't these *Dollarjäger* stay home and mind their own business chasing their dollars? What right did they have in this fray, anyway, *"Elendige Schweinebande,"* he spoke out loud, to better vent his overpowering hate.

But where were the Yankee *Flieger* today? The Baron's "Circus" was up one thousand meters and less than a mile away from the American first line trenches, but still no machine in sight, either American or French. Strange. Quite an unheard of occurrence. Afraid? *"Unsinn,"* he muttered to himself, they were not the sort to be afraid. Von Unterrichter knew that. For the first time he felt a vague sort of uneasiness creeping over him. He could not understand. There was not a *Flieger* anywhere in sight. None on the ground either, as he scanned the vast saucer below him through his *Zeiss*. Was it a new trick, was . . .

Before he finished his train of thought, his engine stopt dead. Cursing volubly he made ready to "bank" his machine in order to volplane down behind his own lines. He congratulated himself that his engines had not stopped later while over the enemy's lines, but his pleasure was short-lived. For he suddenly became aware of the fact that there was a supreme quiet reigning all about him. Why did he not hear the loud roar of the other fourteen engines, now that his own engine was quiet? Looking around he perceived with horror that every one of the fourteen machines of the "Circus" had simultaneously "gone dead," too, *all of them were now volplaning earthward.*

Sick with an unknown terror, von Unterrichter made a clumsy landing in the midst of his other flyers, all of them pale, some shaking, some with a strange animal expression in their eyes. *What unknown, invisible hand had with one stroke disabled the fifteen engines, one thousand meters above the ground?*

"Himmelkreuzdonnerwetter," shrieked von Unterrichter jumping to the ground, near his airdrome. "I . . . I . . . cannot" . . . here his voice broke. For the first time in his life the young Prussian was speechless. He then stamped his foot in a frenzied fury, but finally gave vent to a full round of cursing. At last he collected his senses sufficiently to look for the cause of the mysterious occurrence. It only took five minutes to find it. His mechanician pointed to the magneto.

"Kaput," he said laconically, if not grammatically.

"Auseinander nehmen," commanded the chief.

It took the deft mechanician but a minute to take the magneto apart, and to withdraw the armature. He gave it one look and with a sickly smile uttered:

"*Ausgebrannt, Herr Leutnant.*" *Herr Leutnant* took the armature into his own hand and inspected it critically. Sure enough it was burnt out, if ever there was a burnt out armature. Perhaps fused would be a better term. The armature was beyond repair, a child could see that. He flung it away and went over to the next nearest flyer. But the mechanic had already located the trouble—in the magneto. Burnt out, too!

Von Unterrichter unutterably sick at heart, aimlessly wandered about the other machines. In each case the result was the same: *Every magneto armature of the fifteen flyers was burnt out,* the wires fused together, all insulation gone!

"*Aber so 'was,*" muttered von Unterrichter, looking about him helplessly. It took fully five minutes before it filtered through his thick skull that this disaster that overtook his "circus" could by no means be a coincidence.

"*Verfluchte Amerikaner,*" he said, "probably a new *Teufelmaschine* of Edison!"

But what would the *Kommando* say to this? Instantly he stiffened as he jumped into a waiting automobile, attached to the airdrome.

"*Zum Kommando, schnell*", he ordered the driver as he sank back into his seat. He must report this queer business to headquarters at once. The driver cranked the engine, then cranked it some more. Pfut . . . pfut . . . pfut . . . sputtered the engine asthmatic-like, but it did not start. He tried again. Same result.

"*Donnerwetter nochmal,*" stormed the Baron vexed over the delay, "*was ist den jetzt los,* why in thunder don't you start you miserable dog?" But the engine would not start. The perplexed chauffeur climbed into the seat of the old style car, which still had its faithful spark coils, so necessary to the ignition system. But the spark coil refused to work, although the storage battery was fully charged and all the connections were right. Cautiously he pulled out one of the spark coil units from its box. One look told the story.

"*Ausgebrannt, Herr Leutnant,*" he said weakly, for he had seen the burnt out magneto armatures a few minutes before.

Von Unterrichter, with eyes almost popping out of his head, was struck absolutely speechless for half a minute. "*Heiliger Strohsack,*" he muttered awe-struck, remembering his young sister's favorite expression, whenever something out of the ordinary happened to her. He finally collected himself sufficiently and jumped out of the car.

"*Zum Telefon,*" he muttered to himself. He must report this uncanny occurrence at once to the *Kommando*. Not a second was to be lost. He at last understood that something momentous had happened. He made the airdrome on the run and though it was only 200 yards away he surprised himself at the speed he made. Puffing volubly he arrived at the telephone. He gave the handle several quick turns,[2] grasped the receiver and simultaneously bellowed into the mouthpiece in front of him:

"Hallo, hallo" . . . but he went no further. The receiver flew from his ear, for there had been a loud clattering, rattling, ear-splitting noise in the instrument that almost burst his eardrum. He made a foolish grimace, as he held his ear with his hand. Cautiously he approached the receiver to within a few inches of his other ear and listened. All was quiet, not a sound. Mechanically he unscrewed the receiver cap and looked at the two bobbins. They were charred and black. The telephone was dead.

The instrument slipt from his hand and dangling by its red and purple cord went crashing against the wall of the airdrome, while von Unterrichter limply sank into a chair.

Once more he got up and walked out. He must get in touch with his General at all costs. This was becoming too serious. Ah . . . he had it, the field telegraph. There was one at the other end of the building. He went there as fast as his legs could carry him. He opened the door of the little office, but one look sufficed. The young man in charge of the telegraph sat dejected in a corner, a dumb expression in his eyes. Long purple sparks were playing about the instruments on the table. A child could have seen that it was impossible to either send or receive a telegram under such conditions. . . . Ah! an inspiration. . . .

"Dummkopf," he muttered to himself, "Why didn't I think of it before. *Die Funkenstation!* Surely the wireless must work! Ha, ha, there are no wires there at least to burn out!"

The radio station was over a kilometer away. He knew it well, for he had flown over it a great many times. To get there quick, that was the question. The *Kommando* was at least eight kilometers to the rear, and he knew he could not make that distance on foot very quickly. Ah, yes, there was a horse somewhere around. The cavalry horse was located soon, and as the young airman walked hurriedly about, troubled as he was, he could not help noticing the listless attitude of every man he passed. Men were whispering in a hushed manner, alarm was plainly written on their faces—the fear or the unknown.

Von Unterrichter jumped on to his horse and galloped in the direction of the field radio station. It did not take him long to reach it, and long before he dismounted he could see the bright blue spark of the transmitting station.

"Gott sei Lob," he uttered to himself as he jumped to the ground, "at least that's working."

2. Gernsback: "All German telephones are magneto operated. To call Central you must turn the handle of the ringing magneto."

THE PERVERSITY OF THINGS

Note here the curious mechanism of the Prussian mind.[3] A Prussian officer, the most arrogant, distasteful creature imaginable, is always the great brave hero when he knows that he is fighting with all the advantages on his side. As Heinrich Heine, the poet,—himself a German-hater,—puts it:

> The Germans have no self-respect. They are the only men in the world who, as private soldiers, will stand still while an officer kicks them or bespatters them with mud. They receive the mud with smiles and stand expectantly, cap in hand.

It is the Prussian-German sort of "honor" that makes a Zabern affair possible, where a foul-mouthed young officer, with his sword, beats a helpless, crippled Alsatian cobbler insensible.

A coward at heart, always ready to blaspheme his maker, when things go right, the Prussian quickly turns to his German *Gott,* as soon as things go against him.

Heine,—himself a German, and he ought to know,—will tell you so.

Now it so happened that von Unterrichter had been an expert wireless man before the war, and while he did not know a great deal about electricity, he well knew how to send and receive messages.

He ran to the wagon which carried the mobile radio field apparatus and peremptorily ordered the operator in charge away. *"Aber Herr Leutnant,"* expostulated the thus rudely interrupted man, "I tell you . . ."

"Maul halten," thundered von Unterrichter, with which he sat down, clamping the operator's receiver on his own head.

He pressed the key impulsively, and noted with grim satisfaction that the loud blue spark crashed merrily in the not very up-to-date spark gap.

As he sent out the call mechanically, he wondered vaguely what the matter could be with the government, because it did not even supply a modern, up-to-date *Löschfunkenstrecke*—quenched spark gap—for field use. Things must be pretty bad when the government must economize even a few beggarly pounds of brass, so necessary for a noiseless spark gap.

But he could not give that matter further attention for he had thrown the aerial switch from "sending" to "receiving."

He had strained his ears for a reply from the operator from the *Kommando,* but, as the switch was thrown, instead of a reply there was a loud, constant roar in the receivers, so loud that it was painful. Off came the headgear, while von Unterrichter once more sank into a chair.

He was a pitiful spectacle to look at, the fate of a 20th Century man

3. The following section on the "Prussian mind" was removed from the *Amazing Stories* reprint of this story in July 1926.

flung back a hundred years. His eyes roamed idly about till the distant railroad embankment struck his eye. No train was moving. Everything was at a standstill—how could a train move without a telegraph? How could a train be dispatched—there would be a thousand collisions. He turned to the radio operator, who as yet had not grasped the situation in its entirety.

"Nordlicht, nicht wahr, Herr Leutnant?" he began, thinking no doubt that the phenomenon was an ordinary form of Aurora Borealis,—the northern lights,—in other words, a magnetic storm, that would be over soon.

"Dummes Rindsvieh" . . . snapped the *Herr Leutnant,* who knew better by this time. Indeed he was to know still more at once, for while he was speaking there came to his ear a low dull roar, a sound he had heard once before, far back in 1914 when the Germans had retreated very much in a hurry beyond the Marne.

Panic seized him. Yes the sound was unmistakable. The German army once more was in full retreat—no it was a rout—a panic-stricken rabble that made its way back.

Like lightning the news had spread among the men at the front that uncanny things were afoot, that all communications had been annihilated with one stroke, that no orders could be sent or received except by prehistoric couriers, that the *Grosses Kommando* was cut off from the army, and that in short the German army as far as communication was concerned, had suddenly found itself a century hack.

For what had happened to von Unterrichter that morning, had happened on a large scale not only to every one along the front, but all over Germany as well! Every train, every trolley car, every electric motor or dynamo, every telephone, every telegraph had been put out of commission. With one stroke Germany had been flung back into the days of Napoleon. Every modern industry, every means of traffic— except horse-drawn vehicles—were at a standstill. For days the German retirement went on, till on the fifteenth day, the entire German army had retired behind the natural defenses of the Rhine, the victorious Allies, pressing the fleeing hordes back irresistibly.

And it must have been a bitter pill for the German high-command to swallow when they saw that the Allied fliers were constantly flying behind their own lines and that as the Allies advanced, their automobiles and their trains seemed to run as well as ever behind their own lines. But no German succeeded in flying an aeroplane or in running an automobile. That mysterious force obviously was trained only against them, but was harmless behind the Allied lines. Nor did the Germans find out to this date what caused their undoing.

Peace having not been declared as yet, I cannot, of course, divulge the full details of the scheme of just how the Germans were finally driven across the Rhine. That, of course, is a military secret.

But I am permitted to give an outline of just what happened on that memorable morning, when the German "Kultur" was flung back into the dark ages where it belongs.[4]

But first we must go back to Tesla's laboratory once more, back to that evening when "Why" Sparks first overwhelmed Tesla and his companions with his idea. This is in part what Sparks said:

"Mr. Tesla! In 1898 while you were making your now historic high-frequency experiments in Colorado with your 300-kilowatt generator, you obtained sparks 100 feet in length. The noise of these sparks was like a roaring Niagara, and these spark discharges were the largest and most wonderful produced by man down to this very day. The Primary coil of your oscillator measured 51 *feet in diameter,* while you used 1100 amperes. The voltage probably was over 20 million. Now then, in your book, *High Frequency Currents,* among other things you state that the current which you produced by means of this mammoth electric oscillator was so terrific that its effect was felt 13 miles away. Altho there were no wires between your laboratory and the Colorado Electric Light & Power Co., five miles distant, your *"Wireless" Energy* burnt out several armatures of the large dynamo generators, *simply by long distance induction from your high frequency oscillator.* You subsequently raised such havoc with the Lighting Company's dynamos that you had to modify your experiments, although you were over five miles away from the Lighting Company.[5]

"Now if in 1898, twenty years ago, you could do that, why, WHY cannot we go a step further in 1918, when we have at our command vastly more powerful generators and better machinery. If you can burn out dynamo armatures 13 miles distant with a paltry 300 kilowatts, *why* cannot we burn out every armature within a radius of 500 miles or more."

"The primary coil of your oscillator in 1898 was 51 feet in diameter. WHY *cannot we build a primary 'coil' from the English Channel down to Switzerland, paralleling the entire Western front?* This is not such a foolish, nor such a big undertaking as you might think. My calculations show that if we were to string highly insulated copper wires one-quarter inch thick on telegraph poles behind the front, the problem would become a simple one. Ordinary telegraph poles can be used, and each pole is to carry twenty wires. Beginning three feet above the ground, each wire is spaced two feet distant from the next one. These wires run continuous from the sea to Switzerland. Moreover, every ten miles or so we place a huge 3,000 kilowatt generating plant with

4. Changed in July 1926 reprint to: "when the German army was flung back into the dark ages."

5. Gernsback: "The above occurrences as well as the cited experiments and effects of the Tesla currents are actual facts checked by Mr. Tesla himself, who saw the proof of this story. —*Editor.*"

its necessary spark gaps, condensers, etc. The feed wires from these generating plants then run into the thick wires, strung along the telegraph poles, forming the gigantic Tesla Primary Coil. Of course, you realize that in a scheme of this kind it is not necessary to run the telegraph poles actually parallel with every curve of the actual front. That would be a waste of material. But we will build our line along a huge flat curve which will sometimes come to within one-half mile of the front, and sometimes it will be as much as fifteen miles behind it. The total length of the line I estimate to be about 400 miles. That gives us 40 generating plants or a total power of 120,000 kilowatts! A similar line is built along the Italian front, which is roughly one hundred miles long at present. That gives us another 30,000 kilowatts, bringing the total up to 150,000! Now the important part is to project the resultant force from this huge Tesla primary coil in *one direction only,* namely that facing the enemy. This I find can be readily accomplished by screening the wires on the telegraph poles at the side facing our way as well as by using certain impedance coils. The screen is nothing else but ordinary thin wire netting fastened on a support wire between the telegraph poles. This screen will then act as a sort of electric reflector. So." . . . Sparks demonstrated by means of one of his sketches.

"Everything completed we turn on the high-frequency current into our line from the sea to little Switzerland. Immediately we shoot billions of volts over Germany and Austria, penetrating every corner of the Central empires. *Every closed coil of wire throughout Germany and Austria, be it a dynamo armature, or a telephone receiver coil, will be burnt out, due to the terrific electromotive force set up inductively to our primary current.* In other words every piece of electrical apparatus or machinery *will become the secondary of our Tesla coil, no matter where located.* Moreover the current is to be turned on in the day time only. It is switched off during the night. The night is made use of to advance the telegraph poles over the recaptured land,—new ones can be used with their huge primary coil wires, for I anticipate that the enemy *must* fall back. Turning off the power does not work to our disadvantage, for it is unreasonable to suppose that the Teutons will be able to wind and install new coils and armatures to replace all the millions that were burnt out during the day. Such a thing is impossible. Besides, once we get the Germans moving, it ought to be a simple matter to follow up our advantage, for you must not forget that we will destroy ALL their electrical communications with one stroke. No aeroplane, no automobile, will move throughout the Central States. In other words, we will create a titanic artificial Magnetic Storm, such as the world has never seen. But its effect will be vastly greater and more disastrous than any natural magnetic storm that ever visited this earth. Nor can the Germans safeguard themselves against this electric storm any more than our telegraph companies can when a real magnetic storm sweeps over the earth. Also, every German telegraph or telegraph line in occupied France and Belgium will be our ally! These insulated metallic lines actually help us to "guide" our energy into the very heart of the enemy's countries. The more lines, the better for us, because all lines act as feed wires for our high frequency electrical torrents. . . ."

A few kilometers north of Nancy, in the Department of Meurthe et Moselle, there is a little town by the name of Nomeny. It is a progressive, thrifty little French town of chief importance principally for the reason that here for four years during the great war the French army has been nearer to the German frontier than at any other point, with the exception of that small portion of Alsace actually in the hands of the French.

Nomeny in the military sense is in the Toul Sector, which sector early in 1918 was taken over by the Americans. If you happened to go up in a captive balloon near Nomeny you could see the spires of the Metz Cathedral and the great German fortress, but 16 kilometers away, always presuming that the air was clear and you had a good glass.

On a superb warm summer morning there were queer doings at a certain point in the outskirts of Nomeny. All of a sudden this point seemed to have become the center of interest of the entire French, British, and American armies. Since dawn the military autos of numerous high Allied officers had been arriving while the gray-blue uniforms of the French officers were forever mixing with the business-like khaki of the British and Americans.

The visitors first gave their attention to the camouflaged, odd-looking telegraph poles which resembled huge harps, with the difference that the wires were running horizontally, the "telegraph" line stretching from one end of the horizon to the other. A few hundred yards back of this line there was an old brewery from which ran twenty thick wires, connecting the brewery with the telegraph poles. To this brewery the high officers next strolled. An inspection here revealed a ponderous 3,000 kilowatt generator purring almost silently. On its shining brass plate was the legend: "Made in U. S. A." There was also a huge wheel with large queer round zinc pieces. Attached to the axis of this wheel was a big electric motor, but it was not running now. There were also dozens of huge glass jars on wooden racks lined against the wall. Ponderous copper cables connected the jars with the huge wheel.

One of the French officers, who, previous to the war, had been an enthusiastic Wireless Amateur, was much interested in the huge wheel and the large glass bottles. "Aha," said he, turning to his questioning American confrère, *"l'éclateur rotatif et les bouteilles de Leyde."*

There was little satisfaction in this, but just then a red-haired, tousled young man who seemed to be much at home in the brewery, came over and adjusted something on the huge wheel.

"What do you call all of these do-funnies?" our young officer asked of him, pointing at the mysterious objects.

"Rotary spark gap and Leyden jars," was the laconic reply. The officer nodded. Just then there was a big commotion. The door flew open and a French officer standing at attention shouted impressively:

"Le Président de la République!"

Instantly every man stood erect at attention, hand at the cap. A few seconds later and President Poincaré walked in slowly, at his side General Pétain. It was then five minutes to 10.

President Poincaré was introduced to the red-haired, tousled young man whom he addressed as *Monsieur* Sparks. *Monsieur* Sparks speaking a much dilapidated French, managed, however, to explain to his *excellence* all of the important machinery, thanks to a sleepless night with a French dictionary.

Monsieur Poincaré was much impressed and visibly moved, when a French officer had gone over Sparks' ground, and re-explained the finer details.

The President now takes his stand on an elevated platform near a huge switch which has an ebonite handle about a foot long. He then addresses the distinguished assembly with a short speech, all the while watching a dapper young French officer standing near him, chronometer in hand.

Somewhere a clock begins striking the hour of ten. The President still speaks but finishes a few seconds later. The distinguished assemblage applauds and cheers vociferously, only to be stopped by the dapper young officer who slowly raises his right hand, his eyes glued to the chronometer. Immediate silence prevails, only interrupted by the soft purring of the huge generator. The dapper young officer suddenly sings out:

"Monsieur le Président! A-ten-tion! ALLEZ!!"

The President of the glorious French Republic then shouts dramatically: *"Messieurs . . . le jour de gloire est arrivé . . . VIVE-LA-FRANCE!!"*—and throws in the huge switch with its long ebonite handle.

Instantly the ponderous rotary spark gap begins to revolve with a dizzying speed, while blinding blue-white sparks crash all along the inside circumference with a noise like a hundred cannons set off all at once. The large brewery hall intensifies the earsplitting racket so much that every one is compelled to close his ears with his hands.

Quickly stepping outside the party arrives just in time to see fifteen German airplanes volplaning down and disappearing behind the German lines. A French aerial officer who had observed the German airplanes, drops his glass, steps over to the President, salutes smartly and says impressively:

"Le 'cirque' du Baron d'Unterrichter! Ils sont hors de combat!"

Hors de combat is correct. Von Unterrichter was not to fly again for many a week.

We look around to tell the glad news to General Pétain, but the latter has disappeared into a low brick building where he now sits surrounded by his staff, poring over military maps ornamented with many vari-colored pencil marks, as well as little brightly colored pin flags. Telephone and telegraph instruments are all about the room.

Again the President shakes hands with *Monsieur* Sparks, congratulating him on his achievement. Luncheon is then served in the former office of the brewery, gayly bedecked with the Allied flags along the walls. But even here, far from the titanic rotary spark gap, its crashing sparks are audible. Looking through the window we see a wonderful sight. Although it is broad daylight, the entire queer telegraph line is entirely enveloped in a huge violet spray of electric sparks. It is as if "heat-lightning" were playing continuously about the whole line. No one may venture within fifty feet of the line. It would mean instant death by this man-made lightning.

Luncheon is soon over and more speeches are made. Suddenly the door flings open and General Pétain steps in. One look at his remarkable features, and all talk stops as if by magic. He crosses the room towards the President, salutes and says in a calm voice, though his eyes betray his deep emotion:

*"Monsieur le President,
toute l'armeé Allemande est en retraîte!!"*

And so it was. The greatest and final retreat of the Kaiser's "invincible" hordes was in full swing towards the Rhine.

More congratulations are to be offered to Sparks. A medal, . . . Heavens, where is that young man? But Sparks has slipt over to his machines and is standing in front of the noisy "thunder and lightning" wheel eyeing it enthusiastically.

"Why, oh WHY, do they call you *éclateur!*" he says." "Spark Gap is good enough for me!" "Oh, boy!! But you aren't doing a thing to those *Germins!*"

THE END.

Translation of German and French Terms Used in This Story

GERMAN

Verdammte Yankee Schweinehunde: Damned Yankee Pig-Dogs!

Sie, Müller: You, Müller!

Zu Befehl, Herr Leutnant: As your orders, Lieutenant!

Versammlung, sofort: Assembly, at once!

Dieses Amerikanische Gesindel: This American rabble!

Schiesst die Lumpen zusammen: Shoot the ragamuffins together!

Vorwärts für Gott und Vaterland: Onward, for God and Fatherland!

Dollarjäger: Dollar Chasers

Elendige Schweinbande: Miserable band of pigs.

Unsinn: Nonsense.

Flieger: Flyer (aeroplane).

Himmelkreuzdonnerwetter: A popular German cuss word. Literally it means "sky-cross-thunder." English equivalent is "A thousand thunders."

Kaput: German slang, equivalent to our slang "busted."

Auseinander nehmen: Take it apart!

Ausgebrannt: Burnt-out.

Aber so 'was: Such a thing (of all things).

Verfluchte Amerikaner: Cursed Americans.

Teufelmaschine: Diabolic machine.

Zum Kommando, schnell: Quick, to Headquarters!

Donnerwetter nochmal: By all thunders!

Was ist denn jetzt los? What's up now?

Heiliger Strohsack: Holy bag-of-straw; equivalent to "Holy Gee."

Dummkopf: Blockhead.

Die Funkenstation: The Radio Station.

Gott sei Lob: God be thanked.

Aber Herr Leutnant: But, Lieutenant!

Maul halten: Shut up.

Löschfunkenstrecke: Quenched Spark Gap.

Nordlicht, nicht wahr? Northern lights, is it not?

Dummes Rindsvieh: Stupid piece of cattle.

Grosses Kommando: General Headquarters.

FRENCH

L'éclateur rotatif et les bouteilles de Leyde: Rotary spark gap and Leyden jars.

Le Président de la République: The President of the Republic.

Monsieur le Président! Attention! Allez! Mr. President! Ready! Go!

Messieurs, le jour de gloire est arrivé, vive la France!: Gentlemen, the day of glory has arrived, long live France! (This is from the second verse of the "Marseillaise")

Le "cirque" du Baron d'Unterrichter! Ils sont hors de combat!: Baron von Unterrichter's circus! They are out of the fighting!

Monsieur le Président, toute l'armée Allemande est en retraite: Mr. President, the entire German army is in retreat.

Amateur Radio Restored

Electrical Experimenter, vol. 6, no. 2, June 1919

inning one war seems like a great accomplishment, but winning two wars, one after the other, is a historic occurrence for which we have few counterparts.

When on Nov. 11 the glad news came that the great war had been won, the world once more breathed freely, and so did the American Radio Amateur, who had done his "bit" in bringing the war to a successful termination. But hardly was the armistice signed than the spectre of a new war—to the amateur—crept up over the horizon threatening to take away from him the freedom for which we had fought.

A nation may be opprest by a relentless foe in a similar manner by unjust laws which curtail the freedom of its people. Indeed, when it was proposed last winter to take away from the American Radio Amateur the freedom of the ether, which would have deprived the liberties of over 300,000 young men in this country, war was declared once more. This was war to the knife; it was as short as it was decisive. But Right as usual won over Might, and the American Radio Amateur won *his* war, and the Allies won theirs.

And, as the Allies will win the fruits of their victory in the months to come, so will the amateurs reap the fruit of their victory. Indeed, the reaping has already begun. On April 15th the ban on Radio, at least for *receiving*, was officially taken off, and a mighty shout went up when Radio Amateurs were again permitted to use the ether to their hearts' content. Altho the ban for *sending* has not been removed at this time of writing, the chances are that before the next issue is in your hands the freedom of the ether will be once more restored completely.[1]

As soon as the newspapers published the welcome tidings on April 15th that the ban on receiving was off, hundreds of thousands of amateurs began dusting off their sets and aerials blossomed forth over night by the thousands to resume their former activities once more.

And wonders upon wonders! When we put our sets away two years ago we were accustomed to hear nothing but the crisp dots and dashes in flute-like, staccato sounds coming from the high power stations which we all had learned to love. But the war has changed everything— even radio, for now the *radio telephone* has come into its own.

Where formerly there was nothing but the *tah-de-dah* in our phones,

1. As Gernsback reported three months later, Secretary of the Navy Josephus Daniels attempted to maintain government control over radio activities after the war. Hugo Gernsback, "Government Radio Control—Once More," *Radio Amateur News* 1, no. 3 (September 1919). However, during the 1919 hearings on permanent naval radio control, "two congressmen pointedly criticized [Daniels's] wartime use of navy funds to purchase radio company assets without explicit authorization from Congress. Further discrediting the secretary was testimony from amateur groups and commercial firms, including E. J. Nally's angry retelling of the travails of Pan-American Wireless [the hybrid government/private entity that incorporated all patents, personnel, and equipment from American Marconi and Federal Telegraph companies during the war]. The committee tabled the bill a month later. The ire against Daniels spread to the rest of the House. House Minority Leader James R. Mann publicly called for Daniels's impeachment over the purchase of radio company assets." Jonathan Reed Winkler, *Nexus: Strategic Communications and American Security in World War I* (Cambridge, Mass.: Harvard University Press, 2008), 254.

the ether is now filled with the human voice flung far and broad over the land—nay, over the oceans—and as the months roll by the dots and dashes will grow less and less, and the human voice will come in over our aerials more and more, which is as it should be.

The writer has always contended that wireless telephony was the logical outcome of radio, and in years to come only the commercial high power stations will operate their dots and dashes with their high speed machines where the voice would not be as reliable. But the future of Radio Amateurism in this country is centered upon the radio telephone. While no doubt many of us will still cling to the dots and dashes, the radio telephone will probably soon be used in overwhelming numbers.

What a wonderful world it will be in one or two years hence! You will step out into the star-lit night and myriads of voices—noiselessly and invisibly—will fill the air all over the continent, flung through the ether. You will hear nothing and you will see nothing. When you sit on the sea-shore gazing fascinated at the beauty of the tides, as they follow the moon, all the while the invisible and soundless voices will be all around you—nay, even pass through your very body. But when you step in the humble little amateur radio station and clasp a pair of receivers onto your ears, the ether will be unlocked and all of these myriads of voices will be made audible to your ears. A perfect Babel of voices will greet you, but by means of your tuning devices you will be able to pick out the very voice you wish to listen to, tuning out all the others.

It is a glorious thought that we are fortunate enough to live in an age where such things are possible, and we must prove worthy of our new-gained liberty, now certain.

And the writer regrets to say that before the war we did not realize how fortunate we were, and we did not show by our actions that we appreciated the freedom of the ether.

There was constant bickering and quarreling on the side of the amateur, among themselves as well as with commercial and government stations. Profane talk took place not infrequently thru the ether, and many of us annoyed the commercial as well as the government operators by "hogging" the ether. Hence, it was not surprising that the government tried to pass laws to curb the amateurs and curtail their liberties.

If Radio Amateurism is to prevail in the new era, the amateurs must stop their former nonsense and must settle down strictly to business. They can and should receive all they want to their hearts' content. They can and will listen very shortly to President Wilson's voice when he speaks to us from the "George Washington," where everybody

NAVY DEPARTMENT
Naval Communication Service
Office of the Director.
Washington, April 14, 1919.
Editor Electrical Experimenter
Sir:
The Acting Secretary of the Navy authorizes the announcement that effective April 15, 1919, all restrictions are removed on the use of radio receiving stations other than those used for the reception of commercial radio traffic. This applies to amateur stations, technical and experimental stations at schools and colleges, receiving stations maintained by jewelers or others desirous of receiving time signals, receiving stations maintained by manufacturers of radio apparatus, etc.

The restrictions on transmitting stations of all types are still in effect, as are the restrictions on stations operated regularly for the reception of commercial radio traffic. Both of the above classes of stations will be permitted to resume operation as soon as the President proclaims that a state of peace exists.

Attention is invited to the fact that all licenses for transmitting stations have expired, and that it will be necessary, when peace is declared, for the owners of these stations to apply to the Department of Commerce for new licenses.

Very respectfully,
(Signed) E. B. Woodworth,
Commander, U. S. Navy, Assistant Director Naval Communications.

The Above Is the Text Of The Official Notification Restoring the Freedom of the Ether Once More. The Ban on Sending Is as Yet Not Lifted. When Peace Is Declared, While Technically the Ban for Sending Will be Off, It Should be Remembered That All Sending Licenses Have Been Cancelled During the War and New Licenses Must Be Secured from the Department of Commerce. Address the Office of the Radio Inspector at the Custom House of Your District for Further Particulars.

is free to listen in.[2] Amateurs can listen in to their hearts' delight, whether it be to wireless-telephone or wireless telegraph. But when it comes to sending, we must mend our ways. If we do not, laws will be past [*sic*] sooner or later to curtail our entire liberties, and here the writer desires to make an important suggestion.

There are now a good many Radio Clubs and Associations all over the country. We cannot have too many. The suggestion is that every club and every association should appoint one or a number of expert amateurs to listen in nightly to messages that are violating the common rules, either by profane language, by "hogging" the ether, by Q.R.M. or by willful interference. The offender should be sought out and promptly warned and should he repeat his offense a second time, he should be promptly reported to the Radio Inspector of the district, and the various publications catering to the amateurs should be notified. A sworn statement relating the offense should be submitted to the publications, who will pledge themselves to publish the name of the offender.

The *Electrical Experimenter,* for one, will be glad to publish such names regularly, and it is believed that such measures will do more than anything else to increase the prestige of radio amateurism in the United States, and take it out of the school-boy class. There is little doubt that if we follow such procedures Radio Amateurism in the United States will become one of the great institutions of this country. We must show the world that we amateurs are serious-minded and do not use one of the greatest inventions for a plaything only for just a mere holiday.

The *Electrical Experimenter* thru the RADIO LEAGUE OF AMERICA shall be only too glad to publish letters from amateurs suggesting new ways and means to benefit the cause of Radio Amateurism, our columns being open to all worthy suggestions.

While for the present there is nothing to be feared of adverse legislation, still it is impossible to forecast that such legislation will not be attempted in the future. We have won the freedom of the ether due to the splendid cooperation of Radio Amateurs in this country, and if we have won once we will win again.

The RADIO LEAGUE OF AMERICA is a clearing house for all Radio Amateurs in the United States, and today has more members than any other league, but it has not enough members. It wishes to have every amateur as an enrolled member. The membership of the Radio League, established in 1915, is gratuitous. There are no dues or fees to be paid. All the League wants is the name of every amateur and his address, so that first of all, if a national emergency arises, the Government can rely on the amateurs for quick communication. Also in case

2. The SS *George Washington* was an ocean liner that carried Woodrow Wilson to Europe for the Paris Peace Conference. At the request of the Navy Department, the ship was outfitted with a "radio telephone transmitter," purpose-built by General Electric's Research Laboratory. On July 4, 1919, during its return trip across the Atlantic, Wilson attempted a holiday broadcast address to the troops. The broadcast failed, not for any technical reason but because Wilson stood too far from the microphone. As the story goes, the wireless operators were too nervous to ask him to step closer. See Thomas H. White, "Radiophone Transmitter on the U.S.S. George Washington," *United States Early Radio History,* http://earlyradiohistory.us/1919wsh.htm.

adverse legislation should be attempted, the League wishes to notify every member immediately, as it did last December when the freedom of the ether was threatened.

We ask every amateur who is not a member of the League to sign the blank appended herewith. It costs nothing but a 3-cent stamp to do so. Every enrolled member is entitled to the membership certificate free of charge which is mailed to all members. The membership of the Radio League of America now comprises 21,309 members. If you are not a member at the present time, you should enroll at once by all means.

LONG LIVE THE RADIO AMATEUR!

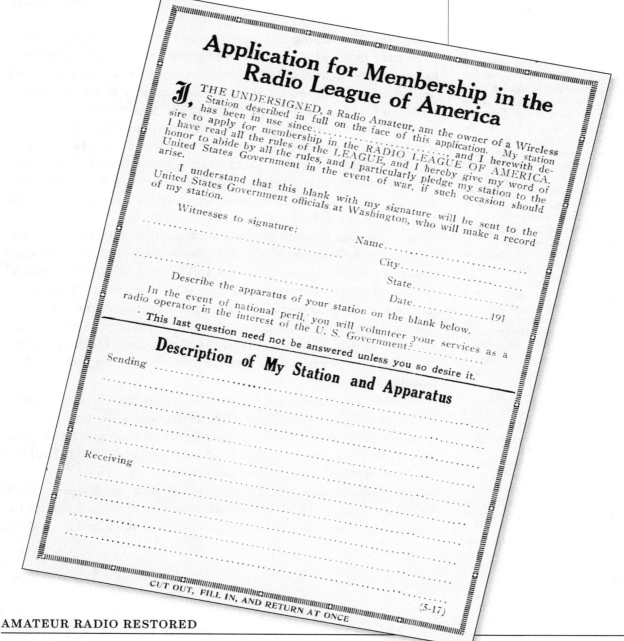

Application for Membership in the Radio League of America

I, THE UNDERSIGNED, a Radio Amateur, am the owner of a Wireless Station described in full on the face of this application. My station has been in use since.................................and I herewith desire to apply for membership in the RADIO LEAGUE OF AMERICA. I have read all the rules of the LEAGUE, and I hereby give my word of honor to abide by all the rules, and I particularly pledge my station to the United States Government in the event of war, if such occasion should arise.

I understand that this blank with my signature will be sent to the United States Government officials at Washington, who will make a record of my station.

Witnesses to signature:

....

....

Name...........................

City...........................

State...........................

Date...........................191

Describe the apparatus of your station on the blank below.

In the event of national peril, you will volunteer your services as a radio operator in the interest of the U. S. Government?...........................

This last question need not be answered unless you so desire it.

Description of My Station and Apparatus

Sending ....

Receiving ....

CUT OUT, FILL IN, AND RETURN AT ONCE

(5-17)

Why "Radio Amateur News" Is Here

Radio Amateur News, vol. 1, no. 1, July 1919

Radio Amateur News with this issue makes its debut to the radio fraternity. At this occasion it may not be amiss to state the pertinent reasons for its existence and why as a matter of fact it just *had* to come.

This magazine is the logical outcome of many attempts to publish a purely Radio periodical, independent thruout and devoted to American Radio Amateurism.[1]

In 1908 I started the first magazine in America in which were publisht many radio articles—*Modern Electrics*. Radio Amateurism being in its infancy then, could not support a purely radio magazine,—so *Modern Electrics* devoted only about one-quarter of its contents to radio. In 1913 I came out with the *Electrical Experimenter*. This magazine has been more prominent than any other on account of its very important radio section. Even during the war—with radio amateurism dead and nearly every radio magazine discontinued—the *Electrical Experimenter,* at a great financial loss, continued publishing radio articles uninterruptedly, month after month, to keep alive the radio spark in the hearts of our amateurs.

But now that the war is won,—now that the amateurs have won *their* war by defeating a proposed new law which would have destroyed American Radio Amateurism—we will witness the most wonderful expansion of the radio arts ever dreamt of. The amateur is here to stay and so is radio in general. I predict an astounding growth of the art during the next ten years. Every other house will have its radiophone, to converse with friends and relatives, for business and for pleasure. Marvelous inventions will be made in Radio during the next decade—unbelievable now.

Because I am a staunch believer in the glorious future of Radio in America, I have launched *Radio Amateur News*. Its first issue will mark the time when amateur radio in America has come into its own again, when it has been re-born greater than ever—a Phenix [*sic*] rising more beautiful than before from his ashes.

I felt that the time was ripe for a purely radio magazine—a 100% radio magazine,—by and for the amateur. I felt that a magazine for the entire radio fraternity, be he scientist, advanced, or junior amateur,

1. *Radio Amateur News* (later abridged to *Radio News*) debuted in 1919 as the standard bearer for a community of amateur experimenters during a period that was about to see a rapid shift in the industry away from independent tinkerers having private conversations, toward powerful corporations that produced consumer-friendly broadcast listening sets. But the magazine wasn't altogether above buying into the broadcast craze of the 1920s. Tim Wu writes that *Radio News* eventually served as one of the first broadcast programming guides in the country's history, publishing lists of each radio station in operation, along with their frequencies and "what one might expect to hear on them—a forerunner of the once hugely profitable *TV Guide.*" These lists of broadcast programming reveal that much of the chatter on the air as early as 1922 came from corporate sources: "many early stations were run by radio manufacturers such as Westinghouse, the pioneer of the ready-to-plug-in model, and RCA, both of which had an obvious interest in promoting the medium. Still many stations were run by amateurs, radio clubs, universities, churches, hotels, poultry farms, newspapers, the U.S. Army and Navy." Tim Wu, *The Master Switch: The Rise and Fall of Information Empires* (New York: Knopf Doubleday Publishing Group, 2010), 39.

Radio News became an increasingly conflicted magazine into the 1920s, as it extolled the virtues of the amateur experimenter while being bankrolled by advertising revenue from RCA. An RCA advertisement appears in *Radio News* as early as January 1922. James Harbord, lieutenant general in the Army during the Civil War and later president of RCA, wrote an editorial in a 1923 issue of the magazine explaining his company's policy on buying and controlling radio patents. James Harbord, "General Harbord Explains the Patent Situation," *Radio News* 5, no. 3 (September 1923): 252.

was badly needed, and that is why you are now reading this the first issue of the newcomer.

And here is the platform upon which *Radio Amateur News* stands. I pledge myself to a strict adherence to every plank:

1st. Only Radio—100% of it—*nothing else.*

2nd. An Organ for and by the amateur. The amateur's likes and wants will always come first in this magazine.

3rd. Absolute Independence. *Radio Amateur News* has only one Boss—its readers. This magazine is not, nor will it ever be affiliated with any stifling, commercial radio interests whatsoever.[2]

4th. Truth—first, last, and always. When you see it in *Radio Amateur News* you may be sure that it is so. Not being affiliated with commercial radio interests, this magazine will have no reason to suppress important articles, discoveries, etc.

5th. *Radio Amateur News* is and will be the sworn enemy of all adverse and unfair radio legislation. Our Washington representative will inform us immediately of any new radio legislative measures. No unfair bill will become a law before all amateurs have had their say.

6th. The uplift of American Radio Amateurism out of the "kid" class, into the serious status to which the art is entitled. Amateur Radio is not a plaything or a sport—it is a useful mind ennobling art—it vanquishes distances, it saves lives, and it will be as necessary as the telephone ten years hence.

7th. Instructive first and last. Up-to-date scientific articles for your instruction will always have first place in *Radio Amateur News.* We shall publish purely scientific articles every month, articles that on account of their length are often crowded out of other publications.

8th. First in print with the News. You will find all important Radio News in this magazine from one to three months ahead of all other publications—always.

Now, my friends, it's up to you how great and how big *Radio Amateur News* shall be. Its future is in your hands. We're off . . . . . .
Three cheers for American Radio Amateurism—
Long live the Radio Amateur.

2. By 1928 the number of radio parts retailers and manufacturers had declined significantly with the rise of more consumer-friendly, complete radio sets. But when *Radio News* began to name specific radio parts and manufacturers in its articles, purportedly to make these endangered resources easier to find, readers cried foul. Many wrote in saying that the magazine's articles seemed far too "inspired" by its advertisers. Gernsback responded, "Against its better judgment, *Radio News,* during 1927, endeavored to better trade conditions by publishing manufacturers' names and giving manufacturers' specifications in its text pages; because it seemed that it would thus be made easier for the reader to buy suitable material than if no names were given and he had to guess them. In doing so, *Radio News* honestly believed that not only would it serve its readers, as it had always done, but the radio trade as well.

"It is, however, admitted now that this policy was wrong and, beginning with this issue, we revert to our former custom of not mentioning any manufacturers' names or trademarks of parts and circuits of any kind in the text pages of *Radio News.* In that respect, we go back to our former practice of the years previous to 1927.

"... At no time, however, did *Radio News* reap any material benefit from specifying manufacturers' names and materials. If any proof is wanted, the constant advertising shrinkage of *Radio News* during 1927 is glowing testimony of this fact. At no time did the trade (particularly the parts manufacturers) really support *Radio News,* in spite of its publishing their names and specifications quite freely." Hugo Gernsback, "Radio Movie," *Science and Invention* 16, no. 7 (November 1928): 622–23.

Grand Opera by Wireless

Radio Amateur News, vol. 1, no. 3, September 1919

A recent newspaper report from Chicago brought the not at all surprising news that grand opera music had been transmitted by wireless telephone for over one hundred miles. Sensitive microphones placed on the stage of the opera house caught the sound waves; the impulses then being stepped up in the usual manner by means of a transformer were led into an amplifying vacuum tube. Here the current was impressed upon the radio telephone transmitter in successive stages and then sent out over the aerial on top of the opera house. Wireless amateurs all about the surrounding country were thus able for the first time to hear grand opera. While this was only an experiment, grand opera by wireless will soon be an accomplished fact.

During the next few years it will be a common enough experience for an amateur to pick up his receivers between eight and eleven o'clock in the evening and listen not only to the voice of such stars as Scotti, Tetrazzini, Mac Cormick, and others, but also to the orchestra music as well, which is picked up by the sensitive transmitter along with the voices of the stars. The surprising thing is that it is not being done now.

This probably is due to the fact that as yet no means has been found to reimburse the opera companies for allowing everyone to listen in. While of course listening to the music is not as satisfying as witnessing the performance in person, still many music enthusiasts would rather stay home listening to the music alone than to witness the performance itself. To your true, dyed-in-the-wool opera fiend the performance is of secondary importance, the music always coming first.

But we must give a thought to the management, which cannot subsist on an empty opera house if everyone could listen in to the actual rendering of the opera without paying for the privilege. Needless to say that the producers would soon find themselves bankrupt. For this reason we cannot expect that grand opera by wireless will be an accomplished fact until some means has been found to reimburse the producers, and, as every wireless man knows, this is very difficult to do. Anyone with suitable radio apparatus can "listen in" to the music without much trouble. No matter on what wave-length the music would be rendered, every wireless man would find a way to listen to it without serious inconvenience.

Probably the only logical way out would be for the management of a grand opera company to advertise in the newspapers, stating that no grand opera via radio would be given unless a certain amount of revenue were guaranteed by radio subscribers before "radio performances" would be given. This would mean that probably ten out of one hundred radio stations, amateurs, and otherwise would pay monthly or yearly dues to sustain the management, which then would not have to care how many were listening in.

This is the only practical solution. As for technical difficulties, there are of course none. All that is necessary for the producing company is to install a high-class wireless telephone outfit which can be bought on the market right now and which is immediately available. The rest is up to the wireless fraternity, which has nothing else to do but listen in.

At the receiving end, the future up-to-date radio opera enthusiast will of course, have a first-class receiving outfit, using vacuum tube amplifiers, and a loud talker. Then it will be a simple matter to listen to Scotti himself, though he be a thousand miles distant. His voice will come out loud and distinct and the amateur's family will be enabled to "listen in" to their heart's content.

There is still another novel scheme recently originated by the writer.

The underlying idea is not only to give grand opera by wireless, listen to the music and to the singers only, *but to actually see the operatic stars on the screen as well.* This is how it can be readily accomplished *by means which are available to-day,* and without the slightest technical difficulty.[1]

Let us say, by way of example, that the opera "Aida" is filmed in its entirety. This may mean a four or five reel feature. The opera will be filmed just like any other photo-play.

Our large illustration shows what happens next. The stars, singers, players, the chorus, orchestra, conductor, etc., are then assembled in a moving picture studio and in front of them is the usual screen. The opera "Aida," which had been filmed before, *is now repeated on the screen* while the entire cast follows the screen picture closely. Each performer, every star, every member of the chorus has his or her own microphone in which he or she sings the regular score, watching closely the film-play as the action is unreeled on the screen. The moving picture opera through the film operator keeps time with the singers, and the singers themselves must keep exact time with the performance as it is unrolled on the screen before their eyes. Inasmuch as the identical cast has been filmed, it will not be difficult for them to keep time with their own performance, as may readily be imagined. *In other words, when Scotti sees his own figure*

1. This was not the first proposal for broadcasting both the sound and image of live opera performances, but it was one of the most fully developed at the time. Thomas Edison, for instance, is quoted in an 1893 *New York Times* article as saying, "My intention is to have such a happy combination of photography and electricity that a man can sit in his own parlor, see depicted upon a curtain the forms of the players in opera upon a distant stage and hear the voices of the singers." See Ken Wlaschin, *Encyclopedia of Opera on Screen: A Guide to More Than 100 Years of Opera Films, Videos, and DVDs* (New Haven, Conn.: Yale University Press, 2004), vii. For a musicological account of more recent initiatives to broadcast opera performances to cinema screens around the world, see James Steichen, "The Metropolitan Opera Goes Public: Peter Gelb and the Institutional Dramaturgy of the Met: 'Live in HD,'" *Music and the Moving Image* 2, no. 2 (July 2009): 24–30.

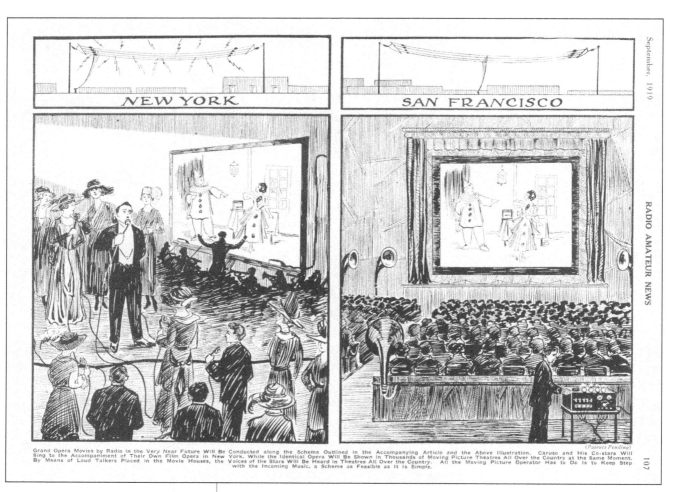

September, 1919

RADIO AMATEUR NEWS

107

(Patents Pending)

Grand Opera Movies by Radio in the Very Near Future Will Be Conducted along the Scheme Outlined in the Accompanying Article and the Above Illustration. Caruso and His Co-stars Will Sing to the Accompaniment of Their Own Film Opera in New York, While the Identical Opera Will Be Shown in Thousands of Moving Picture Theatres All Over the Country at the Same Moment. By Means of Loud Talkers Placed in the Movie Houses, the Voices of the Stars Will Be Heard in Theatres All Over the Country. All the Moving Picture Operator Has to Do Is to Keep Step with the Incoming Music, a Scheme as Feasible as It Is Simple.

FIGURE 1.

appearing on the screen he will know exactly how and when to sing into the microphone in front of him.

All of the microphones go to the wireless telephone station located in the radio room above, and there are, of course, sensitive microphones in the studio which pick up the sounds from the orchestra as well. All sounds are then stepped up through the usual amplifiers and are then fed into the high power vacuum pliatrons, *which finally amplify the original sound several million times.* These impulses are then sent out over the usual aerial located on top of the house and are shot out all over the country instantaneously.

Five hundred to 1,000 miles away—and for that matter all over the country—every moving picture house will have been supplied with the identical film at the stated performance, it having been announced days ahead that the grand opera "Aida" will be given at such and such an hour.

Of course, where the distances are large, the hour of rendering the opera will vary. Thus, for instance, if Scotti were singing in New York and a performance would start at eight o'clock in the evening, New York time, it would start in San Francisco at four o'clock in the afternoon, as a matinee, due to the difference of time. Inasmuch as such performance would probably only be held once a month, people would not mind the inconvenience due to slight difference of time.

Every moving picture house will have its receiving apparatus with its usual amplifiers and anywhere from six to one dozen loud talkers scattered through the house. Exactly at the stated time the moving picture operator will begin grinding away—the opera has begun. Simultaneously the distant orchestra will begin playing, filling the house with music.

When the actual performance begins, it will be an easy matter for the operator to keep time with the incoming music. All he needs to do is to grind faster or slower, and inasmuch as Scotti with his performers in New York is watching the identical film, the distant operator will have no trouble in having the music keep time with his film. If he finds that he runs ahead for one second, he can readily slow up the next and vice versa. With a little practice it will be easy for the distant operator to time himself perfectly, thus giving the patrons of his house an ideal performance.

From a financial standpoint it would be good business for the opera company, as well as for the moving picture house, *both of which would thus derive a new income running into the hundreds of thousands with hardly any expense whatsoever.* The grand opera with an outlay of from one thousand to three thousand dollars could buy its high power radio telephone outfit, while every live picture house throughout the country would be able with an expenditure of less than five hundred dollars to buy its necessary radio telephone equipment *and this cost would only be initial,* because nothing except burnt-out vacuum tubes need be replaced and there is practically no cost of upkeep.

The writer confidently expects that this scheme will be in use throughout the country very shortly.

The Future of Radio

Radio Amateur News, vol. 1, no. 4, October 1919

From time immemorial, crass ignorance and great scientific discoveries have cheerfully trotted side by side. Whether it was Columbus discovering a new world, or whether it was Galileo who maintained that the world was turning around the sun and not *vice versa*—a stupid and narrow humanity was ever ready to step in and command a threatening *Halt!* to scientific exploits.

It is no different with radio. No sooner has the new art demonstrated its inconceivable boon to the world that some well-meaning but misguided official steps up and frantically tries to shackle it down, hands, body, and feet.

We ask ourselves with horror what would have happened to our telephones if our Government had taken control of them in the early eighties, as was the case in most European countries. Today there are more telephones in New York City than in France and Belgium combined. A single New York building—the Hudson Terminal building—has more telephones than all of Greece!

Now Europe needs the telephone no less than America, but it has been proved time and again that it is the Government control which retards development. As soon as the Government steps in, competition as well as the natural development ceases—the art decays.[1]

If this is true of the telephone and telegraph, how much truer is it of Radio? And particularly Radio in the United States—the land of enormous distances, where Radio will be more necessary within twenty-five years than the wire telephone is to-day.

We are certain that if the men who now advocate Government Radio control were possest of but a little vision as to the marvelous future and the possibilities of Radio, they would recoil with horror at their preposterous suggestions.[2]

At the present the radio art comprises but two branches: Radio telegraphy and telephony. Before the war the latter was only a laboratory experiment. To-day sets are being sold at a low price that plug into your lamp socket and you can then talk to your friend thirty miles away—be he in the house, his auto, his yacht, or in his airship. Soon every farmhouse will boast its radiophone. Every limousine will be equipt with it. How would these developments fare under Government control?

But these two applications are but a small part of the whole art of

---

1. In Europe, a similar model of state ownership emerged for radio with the explosion of a listening public over the next decade. In a 1927 *New York Times* editorial response to H. G. Wells's views on the poor quality of broadcast content, Gernsback not only accuses Wells of a Eurocentric elitism but also argues that the situation Wells describes in Europe—an emphasis on trivial entertainment over worthwhile news and information—is a direct product of the state ownership of broadcast stations: "In America no fees are paid by listeners. In Europe fees vary from 50 cents upward a month. Practically all European broadcasting is Government controlled, which probably accounts for the fact that it has never attained the high plane it has reached in America. Furthermore, it is conceded by authorities that European broadcasting is at least two years behind American broadcasting." The editorial is a simultaneous defense of the lowbrow ("Mr. Wells . . . evidently hankers to listen constantly to the great, when a simple mathematical calculation would show that this would not be possible. There are not enough great people in this world.") and defense of privatized news and entertainment, since the American model seeks to provide the widest possible audience with diverse programming that is well-funded by advertising. Not only do American broadcast listeners hear the Moscow Art Orchestra's *Pagliacci*, visiting performers from the Metropolitan Opera, and personal addresses from President Coolidge, but they can also tune in to entertainment from comedic acts and popular bands. Hugo Gernsback, "Wellsian Opinion of Radio Tinged with Provincialism," *New York Times*, April 11, 1927.

2. Gernsback is referring to Secretary of the Navy Josephus Daniels's suggestions to Congress for a bill keeping radio broadcasting under the control of the Navy. See also **Amateur Radio Restored** in this book.

Radio. Take for instance *Distant Radio Control*. Switches can now be thrown by wireless a hundred miles away. Alarm horns can be sounded without wires from five to fifty miles distant.

Then we have the *Radio Telautomata,* as first invented by Nikola Tesla. This famous inventor has worked out plans whereby it is possible to send a ship across the Atlantic without a human being on board. The entire control is by radio; and the ship is guided unerringly to its harbor by a man who sits in an office building in New York. The ship automatically discloses its position hourly to its New York control office by means of the Radio Compass.

But one of the most important branches of the new art is undoubtedly *Radio Power Transmission*. Nikola Tesla, the inventor of the system, already demonstrated the feasibility of this in his famous Colorado experiments in 1898. He was able to light lamps hundreds of feet away without the use of wires, using only a ground connection. To be sure no Hertzian waves were used in these experiments, but the transmission of energy was accomplished wirelessly.

In the face of these developments in the Radio Act, who can doubt that the future will bring forth still more wonderful achievements, not to be imagined to-day by the most fervid imagination.

And now Washington officials once more threaten this art—a distinctly American art—the eighth wonder of the modern world. But the country at large wants no further harmful radio legislation, we are certain. On page 190 we found a few voices. Radio enthusiasts should read particularly Mr. Manderville's letter to his Senator—and follow suit. Let our slogan be: *"Hands off Radio."*[3]

Radio Restrictions Off

Washington, Sept. 27—Effective Oct. 1, all restrictions on amateur radio stations are removed. This applies to amateur stations, technical and experimental stations at schools and colleges, and to all other stations except those used for the purpose of transmitting or receiving commercial traffic of any character. These restrictions on stations handling commercial traffic will remain in effect until the President proclaims that a state of peace exists.

Attention is invited to the fact that all licenses for transmitting stations have expired and that it will be necessary for the amateurs to apply to the Department of Commerce for new licenses. So far as amateurs are concerned, radio resumes its pre-war status under the Department of Commerce.

3. Manderville's letter to his senator reads, "The writer wishes to appeal to you to prevent, if possible, the control of all wireless telegraphic communication being placed in the hands of the Navy Department. He asks this in a dual capacity; first, as a private citizen who is a wireless amateur and vitally interested in wireless telegraphy, both as an electrical engineer and as an experimenter, and, second, as chief engineer of the T. W. Phillips Gas & Oil company, a public service corporation engaged in supplying gas to 20,000 consumers in Western Pennsylvania. . . . Navy control of wireless telegraphy spells death to both amateur and business use of this method of communication. Another point is that all wireless installations now are thoroly [sic] under government control in the hands of efficient officers of the Department of Commerce and there is no good reason for any change. . . . In closing I wish to point out that all the great advances in the use of wireless telegraphy and telephony have been made by amateurs or commercial companies."

Thomas A. Edison
Speaks to You

The Electrical Experimenter, vol. 7, no. 8, December 1919

This is the first interview which Mr. Edison has given out for some years past.

Mr. Edison, who, as is well known, was elected Chairman of the Navy Consulting Board at the outbreak of the world war, was taken up with important duties, refusing to see all visitors. Even several years before this, no general interviews were given out. In this story are covered many points of interest not only to all experimenters and the man interested in science, but to the world at large. Much that is new has been presented here, and it will be noted with satisfaction by all that at the age of seventy-three, Mr. Edison's mind is as keen and clear as ever. We are certain our readers will appreciate this important article. Nearly all of the photographs and illustrations appearing in this story have never been publisht.

Mr. Edison having kindly consented to speak to the readers of the *Electrical Experimenter,* an interview with the illustrious inventor had been arranged for during the latter part of October of this year.

This interview, by the way, has some history attached to it. During the early part of 1917 a similar appointment had been made to interview Mr. Edison on the same subject. But just then the great war broke out and Mr. Edison, who, as is well known, was immediately appointed the head of the Naval Consulting Board, broke off all engagements, devoting himself day and night to the welfare of his country. For this reason the interview only took place a few weeks ago.[1]

I Arrive at Mr. Edison's Laboratory

I arrived at West Orange on a crisp October morning and was soon in the little gate house which protects Mr. Edison from a curious public. Plain and modest as it is, the little red house has past [*sic*] thru its gates hundreds and thousands of the world's most famous men and dignitaries. Few such modest little houses, if any, have held under their roofs such an array of famous people who have come to pay homage to one of the greatest inventors the world has ever known.

1. This interview was the product of a string of correspondence between Gernsback and Edison (or more accurately, his assistants) that in fact dated back to 1908. With the launch of his first magazine, Gernsback sent a personal notice to Edison with a copy of the debut issue enclosed, saying, "The writer believes you to be interested in the propagation of the electrical arts among the Young and as *Modern Electrics* is devoted exclusively to young people the writer hopes that this step will find approval by the Master of Electricity" (March 28, 1909). Gernsback would request articles or interviews from Edison periodically afterward, especially in preparation for the May 1917 "Edison Number" of *The Electrical Experimenter.* William H. Meadowcroft, Edison's assistant, responded with regret that, due to the war effort, he was "busy on Government work and is working about 20 hours a day just now. He will not make any appointments at all for the present, but let me suggest that you call me up twice a week, and I will try and get the interview for you specially" (February 16, 1917). After the Great War's end, Edison finally agreed to an interview in July 1919, upon return from one of his camping trips with the self-styled "Four Vagabonds," who included Henry Ford, the naturalist John Burroughs, and tire magnate Harvey Firestone.

This hagiographic profile followed in the wake of Nikola Tesla's autobiographical fragments, "My Inventions," which ran throughout 1919 in *Electrical Experimenter.* After years of waiting, Gernsback's visit to West Orange took place in October 1919. Edison saved a clipping of the published article in his scrapbooks.

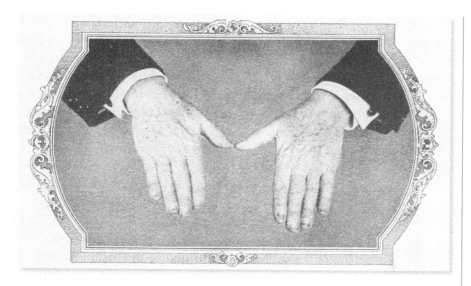

In this little gate house is located the famous time clock on which Mr. Edison rings in his time and rings out every day of the year, many holidays included. An inspection of his week's time card revealed that Mr. Edison had invariably been at the laboratory before 8 o'clock in the morning and had worked as many as eighteen hours for three days at a stretch. Only once did he have a twelve-hour day. Right then and there I wondered how Mr. Edison felt about the now so popular eight or six hour day, and I meant to ask him about it, but we became so engrossed in other more important questions which are moving the world, that altho we touched upon this subject, the eight-hour day question was never broached by me.

Thomas Edison and Henry Ford on a "Vagabonds" Camping Trip, 1923. From the Collections of The Henry Ford. Gift of Ford Motor Company.

FIGURE 2.

After passing thru the gate house, I made my way to Mr. Edison's library, where I was welcomed by Mr. W. H. Meadowcroft, his trusted and capable friend and secretary. While waiting for Mr. Edison, who was just then engaged with some chemical experiments, Mr. Meadowcroft pointed out all the interesting objects of Mr. Edison's library. This library is a huge affair and, besides containing electrical, chemical, and physical reference works printed in almost any imaginable language with English, French, and German predominating, many other curiosities are to be found here. There are many dozens of autographed photographs of famous men hanging about the walls, as are famous historic photographs portraying this or that view of an important phase of Mr. Edison's great discoveries, such as the electrical traction, the electric light, the phonograph, the moving picture, and many others. A huge white marble statue immediately catches the eye. Mr. Edison brought this from Paris at the time of the World's Exhibition in 1889, it having caught his fancy. The marble figure represents a boy seated upon a broken gas lantern, holding aloft triumphantly an electric light. Another object of recent dating is a solid cubic foot of copper weighing several hundred pounds mounted upon a mahogany pedestal. This solid piece of copper, made by Tiffany and suitably engraved, was presented to Mr. Edison by the copper interests of this country as a tribute to the great inventor. This symbolic gift can be better understood when it is realized that fully 50 percent of most of this copper is used for lighting purposes, which art was founded by Mr. Edison.

In this library I also inspected the famous "bed" upon which Mr. Edison catches a little sleep when he is engaged in day and night work at the laboratory. This cot is a very prosaic affair, and is located in the corner of the library amidst books and other curios. It is a bed in name only, for it is comprised of nothing but a mattress, pillow, and a blanket.

On one of the walls, we also find a complete history of Mr. Edison's Alkaline Storage Battery, exhibited on a large wall board. This is graphically shown by displaying every part, chemicals included, that go into the making of this famous non-lead battery. The thing that most interested me in Mr. Edison's library, however, were his notes, and right here we find one of the pregnant reasons for Mr. Edison's great success.

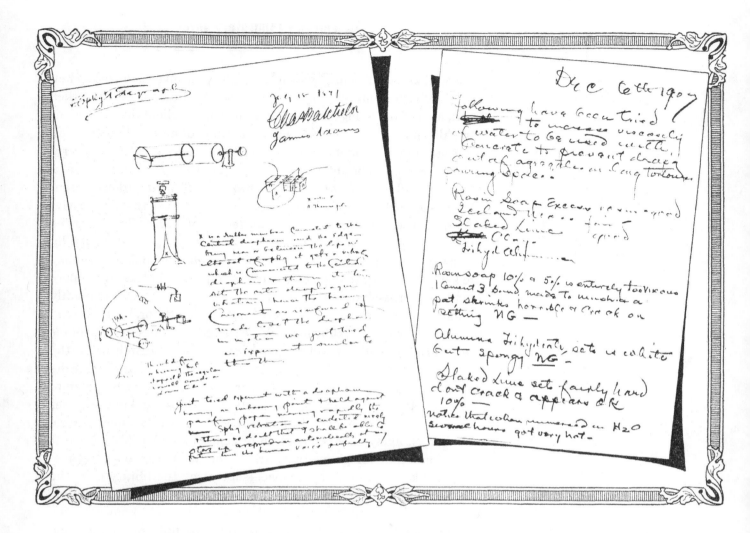

FIGURE 3. The Illustration [above] Shows a Photographic Reproduction From One of Mr. Edison's Note Books. It Is Nothing Less Than the Now Historic Proof of the Invention of the Phonograph. It Shows the Conception of the Idea Plainly. On July 18, 1871, Mr. Edison Was Making Some Experiments Which Had Nothing To Do with the Phonograph. As a Matter of Fact It Seemed To Be More or Less Nebulous in His Own Mind, for at the Top of the Page He Wrote "Speaking Telegraph." The Part of Interest to History Is Found in the Foot Note of the Page, Which Reads as Follows: "Just Tried Experiment with a Diaphragm Having an Embossing Point and Held Against Paraffine Paper Moving Rapidly. The Speaking Vibrations Are Indented Nicely and There Is No Doubt I Shall Be Able to Start Up and Reproduce Automatically at Any Future Time the Human Voice Perfectly." Note Also the Two Witnesses' Names Under the Date, at Top of Page. The Other Page Shows a More Recent Sample of Mr. Edison's Handwriting. It Is From One of His Memorandum Books and Is Dated December 6, 1907. Note How Much Attention Mr. Edison Pays to Record Even Trivial Experiments, Also His Characteristic "N. G." In Two Places. The Changed Handwriting After an Elapsed Time of Thirty-Six Years Is of Great Interest.

Mr. Edison's Laboratory Notes

As is well known, few inventors have made their inventions pay. They are usually "inventors"—and that is all. Mr. Edison, on the other hand, aside from being an inventor, is also an excellent business man. Mr. Edison is systematic and knows the value of notes. He won many a patent suit on account of his notes written dozens of years ago. Early in his career it was brought home to him that when you have an idea, a record should be made *at once!* This was so engrained into his system that it is almost an impossibility for him to make even the most trivial experiment without making a careful record of it. Mr. Edison writes his notes in pencil, and at the end of the day the notes are carefully put away. Each month these notes are bound into a book, where they are kept for further reference. The office staff card-indexes his experiments and cross-indexes them, so at any time Mr. Edison can readily find any one experiment he made during the month, or, for that matter, any experiment that he made 25 years ago, if necessary. The notes are invariably made on yellow paper on ordinary scratch pads, the pages measuring about four by six inches. Being systematic, Mr. Edison always uses the same size pad year after year, and it is therefore refreshing to see the same sized sheets, and the same kind of paper made into bound volumes. While I was still inspecting some of these note books, many of which contain priceless data, Mr. Meadowcroft was informed by telephone that Mr. Edison was ready to receive me.

We walked across an open space separating the library from the chemical laboratory, and Mr. Meadowcroft explained that the building in which Mr. Edison was working that day was known as the chemical laboratory. Mr. Edison, however, does not work in the same laboratory all of the time, but for the past few years he has worked more in the chemical laboratory than in any of the others, which are located in different buildings about the great Edison works.

I Meet the Great Inventor

Together we entered the laboratory, the first impression being mostly green fumes. A half dozen of Mr. Edison's assistants were to be seen busily engaged in performing various chemical experiments under the direction of their chief. Mr. Edison himself was nowhere to be seen. Advancing to the rear of the laboratory, I finally discovered Mr. Edison sitting at a little table busily engaged in writing notes upon a yellow pad, using a small pencil. To the right of Mr. Edison on the table was a big chemical dish containing some green solution, presumably copper nitrate. The table, which was rather small, measured perhaps four feet in length by three feet in width. Mr. Edison was seated on a simple wooden chair. Altho the inventor is 73 years old at the present time, I was surprised at his vigorousness and apparent strength. The first impression at close quarters is a kindly face of ruddy complexion, from which peer two light blue-gray eyes of unusual intelligence. There is an enormous broad forehead, over which falls some silver white hair, giving the characteristic Edison curl. The jaw as well as the ears are well formed, both denoting character and a strong will. Like all great men, Edison has a big and massive nose, which denotes the thinker and philosopher. The lips are rather thin, around which a smile is constantly hovering.

Mr. Edison rose and we shook hands. I was startled somewhat, for I expected a large, hard hand. On the contrary it was as small and soft as a woman's hand, and white except for green stains upon it caused by chemicals. There is perhaps no more famous right hand in the world than Edison's. *If the world were called upon to make an inventory of what Mr. Edison's hands have actually wrought in enriching this planet, there would not be gold enough to pay him.*[2] This is not a mere figure of speech or written in order to make this review grandiloquent. It is the unvarnished truth, as anyone with a clear mind and a pencil can easily figure out to his satisfaction. There certainly has never been one man since the dawn of history who has contributed so much to the world's progress as has Mr. Edison. While of course the mind is supreme, Mr. Edison's hands are the tools that achieved his success, and that is why I place so much stress upon this phrase.

Mr. Edison spoke and welcomed me—a curiously high-pitched voice, unusual for one who has never heard it, but characteristic of the famous man.

Mr. Edison Answers Some Important Questions

Mr. Edison in late years has become hard of hearing, and it is necessary to speak quite loud in order that he may understand you. There being no second chair around, and Mr. Edison having seated himself again, I leaned against the laboratory table, being careful not to sit in the chemical dish, and taking other precautions not to disturb anything. After a few pleasant remarks on both sides, I immediately launched into the object of my visit, namely, to put up to Mr. Edison

2. Edison's hands were also the subject of a bronze sculpture made by Rupert Schmid for the 1884 Philadelphia International Electrical Exhibition, depicting him holding two Bunsen cells at the moment he "made the discovery of the electric light." The sculpture was ridiculed in the pages of *The Electrical World* due to the fact that the principles of electric lighting were successfully demonstrated as early as 1800. "Rewriting History in Art," *The Electrical World*, June 1884: 185. Paul B. Israel et al., eds., *The Papers of Thomas A. Edison, Vol. 7: Losses and Loyalties, April 1883–December 1884* (Baltimore: The Johns Hopkins University Press, 2011), 590.

The photograph of Edison's hands printed here, taken by Gernsback, was republished in the *New York Times* twelve years later on the occasion of Edison's death. "10,000 Mourners Pass Edison's Bier in Day; Nation Plans Tribute at Burial Tomorrow: 10,000 PAY HOMAGE AT EDISON'S BIER Clock Stopped After He Died. Reminders of Latest Labors. Inventor's Religious Views. 2,000 Employes Pass Bier. THRONGS AT LLEWELLYN PARK PAY HOMAGE TO INVENTOR." *New York Times*, October 20, 1931: 1.

certain questions which had been in my mind, and which I knew would be of great interest to our rising generation and the world at large. I said to Mr. Edison:

"Do you believe that the young man who embarks on his electrical studies today, has the same chances and opportunities which you had at the beginning of your career?"

Mr. Edison was very emphatic when he replied: "He has far greater opportunities than I had—infinitely more. There is absolutely no comparison to be made, for the world has grown larger, and therefore the opportunities have multiplied."

This prompted my second question: "In what branch of electricity, in your opinion, can the young man of today accomplish most—where is he most desired?"

Mr. Edison looked squarely at me and said: "There are thousands of men wanted in every branch of electrical engineering and hundreds of new branches based on new discoveries are being created continuously. The field is being enlarged every day and keeps growing. It would be difficult to pick out any one branch, since all are still in their infancy, and all have wonderful futures."

My next question was: "What training should the young man undergo? Should he acquire practical experience first in the shop or laboratory; or should he take a correspondence course or go to college; or all two or three?"

After a few seconds reflection, Mr. Edison replied: "I think one of the correspondence schools at Scranton, Pa., is the more available for most young men. After passing an examination, I think he should specialize in one branch and study it real hard. Then he should get busy and get a position. Any boy can be a success, providing he is willing to pay the price, which is continuous hard work. The trouble with a large number of young men today who are given positions is that they refuse to pay the price, and thus become a drag on the industry. Our young men today do not wish to work as much and as hard as they did when I was a boy. They want shorter hours and more pay, and too many amusements, the same as all workmen do; but I think this condition will be rectified when people come to their senses. It certainly cannot last."

"Be Sure Your Idea Works, before Patenting It," Says Mr. Edison

"Mr. Edison, what constructive advice can you give to our young and rising inventors? Is it worth while to patent every idea, or only certain ideas? How can the young inventor differentiate between good ones and bad?"

After a few seconds of meditation Mr. Edison answered: "I suggest that if the young inventor has an idea he had better reduce it to actual practise and be sure that it works before applying for a patent. Ideas are easy, but working them into commercial shape is generally a long, tedious, and expensive job. After successful operation and the results warrant it, a search of the United States Patent Office should be made to learn if it has not been previously invented or patented by others. Here is where the young inventor will have his greatest disappointment. He will find many a time and, as a matter of fact, in a majority of cases, that the idea has been patented already in one form or another. But disappointments show the salt of the inventor. Only by such disappointments can he triumph finally."

This brought about my next question: "You have patented over one thousand inventions, Mr. Edison. How many of these have been actually worked?"

Mr. Edison Has 1400 Patents

Mr. Edison thought for a while—"Of the fourteen hundred patents which I have obtained, about four hundred were actually worked. This figure may be taken as proportionate for inventors. It is seldom that an inventor makes a success of his first invention. Usually he finds that altho he obtains a patent, for some reason or other the idea did not prove to be successful commercially, or could not be exploited otherwise. I have made a rule in my later years, not to patent anything for which I knew there was no actual demand. Merely collecting patents is a waste of time, money, and energy."

This answer of Mr. Edison's was of more than passing interest to me, and at the spur of the moment, I sang out: "Which is your pet invention?"

Mr. Edison smiled broadly, and it seemed to give him much pleasure when he said: "My pet invention I think is the phonograph first—and moving pictures second. Somehow or other these two inventions have taken hold of me more than my other ones, as I have probably spent more time upon them than upon any of the others." This naturally brought me to the next question, which was, "Are your inventions perfected first in your mind or perfected in your models? Or by actual experiments?"

Without hesitation Mr. Edison continued: "I always start out with a definite idea of accomplishing a certain result. I collect all the data possible, both scientific, commercial, and otherwise. I then proceed to sketch out every possible and probable way of attaining results and carry it to success by experimenting. In other words, first plan—then act. I usually find that the first model is not at all what I had in mind when I conceived the idea first. Any inventor knows this of course. I have found it necessary for this reason to build many models, and only in the exceptional case is the first model a success."

Mr. Edison Tells What World Most Needs

Knowing that Mr. Edison was perhaps one of the greatest authorities on what the world needs most to-day, I asked him: "What inventions does the world need most to-day?"[3]

Becoming reflective, Mr. Edison thought for a few seconds and answered: "Automatic machinery, and systems for the quantity production of one-family houses so cheaply that every man can possess his own home. These two are the world's greatest and most pressing problems today. Take a city like New York, for instance. Conditions there are indescribable. There are too many people in New York at present, and but little new building can be done there. I am in great favor of a law being passed that no additional factories should be built in the city of New York after 1925. This would mean that the housing as well as transportation facilities would not be continually overtaxed as they are now. In other words, the city should be forced to spread out either towards Long Island or to the north."

At this juncture I mentioned to Mr. Edison a plan that was advanced some years ago, whereby it was advocated to build the so-called road house, i.e., a city running thru the country by having a single line of houses built one next to the other with a subway underneath. This would give us a city as well as country at the same time. There would be a sidewalk and roads on each side of the house running continuously without interruption and paralleling the houses.[4]

"I do not think this idea is very practical," vouchsafed Mr. Edison. "I think it would be too expensive and would make the traveling distances uncomfortable."

While discussing the printers' strike, which just then had started in New York, paralyzing the entire printing industry, I put the question: "What known substitute is there for white print paper when our raw materials give out during the next twenty-five years?"

Mr. Edison's answer was surprising: "Print paper will never give

3. Edison was a magnet for speculations about the future, with fictional "Edisons" starring in Villiers de l'Isle-Adam's *L'Eve Future* (1886), a sequel to Wells's *War of the Worlds* titled *Edison's Conquest of Mars* (1898) by Garrett P. Serviss, and the Tom Edison Jr. dime novel series created by Edward Stratemeyer that ran throughout the 1890s. But Edison also actively collaborated on some of his own speculative fiction. Paul Israel has found handwritten notes Edison shared with the journalist George Parsons Lathrop, a son-in-law of Nathaniel Hawthorne, in preparation for a novel they were planning to co-write titled *Progress*. Although the novel was never published, it appeared in truncated, serialized form in the *Milwaukee Sentinel* as *In the Deep of Time*. "Invention, after all, is about imagining the future, and many of Edison's ideas for the novel had been the subject of his experiments," writes Israel. In the serial stories that were published, Edison's notes for future advances like photography in the dark, generating electricity from coal, and the manufacture of artificial wood, diamonds, and leather were wrapped in a narrative by Lathrop about the discovery of a Darwinian society working in the Amazon to breed an intelligent species of apes that could walk upright and speak English. Paul Israel, *Edison: A Life of Invention* (New York: John Wiley & Sons, 1998), 365–68.

4. Gernsback may be referring to Edgar Chambless's utopian proposal for Roadtown, a linear city whose goods and people are arranged in a single line. Edgar Chambless, *Roadtown* (New York: Roadtown Press, 1910).

out as long as wood grows in the Amazon and Congo river basins. It is simply a matter of transportation, and that, I believe, will soon be solved, as soon as the world is upon a peace basis once more."

I have always had a pet idea on the subject of cold light, so I ventured my next question. "Over 99 per cent of the energy is lost today in useless heat in our incandescent lamps. How near are we to 'cold light,' and do you think it will be invented at all?"

"I think we are slowly advancing in increasing the efficiency of light production," replied the inventor. "Any moment a discovery is liable to be made that will advance the efficiency of our present lighting methods enormously. The time is surely coming when 'cold light' will be a matter of fact. What shape this invention will take, it is impossible to predict today."

"On which of our dormant and unworked sources of energy should our coming generation work most intensely, Mr. Edison?" I asked. "In your mind, is the exploitation of the following sources of energy chimerical or are they within the realm of possibility from the standpoint of modern electrical engineering:

Power derived from the earth's internal heat.

Power derived from the earth's atmosphere.

Power derived from the tides. Power derived from the sun's heat."

"Utilize Earth's Natural Volcanic Heat," Says Mr. Edison

"Volcanic power to the extent of 5,000 H.P. is utilized already in Italy, and 20,000 H.P. more is being arranged for," explained Mr. Edison. "Italy probably has more in her volcanic regions to work all her machinery and heat every house, carry on every metallurgical process, and in fact make coal unnecessary in that country. My impression is that in Nevada and the Yellowstone region there is available volcanic energy greater than that given by all the coal mined in the United States. 'As to solar energy' we are getting there step by step. It is a long and weary road we have to travel, but we are making it slowly. I am an urgent advocate of water power. We are using already too much coal without adequate returns. Water power in the United States is not at all developed as it should be, and I see a great future in its proper development. I have advocated many times that the coal should be burnt at the mine instead of shipt by cars over long hauls. Electric power can be sent much cheaper thru electric wires than over the railroads; in other words, first hauling the coal which is then burnt at the destination."

This prompted my next question: "What are your ideas, Mr. Edison, as to atomic energy?"

Mr. Edison smiled broadly and, with a twinkle in his eye, said: "You know, Mr. Gernsback, I am an inventor, and as such I do not concern myself overmuch with philosophical research, and altho I have my own ideas on atomic energy I am not at present making them public."

My next question was: "What shall America do to prevent Germany from flooding the world with its cheap goods, and winning the war commercially twenty years hence?"

Here, too, Mr. Edison's reply was surprising as well as illuminating: "Germany never has and never will flood the United States with cheap goods or undersell us if we make up our minds to beat her at that game. Out of thousands of articles, she is only efficient in two, to wit: chemical dyes and toys. This is due to our indifference to going into these lines of manufacture. I am happy to note, however, that American manufacturers are beginning to see the light, and are protecting themselves adequately."

We then discuss various other subjects, and it soon became apparent that Mr. Edison thought that every prophet is honored save in his own country. Mr. Edison was of the opinion that before the war, and particularly during the war, American inventors had not received their due credit, most of the fame having gone across the water. Mr. Edison felt particularly strong about a recent patent decision, where the honors of the vacuum tube used for radio purposes went to an English inventor. It is an incontrovertible fact that the "Edison effect" was known years before the Fleming valve was discovered, having been publisht in American and foreign scientific papers. Mr. Edison was certainly right in his contention that the honors for the invention of the vacuum tube should go to America, and there seems to be no doubt as to this.[5]

My final question to Mr. Edison was, "What is your hobby, and how do you relax from your work?"

Great Inventor's Hobby Is "Experimenting"

"Just now my hobby is 'experimenting.' I like experimenting better than anything that I know of. As for my relaxation, I like to camp out in the mountains, which I do every summer. This makes me fit for another winter's hard work."

Mr. Meadowcroft by this time was beginning to look at his watch, which I took for a gentle hint, and shaking hands with Mr. Edison, I took my leave.

5. The Edison effect was an experimental precursor to the vacuum tube. "Working with his carbon filament [incandescent] lamps in the early 1880s, Edison noticed that after a period of use the inner surface of the glass bulb became progressively darkened, and that, when this happened, there was always a thin and lighter streak on one side of the bulb in line with the plane of the filament. It was as if particles of some kind were being emitted from one leg of the filament—particles that darkened the glass except where they were intercepted by the other leg. Edison's effect attracted attention among scientists partly because it linked the new technology of incandescent lighting with a continuing line of scientific research into the conductivity of rarefied gases. This had origins that long preceded Edison." Hugh G. J. Aitken, *The Continuous Wave: Technology and American Radio, 1900–1932* (Princeton, N.J.: Princeton University Press, 1985), 205–9.

True to form, Edison patented the "effect" without knowing what he had, explaining that the effect could be used to measure the flow of electrical current. It wasn't until John Ambrose Fleming showed that this setup could be used to create a detector more sensitive than the coherer that the practical value of the Edison effect was understood.

While shaking hands I was again imprest with Mr. Edison's hand, and I subsequently made a special request of Mr. Meadowcroft to let me have a photograph of the great inventor's hands for publication. I was much astonished to learn that no photograph of Mr. Edison's hands existed, none having ever been taken, the inventor feeling rather sensitive about this. I had seen many sketches of Mr. Edison's hands, but I only then remembered never having seen an actual photograph. It took several weeks to secure permission from Mr. Edison, but finally the photograph of his hands was taken, and it is here presented to the readers of the Electrical Experimenter for the first time.

I also made another discovery, namely, that there was no oil painting in existence of Mr. Edison. True, several of these had been made by certain artists after Mr. Edison had patiently sat for them, but he was more or less displeased with the result, and on one occasion did not hesitate to put his foot thru one of them. After securing Mr. Edison's permission, I charged Mr. Howard V. Brown with the delicate mission of making an oil painting of the famous man. It is reproduced on the front cover of this magazine in full colors. It is the only oil painting in existence of Mr. Edison today, and the inventor, who inspected it, was very much pleased with it, declaring it a perfect likeness.

As I past out of the laboratory I caught a last glimpse of Mr. Edison. He had risen from his chair, making his way to a little room containing delicate scales and apparatus. The tall, white-haired figure, somewhat stooped under its 73 years' load, clad in a duster, bespotted with chemicals, slowly faded out of view into the adjoining room.

December 1919 cover
of *Electrical Experimenter.*

Interplanetarian Wireless

Radio Amateur News, vol. 1, no. 8, February 1920

nce again interplanetarian radio has come to the foreground. For the past few weeks the press has been full with all sorts of talk about radio from Mars and radio from Venus. Even the poor old moon has not escaped.[1]

Signor Marconi recently announced from London:

> We occasionally get very queer sounds and indications, which might come from somewhere outside the earth. We have had them both in England and in America. The Morse signal letters occur with much greater frequency than others, but we have never yet picked up anything that could be translated into a definite message. The fact that the signals have occurred simultaneously at New York and London with identical intensity seems to indicate that they must have originated at a very great distance. We have not yet the slightest proof of their origin. They are sounds. They may be signals. We do not know. They are not static and we have nothing to guide us at present as to how the signals are caused.
>
> We do not get them unless we set up a special wave length, very much greater than the wave length ordinarily used. Sometimes there may be a long wait before we hear anything, or we may hear these sounds in twenty minutes or half an hour. They occur when we are *using a wave length of approximately 100 kilometers,* which is three or four times the length used for commercial purposes.
>
> They might conceivably be due to some natural disturbance at a great distance, for instance, an eruption of the sun causing electrical disturbances.

Asked whether attempts were possibly being made by another planet to communicate, Signor Marconi said:

> I would not rule out the possibility of this, but there is no proof. We must investigate the matter much more thoroly before we venture upon a definite explanation.

He added that the mysterious sounds are not confined to any particular diurnal period. "They are equally frequent by day and night," he said.

Since Marconi made this announcement a great controversy has raged among scientists and would-be scientists. Many interesting things have been printed among the wagon-loads of pure rubbish, that convulse one with its unintentional humor.

1. As early as 1896, Nikola Tesla speculated that disturbances in his equipment during the famous experiments with wireless power transmission could be of extraterrestrial origin. "It was some time afterward when the thought flashed upon my mind that the disturbances I had observed might be due to an intelligent control. Although I could not decipher their meaning, it was impossible for me to think of them as having been entirely accidental. The feeling is constantly growing on me that I had been the first to hear the greeting of one planet to another." Nikola Tesla, "Talking with the Planets," *Collier's Weekly* 26, no. 19 (February 1901): 4–5. For a complete bibliography of Tesla's comments on the possibility of extraterrestrial communication, see Steven J. Dick, *The Biological Universe: The Twentieth Century Extraterrestrial Life Debate and the Limits of Science* (Cambridge: Cambridge University Press, 1996), 401n5. The recent flurry of talk mentioned here most likely refers to a January 27 announcement from Guglielmo Marconi that he had received "very queer sounds and indications, which might come from somewhere outside the Earth." The *New York Times* followed up on this story almost every day for weeks after. "Marconi Still at Sea on Mysterious Sounds," *New York Times*, January 27, 1920: 7.

Scientists, as a rule, are the most one-sided folk on the face of the globe. It is seldom that you find an expert on astronomy who is at the same time an expert on radio or in physics.

To give but one ludicrous example. Let us only quote Professor Harold Jacoby, the eminent head of the Department of Astronomy of Columbia University of New York. Says the Professor:

> It is highly improbable that the people of another planet, if there are any such, *would be acquainted with the Morse code, which is a complicated system of dashes and dots based on our alphabet. It was invented by Morse and cannot be regarded as universal among civilized peoples.*
>
> If the people of another planet were seeking to signal us they would probably select a system of signals which would be understood on any planet where civilization exists. Such a system would much more *probably be based upon number than upon letters of the alphabet, for the people of different planets would be no more likely* to have the same alphabet than different peoples here on earth are to have the same language.

The italics are ours. Evidently the worthy Professor imagines that the Martians would drop wooden or steel numbers over our aerials. We might ask WHAT else the Martians could use besides dots and dashes. A radio telegraph message cannot by any conceivable means be made up of any other code except either dots or either dashes, or else a combination of the two. The dots and dashes may be high or low buzzes, whistlings, flute-like tones or any other form of sound. BUT there must be dots or dashes or both.[2] There is only one alternative

2. One proposal for "building up a common language with our new correspondents" was put forward by H. W. and C. Wells Nieman in the pages of *Scientific American* the month after Gernsback's article was published. Analogizing Native American beadwork to Morse code ticker tape, the scheme involves sending a message that can be reconstructed on the receiving end line by line as an image, complete with the Morse code word for the depicted object as a caption, "As an intelligent being you must be familiar with the principles of pictorial representation, and while this may not conform to the Martian standard of art its intention is unmistakable; your knowledge of gravity, also, will enable you to place it right side up according to the horizontal line." H. W.

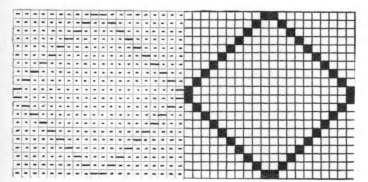

From *Scientific American,* March 20, 1920. Courtesy of HathiTrust

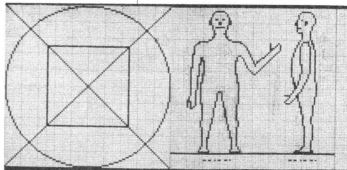

From *Scientific American,* March 20, 1920. Courtesy of HathiTrust.

Nieman and C. Wells Nieman, "What Shall We Say to Mars?" *Scientific American* 122, no. 12 (March 1920): 298, http://www.scientificamerican.com/article/what-shall-we-say-to-mars/.

Recourse to Native American weaving techniques in understanding new media technologies was later made in the context of integrated circuits. Lisa Nakamura discusses how Fairchild Semiconductor built a factory on a Navajo reservation in 1965, purportedly for the community's "manual dexterity, and affective

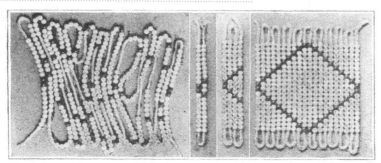

From *Scientific American,* March 20, 1920. Courtesy of HathiTrust.

investment in native material craft." See Lisa Nakamura, "Indigenous Circuits: Navajo Women and the Racialization of Early Electronic Manufacture," *American Quarterly* 66, no. 4 (2014): 919–41, doi:10.1353/aq.2014.0070.

For an account of indigenous North American recording systems that complicates commonly accepted distinctions in the history of writing technology, see Angela M. Haas, "Wampum as Hypertext: An American Indian Intellectual Tradition of Multimedia Theory and Practice," *Studies in American Indian Literatures, Series 2* 19, no. 4 (December 2007): 77–100, http://www.jstor.org/stable/20737390.

3. Despite his skepticism about extraterrestrial wireless messages, Arrhenius was a proponent of the theory that—due to its observed heavy cloud cover—the planet Venus was home to a lush, tropical biosphere. In his lectures on the evolution of the planets, *The Destinies of the Stars* (1918), he claims, "A very great part of the surface of Venus is no doubt covered with swamps, corresponding to those on the Earth in which the coal deposits were formed, except that they are about 30 degrees C (54 F) warmer. . . . The temperature on Venus is not so high as to prevent a luxuriant vegetation. The constantly uniform climatic conditions which exist everywhere result in an entire absence of adaptation to changing exterior conditions. Only low forms of life are therefore represented, mostly no doubt belonging to the vegetable kingdom; and the organisms are nearly of the same kind all over the planet. The vegetative processes are greatly accelerated by the high temperatures." Svante Arrhenius, *The Destinies of the Stars*, trans. J. E. Fries (New York: G. P. Putnam's Sons, 1918), 251–53.

4. By the time Karl Jansky, an engineer with Bell Telephone Laboratories, discovered in 1933 that natural radio waves emanate from celestial bodies, giving birth to the field of radio astronomy, the fervor over interplanetary communication had died down significantly. See K. G. Jansky, "Electrical Disturbances Apparently of Extraterrestrial Origin," *Proceedings of the Institute of Radio Engineers* 21, no. 10 (October 1933): 1387–98. Karl G. Jansky, "Radio Waves from Outside the Solar System," *Nature* 132 (July 1933): 66.

Steven J. Dick writes that no one "considered Jansky's discovery as interstellar signaling, since no regularity was detected in the signal. Gone were the days of Marconi when any mysterious transmission could be so interpreted, perhaps

and that is the voice.—Radio Telephony in other words. We will return to this later. Other eminent scientists such as Dr. Greenleaf W. Pickard, John Hays Hammond, Jr., Professor Svante Arrhenius of the Chair of Physics of the Stockholm Technical Institute, seem to think that the mysterious signals are caused by the sun.[3] So thinks Dr. Charles P. Steinmetz, adding that interplanetarian wireless "must be regarded as a wild dream."

Other eminent scientists such as Nikola Tesla, Thomas A. Edison, and of course Marconi think it not impossible that the signals are coming from some planet such as Mars or Venus. Indeed the believers in this theory are far more numerous than the unbelievers.

In his former editorials the writer has often dwelled upon interplanetarian communication, and he is of the firm opinion that if communication is ever establisht between the earth and the outside world, it will of course be by the agency of Radio.

Let us now analyze the situation and draw a logical conclusion from the facts on hand.

The most important fact—entirely overlooked by the press and all would-be scientists—is found in that one line of Marconi's statement:

The signals occur at a wavelength of approximately 100 kilometers.

That is a fact of tremendous importance.

When radio first was invented some twenty odd years ago, we used but trifling wave lengths from less than 500 meters upward. It was soon found out that to bridge great distances, such as sending across the ocean and further, much longer wave lengths were absolutely necessary. In other words, *the greater the distance we wish to cover the greater the wave-length we require.* Thus today our great transoceanic radio stations operate on wavelengths from 6,000 to 16,000 meters (6 to 16 kilometers). This gives an average of about 10,000 meters.

Let us now assume that the average distance these stations cover is 5,000 miles. That gives us 2,000 meters wave-length for every 1,000 miles we cover.

If now one of our radio engineers were to build a station that could transmit from the earth to Mars—a distance ranging from 60 million miles down to 35 million miles during opposition—he would certainly adapt his wavelength to the distance. A simple calculation—based upon terrestrial standards—reveals then that the necessary wavelength would be at least 30,000 *kilometers,* an unheard of figure compared to our 6 or 12 kilometer pigmy wavelengths.[4]

But we must not forget that the human mind is unused to apply this terrestrial yardstick to celestial distances.

Our conclusion must then be that *if any extra-terrestrial messages are picked up by us they will be received on wavelength at least above 20,000 kilometers.* This would make Marconi's 100 kilometers look quite sick!

Here then is a chance for our earnest investigator to get busy at once. Tuning coils or concentrated inductances to tune up to 30,000 kilometers (30,000,000 meters) can be assembled for less than $100 today, if used in connection with a *very* large aerial. If there are any extra-terrestrial messages, I predict quick confirmation of them.

On the other hand, the intelligent creatures who know how to send radio messages across a chasm of 50 million miles, admittedly know a few things about the game themselves. They may have sent these messages for centuries upon centuries, waiting for us to grow up and finally hear them. Thus the Martians and probably the Venerians are surely on an infinitely higher plane of civilization than ourselves. And we with our twenty odd years' experience in radio—don't we really look foolish in the extreme?[5]

But to make the point clear, if we *do* get messages, they probably will not all be in dots *and* dashes. They may come in musical notes, the same as we transmit radio band music from a phonograph over hundreds of miles, or they may come in the actual voice (providing these beings talk like we do), in other words, radio telephone messages. But time will tell. In the meanwhile we shall wait a bit longer.

because the likelihood of intelligence on Mars had declined and the planet in any case was not at one of its favorable oppositions. But the lure of discussing interplanetary communication at least in theoretical terms was not gone." Steven J. Dick, *The Biological Universe* (Cambridge: Cambridge University Press, 1996), 411.

5. The then-dominant idea that Martian civilization was much older than humanity in part stemmed from the Kant-Laplace nebular hypothesis of solar system formation. Developed independently by Immanuel Kant in 1755 and Pierre Simon Laplace in 1796, the theory was that the sun and its planets formed out of a rotating nebula that gradually flattened and compacted under the force of gravity. In this model, because Mars is smaller than the Earth, it cooled out of the cloud of materials surrounding the young sun faster and therefore began its ecological evolution much earlier. See Robert Markley, *Dying Planet: Mars in Science and the Imagination* (Durham, N.C.: Duke University Press, 2005), 65.

The Physiophone

Music for the Deaf

The Electrical Experimenter, vol. 7, no. 12, April 1920

When I was sixteen I secured an old-fashioned Pathé phonograph of the cylindrical record type. You know the kind that was in vogue years ago. Being much interested in electricity in those days, the thought occurred to me, the same as it occurred to thousands of others, namely, why not transmit the music electrically by putting a sensitive microphone somewhere on the phonograph and thus get the music at a distance.

No sooner said than done. An old-fashioned Hughes microphone was constructed by means of three little carbon rods, and this miniature microphone was attached to the sound box. The microphone was in series with a battery and the primary of an ordinary telephone induction coil. The music transmission was excellent and phonograph music was transmitted over a distance of three hundred yards on my father's estate. These experiments created quite a sensation in those days, and my friends, all electrical "bugs," were much elated and pleased with the stunt.

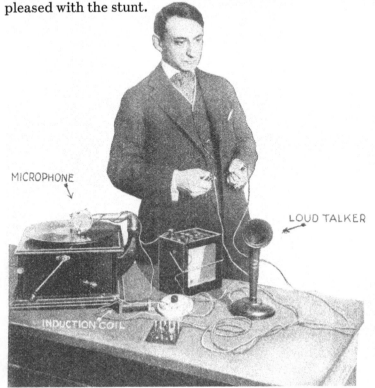

MICROPHONE

LOUD TALKER

INDUCTION COIL

Mr. Gernsback Demonstrating the Physiophone. The Photograph Shows How the Instruments Are Connected. By Means of the Double Pole Switch, the Music Is Reproduced by a Loud Talker. Then the Switch is Thrown, and Physiological Music Is Had.

One evening I accidentally touched the two wires of the secondary terminals of the telephone coil and was quite surprised to get a smart and disagreeable shock. That was in 1900. The early experiments were soon forgotten, but in 1917, while editing an article in this magazine, where a young man had re-discovered the ancient experiment, I thought of that shock, and I understood immediately what that shock really was. I accordingly set to work and immediately built a new transmitter which was attached to a Victor phonograph sound box, and which is shown in Fig. 4. The connections are shown from which it will be seen that the microphone is in series with the 6-volt storage battery and the primary of an induction coil such as is used in telephone work.

The writer used a regulation sound box merely by making a microphone out of it and substituting a carbon diafram for the mica-diafram. The space between the back carbon and the carbon diafram is filled out with polisht carbon grains. The mechanical suspension of the carbon diafram must be the same as the one for the mica-diafram. In other words, the vibrations of the phonograph needle must be faithfully past onto the carbon diafram, the same as is the case with the mica type. When the connections are correctly made and two handles are now attached to the secondary of the induction coil, and these graspt in your hands, the rhythm of the music will be felt faithfully and with astonishing fidelity. What we do feel is sound vibration translated into electrical impulses which in turn are felt physiologically by the human nerves. It is surprising how well this translated music is communicated to the nervous system of the human being, and with a little practise it becomes possible to recognize the different tunes merely by the variations of the little tingling shocks.[1]

1. Gernsback had an abiding interest in the transposition of the senses and the reproduction of sensory experience. See, for instance, After Television—What? in this book. In a 1921 editorial setting up an article by Edwin Haynes on "A New Color-Music Instrument" designed by William Maulsby Thomas in Los Angeles, Gernsback writes: "An interesting feature of the unison color-music instrument is that it gives each note of the musical scale a distinctive geometric figure and definite color, so that when an orchestral performance is translated and projected upon a screen the several geometric figures assemble themselves into a sequence of symmetrical multicolored forms, of great variety and beauty."

Gernsback speculates that all of the senses may in fact be translatable, not just from sound to sight or touch: "Will we stop here? Not at all. We can still transpose the ear for the touch. The writer recently showed this in his Physiophone where ordinary phonograph music was transformed into electrical impulses, which latter were felt by the hands. There was of course here no audible music—nothing was heard—but the rhythm was faithfully preserved. Thus it was possible with but little practice to recognize

Double Barreled Music. Showing How an Audience Can Enjoy the Music Orally As Well As Physiologically. In Other Words You Hear the Music and You Feel It As Well. A Brand New Source of Enjoyment.

different pieces of music, were they a march or a waltz.

"How about the sense of smell? Can we transpose music into smells, or rather odors, scents, or perfumes? To be sure it is possible, but it will be difficult to convey a sonata of odors to our audience. But with pipes scattered thru an auditorium and powerful blowers as the "odor-organ," it seems not so difficult. The trouble will be to find the proper odor for each note. Thus, to decide if A-flat is represented by jasmine, or violet, or attar of roses, may presumably be left to the future poet-musician.

"This leaves us with the last of our senses—taste. Can we transpose music into taste? Why not? Our ever-ready servant electricity may solve the problem. We can already taste radio messages, so why not music? We know that placing two wires from a dry cell on the tongue gives us a sour sensation—taste. By using different metals, different tastes are had. Thus copper gives an acrid-metallic taste, silver a clean-sour taste, etc. Suppose we make a "comb" of many metals and conductors to each of which a wire is attached. The comb to be placed into the mouth so that it lies on the tongue. By a little experimenting we will readily transpose music into gustatory sensations—and not to make a pun—we will now be able to suit every taste!" Hugo Gernsback, "Innovations in Sensations," *Science and Invention* 9, no. 8 (December 1921): 690.

An apparatus designed by William Maulsby Thomas for translating sound into light and color patterns, allowing deaf people to hear visually. From the December 1921 *Science and Invention.*

Different records were tried in 1917, but just then the United States entered into the war, and the experiments came to a sudden end. Recently, however, they were taken up again with the following results. Improvements were made on the microphone and a great many new types were tried out, because the original type was not entirely satisfactory. Later experiments, however, proved to me that 1917 type in principle was probably the best that could be produced. From some ten or twelve types which were evolved by me then, a few are shown here. Many different records were tried, and it seemed to be readily establisht that the different tunes, the different musical instruments,

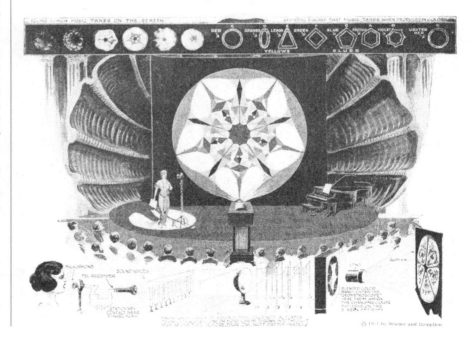

as well as voices, could be readily differentiated physiologically *without listening at all to the music,* or without hearing any sound whatsoever. Of course, it goes without saying that such experiments must be made with the phonograph in a different part of the building with the handles so far away that the music from the phonograph cannot be heard at all. When using the microphone type as shown in Fig. 4, the sound near the phonograph is still audible, altho not anywhere near as loud as if the original mica-diafram were used.

For this reason the handles must be located in another room so that whatever music leaves the phonograph cannot be heard. I used a double-throw switch and a loud talker of the type commercially sold, and invited visitors to first listen to the phonograph record by means of the loud talker. Then the switch was thrown, and of course no sound was heard at all. By grasping the handles, the visitors could readily follow the rhythm of the record, and right here a curious thing happens.

Some people, altho musically inclined, have trouble in following the music, while others immediately recognize the different strains and have no trouble to sing or whistle with this physiological music. There seems to be a difference in the nervous system of individuals, and some people can more readily translate the rhythm than others, altho not necessarily more musically inclined. Other tests were made paralleling the idea *by using no phonograph at all,* but simply a microphone into which a person spoke. By having the experimenter count from one to a hundred, the other party at the far end could feel the voice impulses and after a while managed to "understand" the voice by physiological impulses.

In Fig. 1 is shown another type of transmitter attacht in front of the sound box in the phonograph. It consists simply of three transmitters connected in parallel, the connections being the same as usual. This, however, did not work as well, and very little could be felt at the secondary end.

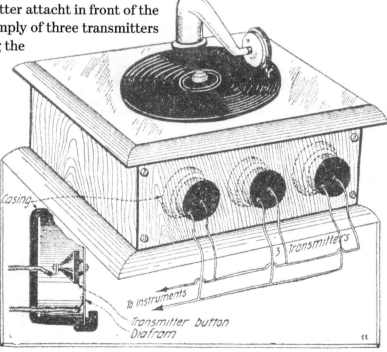

FIGURE 1. Experiment to Transmit Sound Impulses Into Electrical Impulses. This Scheme Works Very Well to Transmit Music.

In Fig. 2 is shown two transmitters, because the writer at one time when working with a single transmitter on a disc type phonograph did not receive all the impulses, that is to say, only one side of the lateral cut vibrated the diafram.

The same idea is shown in Fig. 3 where two transmitters were simply attached to the needle holder, the rubber band serving simply to feather the action, but the results in neither case were good, and the original 1917 type transmitter with certain refinements has been found to work best in all respects.

What purpose is accomplisht by these experiments? Ordinary human beings certainly do not require the translation of musical impulses into their nervous system, *but for the deaf a vast and important field has been opened.* Here we have a means of translating music into the nervous system of a deaf person who has not the slightest conception of music. Of course, it should be understood right here that I do not mean to convey the idea that a deaf person will actually "hear" the music. What he does get, however, is the rhythm, and he certainly gets this very definitely as has been actually demonstrated by tests with deaf persons upon which the writer experimented. It has been found that a deaf person can readily understand the different musical pieces, and can even recognize different musical instruments with very little practise.

FIGURE 2. Two Transmitter Buttons Connected in Parallel, Attached to Phonograph Arm. This Scheme Works Fairly Well for Transmission of Music, But Not so Well in Connection With the "Physiophone."

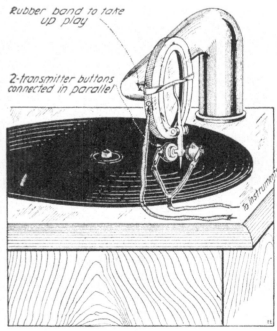

FIGURE 3. Using Two Transmitter Buttons in Parallel. Note Feathering Arrangement.

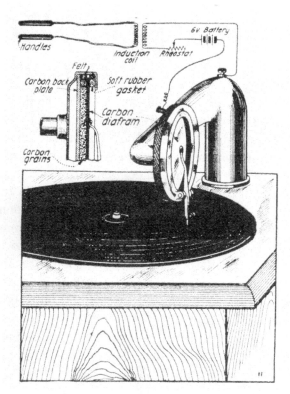

FIGURE 4. Schematic Illustration of the Components of the Physiophone.

Of course, the deaf person must learn the same as any other human being, for it is a well-known fact that if a person totally deaf were restored his hearing, he would not be able to understand what you said to him for some time to come. He would have to learn and judge the sound just the same as a child. The same is the case with a totally blind person who has never seen daylight. Experiments made with such people invariably prove that as soon as the eyesight is restored, they do not see in the sense that the normal person does. They have not the slightest conception of perspective and will invariably put the hands out before the eyes in order to judge the distance. The far away mountain will look to them just as near as a wall three feet away, and it is only gradually that a previously blind person learns to judge the distance, i.e., by experience.

The same is the case with the deaf person and physiological music. He must first learn the rhythm of the music, and after a while it will become easy for him to understand what the music or even the human voice is like if introduced to him by means of the Physiophone.

I even go further and predict that sooner or later a totally deaf person will carry around with him an apparatus along the line of the "deaf-phone" type, the apparatus consisting of a sensitive microphone such as is used with the deaf-phones now. The microphone will be in series with a portable battery and telephone coil, while the secondaries of this coil will go to metallic wrist bands of the wearer. Then when he is spoken to he will soon be able to understand the meanings of the small electrical shocks which are the result of the human voice impinging upon the sensitive microphone.

Coming back to our phonograph experiments, the deaf person can now enjoy dancing when he certainly could not do before, for he had no music or rhythm to dance by. One of our illustrations shows how deaf persons will now be able to dance by means of phonograph music and the Physiophone. The revolving wheel below the ceiling in the room or dance hall has flexible wires hanging down, which in turn connect with the secondaries of the telephone coils. These are connected with the microphone and battery near the phonograph in the manner stated above. Wires hanging down from this wheel connect to metal wrist pieces attached to each one of the dancers. As soon as the music starts, the dancers become conscious of the music and begin to dance, enjoying the dancing, perhaps, even more than normal human beings, due to the electrifying effect of the current. As the dancers progress, the overhead wheel keeps turning so that the wires do not become entangled.

As for the normal human being he can also enjoy the physiological effect of the electrical current for the principal reason that Faradization is healthy and invigorating and entirely harmless. Another one of our illustrations shows what can be accomplisht along these lines for the normal human being. Picture a concert hall, along the stage of which a battery of sensitive microphones is located. The current is led to the primary of a large induction coil in series with the battery. Secondary leads go to metallic handles hung at the back of each orchestra chair thruout the theatre. As soon as the music starts, every one in the audience can grasp the handles, and they will now experience the novel effect of *not only hearing the music but also feeling it in a very invigorating manner.* This thing has already been tried out by the writer in a small way. It is perfectly possible to put your body in series with a loud-talker telephone. Of course, the music coming out of the horn will now be weakened, but if your hands are wet or moist, the weakening of the loud-talker music is not too great. You are now enabled to *hear* the music as well as *feel it.* It will be noted in one of the photographs that a rheostat is used, which is quite necessary to reduce the current. Some phonograph records as for instance band music are so loud and act so powerfully upon the microphone diafram that it is quite impossible to hold onto the handles comfortably. It should be noted that on some phonograph records not all the sounds can be transmitted. In other words, very high sounds do not make themselves felt at all, and some such records as the flute or other soft musical instruments do not lend themselves well for physiological effects. Records containing xylophone, piccolo, band music and the ones with singing voices are probably the best.

Science and Invention

Science and Invention, vol. 8, no. 4, August 1920

The word *Science,* from the Latin *scientia,* meaning knowledge, is closely related to *Invention,* which, derived from the Latin *inventio,* means, finding out. There is little in Science that did not at one time require some inventive powers, while conversely most of the world's inventions are based upon one or more of the sciences.

But "invention" antedates "science" by thousands of years. When our prehistoric man first fashioned his crude hammer by binding a stone to a stick, by means of reeds, he had made a basic invention in every sense of the word. And when he first applied his stick to a huge boulder he wisht to move, then placing a smaller stone under the stick—he had made another notable basic invention—the lever.

In fact, both of these basic inventions are discoveries, and if they were first made today, would be patentable. Right here we may state that in patent law "discovery" and "invention" are held to be synonymous, tho popularly an "invention" designates one that is new and useful as well as patentable.[1]

Science, or rather the sciences, on the other hand first came into being with the ancient Greeks. Of course, some sciences existed before the Greeks, but they were not recognized as such. At least there is no record of any sciences classified as such by the Phoenicians or the old Egyptians. Even in Grecian times there were comparatively little sciences. Thus the Platonists had their sciences divided into dialects, physics, and ethics.

Even in comparatively modern times there seems to be little agreement as to what the sciences really comprise. Thus Bacon in 1605 has history, poesy, and philosophy as his sciences. As late as 1830 Comte classifies the sciences into six parts in their following orders: Mathematics, astronomy, physics, chemistry, physiology, and sociology.

Even today there exists no classification of the sciences that would be acceptable to all of our great thinkers.

The general public and "the man in the street" possibly come nearer the actual definition of "science" than most of our philosophers. To the public, the arts, discoveries, inventions—all fall under the term science. Anything under the sun nowadays becomes a "science"—be it the science of cooking, the science of darning socks, or the science of cleaning streets.[2]

1. This issue marks a change in the title of the magazine from *Electrical Experimenter* to *Science and Invention,* though a shift in its content had been under way for over a year from specialized articles for tinkerers to general interest pieces. Thomas Edison, when solicited for comment on the name change by Gernsback, replied in a letter, "Change of name from Electrical Experimenter to Science and Invention better indicates the proper sphere of your journal. Your field is not unlimited and your journal will be of great value in the advancement of invention and individual applications of science." Thomas A. Edison, "Telegram to Editor, Electrical Experimenter," July 1920.

2. A "Publisher's Announcement" prepared the *Electrical Experimenter*'s roughly 200,000 readers for the change in title and content in the previous month's issue: "your average experimenter wants more than experiments. He wants to know the latest word in science, the newest invention, the latest developments in the realm of human endeavor. He wants to know what the scientists and his fellow workers are doing the world over, and he wants these facts in plain English, adequately illustrated."

The new format would address a growing readership with more general interests: "The business man, the manufacturer, the doctor, the professor, the student, and countless others found in the *Electrical Experimenter* an intellectual gold mine, second to none. The *Electrical Experimenter* always has and always will appeal to the thinking class. . . . We kept all of our old friends and supporters—the ones who buy the *Electrical Experimenter* mostly for the experimental section, and we added besides many thousands of new readers who derived their greatest pleasure from the other 'general science' departments. But we are out for the half million mark, because we know from past experience that the greater the circulation the better we can make the magazine, the more text we can give, the better we can satisfy *all* readers."

The myriad of inventions and discoveries all tend to make the world more "scientific" and whether we like it or not, one science or another creeps into every one of our homes. We are surrounded with science all day long as well as during the night. Science does this thing for us, and makes us do that. There is no escaping it and the general public has awakened to the fact but yesterday, that science no longer is the sombre book closed with seven seals. Quite the contrary, it is the public that popularizes science—not our scientists. Just at present, for instance, educational scientific films are all the rage and the public clamors for more and heartily applauds them.

But our *real* scientists are as backward as in Galileo's times. The public applauds and instantly believes in anything new that is scientific, whereas the true scientist scoffs and jeers, just as he did in Galileo's times when that worthy stoutly maintained that the earth moved and did not stand still.

Then as now they burn our great discoverers and our great scientists at the stake. Only today the stake is moral and the fire derision.

It matters little that Jules Verne or Nikola Tesla are a hundred years ahead of the times—the scientists scoff and laugh unbelievingly.

But happily, the great public today appreciates the "fantastic dreamer," because it knows from experience that these "fantastic dreams" have a habit of coming true on the morrow.

An American Jules Verne

Science and Invention, vol. 8, no. 6, October 1920

Altho not generally known, there lived some forty years ago a personage known to thousands of readers by the *nom de plume* of "Noname." These were the days of the nickel novels when the Nick Carter series, Jesse James stories, and Old King Brady were the talk of the day.[1] For a nickel you bought a complete 32-page novel on closely printed pages in small type. These novels were always hair-raisers in more than one respect, and many of them have gone down as classics.

Lu Senarens, known to tens of thousands as "Noname," was perhaps the most prolific of these writers, and one of the most prophetic.[2] Not only did he turn out a host of these wonderful stories, but he wrote over one thousand of them, each one containing from 35,000 to 50,000 words.[3] Each of these stories were compete and had no continuations. The hero of most of the stories was Frank Reade, Jr., "the boy inventor," who supposedly invented all the many marvelous scientific wonders of that day.

1. Now collectively referred to as "dime novels," these late nineteenth-century saddle-stitched pamphlets actually cost five or six cents. Nick Carter and Old King Brady were popular private detective characters, the former created by John Russell Coryell (1851–1924) and first appearing in the *New York Weekly* in 1886, the latter by Francis Worcester Doughty (1850–1917) and first appearing in the *New York Detective Library* in 1885. Fictionalized versions of Jesse James, the notorious frontier outlaw, were omnipresent in dime novel westerns. For a comprehensive catalogue of dime novel characters, authors, and publications, see J. Randolph Cox, *The Dime Novel Companion: A Source Book* (Westport, Conn.: Greenwood, 2000).

2. Luis Philip Senarens (1863–1939), a Brooklyn-based Cuban American, was one of the few dime novel authors to be known by his given name in addition to his pseudonym. Considering Samuel R. Delany's comments on what we lose along with authorial identities in the history of science fiction, Senarens's background is especially interesting: "I believe I first heard Harlan Ellison make the point that we know of dozens upon dozens of early pulp writers only as names: They conducted their careers entirely by mail—in a field and during an era when pen names were the rule rather than the exception. Among the 'Remington C. Scotts' and the 'Frank P. Joneses' who litter the contents pages of the early pulps, we simply have no way of knowing if one, three, or seven of them—or even many more—were blacks, Hispanics, women, Native Americans, Asians, or whatever. Writing is like that." Samuel R. Delany, "Racism and Science Fiction," in *Dark Matter: A Century of Speculative Fiction from the African Diaspora* (New York: Warner, 2000).

In a way, Senarens proves the exception to this rule. Much like Lisa Yaszek's work on the countless women authors (many publishing under male pen names) for whom writing science fiction in postwar suburban America was "virtually the only vehicle of political dissent," Senarens as a Cuban American author is an example of possibly many other writers of marginalized backgrounds that participated as editors, readers, and storytellers in the otherwise suffocatingly white male world of dime novel—and later of pulp—fiction. Lisa Yaszek, *Galactic Suburbia: Recovering Women's Science Fiction* (Columbus: Ohio State University Press, 2008), 3, quoting Judith Merril.

Unfortunately, Senarens's work is in fact a distillation of the absolute worst tendencies of the dime novels' virulent racism and gleeful violence. E. F. Bleiler's catalogue barely scratches the surface: "sadism, a rancorous disparaging of all races with the exception of WASPS, factual ignorance, coarse imperialist clichés about Manifest Destiny and the right of the White Man to rule the world." Everett F. Bleiler, John Eggeling, and John Clute, "Senarens, Luis Philip," in *The Encyclopedia of Science Fiction*, ed. John Clute, David Langford, Peter Nicholls, and Graham Sleight (London: Gollancz, 2014).

Of course, as a genre writer, one's work must inherently fit within a given paradigm. But take, for instance, Senarens's first Reade title, *Frank Reade, Jr. and His Steam Wonder* (1882), in which the teen-aged protagonist and his friends (including Pomp, "the darkey who had accompanied Frank Reade, Sr., on so many of his wild escapades") massacre two hundred Native Americans in minutes, circling around the group in the Steam Wonder, a tank armed with repeating rifles and cannons that shoot boiling water. When it is done, they "rejoice":

> "Hanged if that doesn't clear the field!" cried Sam Watson, in the greatest glee. "The whole gang is busted!"
>
> "Yes," remarked Jack, "we can lick a thousand as well as a hundred, for they can't get in, you know."
>
> "Why in thunder and chain lightning don't Uncle Sam buy a lot of these machines, and run 'em out here?" demanded Watson. "They'd clear the plains of red-skins so quick it would make their heads swim."

Senarens, Luis. *Frank Reade, Jr., and His Steam Wonder.* Vol. 1. Frank Reade Library 20 (New York: Frank Tousey, 1893).

The dime novel paradigm is a precursor that many histories of science fiction would gladly speed past. But one wonders the degree to which this predecessor is still with the genre, considering just how many of us have played out the exact tactics of Senarens's Steam Wonder tank scenario in SF-themed first-person-shooter video games.

3. Brooks Landon writes that Luis Senarens "wrote between 1,500 and 2,000 dime novel stories during his prolific career," a number that Sam Moskowitz estimates to make up "more than 75 per cent of all the hundreds of prophetic dime novels written during that period . . . the work of a single man." In 1911, Senarens transitioned to the film world, writing silent film treatments and editing "one of

Naturally, these were the days before the trolley car, the telephone, the submarine, the aeroplane, and many another modern invention. Mr. Senarens, a true genius, the same as Jules Verne, had one of the most fertile imaginations. He was not a technically-trained man nor even an engineer. His scientific knowledge was obtained solely from reading books and other scientific publications. His inventions were of course nothing but pure fiction and existed only on paper.[4]

Lu Senarens "invented" some of the most astonishing submarines, airships, war machines, and other world beaters, but the strangest part about it is that in those days no one believed that either the submarine or the aeroplane would ever be actually invented. As a matter of fact, they were considered as physically impossible, not only by Senarens himself, but by leading scientists who lived in the early 70's and 80's. It is not generally known that Senarens corresponded

the earlier film fan magazines, *Moving Picture Stories.*" Brooks Landon, *Science Fiction after 1900: From the Steam Man to the Stars* (London: Routledge, 2002), 45. Sam Moskowitz, *Explorers of the Infinite: The Shapers of Science Fiction* (New York: The World Publishing Company, 1963), 108, 124.

4. In their premature eulogy to Senarens in June 1928 (Senarens actually died in December 1939 at the age of 76), *Amazing Stories* collapsed its celebration of his almost machine-like productivity as a writer with the number of technologies he anticipated. Senarens wasn't just producing stories, he was creating prototypes: "The centennial of the birth of Jules Verne is but a few weeks back of us, and it seems fitting to show at this time that we, too, had a Jules Verne, a man whose industry in turning out reams of copy was as remarkable, as was his ingenuity in evolving the strange machines, prototypes of so much of the present, out of his imagination, though he died unheralded and practically unknown." Quoted in Moskowitz, *Explorers of the Infinite,* 109.

The illustration contains the following captions and embedded text:

FRANK READE JR.'S

FRANK READE JR.

Frank Reade Jr.'s "Sea Serpent;" Or, The Search for Sunken Gold

THE BOYS' STAR LIBRARY

Jack Wright And His New Electric Horse

FRANK READE JR.'S

Reade Jr.'s New Electric Terror; the "Thunderer;"

Frank Reade, Jr., And His Greyhound of the Air

(Continued on page 665)

regularly with Jules Verne, who encouraged the American writer and read his stories as well.[5]

The illustrations which we are publishing herewith give but a faint idea of the stories which appeared in the early 80's, almost forty years ago. We have among others the "Electric Tricycle"; then we have the "Steam Man of the Plains; or Terror of the West," a most marvelous imaginative piece of machinery which was supposed to take the place of a horse and could draw an iron-clad wagon over the prairies. The "Sea Serpent," as its name implies, was a most ambitious submarine; then we have the "Electrical Horse," an adjunct to the steam man which is propelled similarly. The "Electric Thunderer" (armored war car) is suggestive of the recent modern war machines and quite as formidable. The "Electric Submarine Boat," which may be noted, digs its path under an ice field and is worth our attention, because only three years ago Simon Lake, the inventor of the modern submarine, proposed the identical submarine to travel to the North Pole underneath the ice. The "Greyhound of the Air" approaches the modern

5. According to Sam Moskowitz, Verne first wrote a letter to Senarens in 1879, praising the author's works: "Verne did not know who 'Noname' was, but addressed his letter to the publisher, who forwarded it to Senarens. The youthful author was immensely flattered and grateful that so important a literary figure as Jules Verne should condescend to write him, but he need not have been. Verne was doing no more than acknowledging a debt, since he had just finished lifting the basic idea for the Frank Reade series *in toto* and incorporating it in his then current novel, *The Steam House*. Taking the idea of a steam man and a steam horse one step further, Jules Verne used as the basis of his story a steam elephant which carried hunters in India. . . . Everywhere in Europe and America writers were borrowing from Verne. Now, sometimes, he borrowed in return. In the final reckoning he gave far more than he took." Ibid., 116.

airships, while the "Deep Sea Diving Bell" is only little different from those actually used today.

Note particularly the "Electric Air Monitor" engaged in an aerial bombardment. If you substitute the aerial monsters which were in use during the war and which bombarded the various cities, you have here a prophecy that is not far wrong. One of the most curious inventions of Mr. Senarens is undoubtedly the "Flying Submarine" as shown in one of our illustrations. This was a flying machine and submarine combined, and altho we have nothing like it in existence today, who dare say that it will not be in use at not a too distant date? Then again we had the "Electric Warrior," a war chariot used against Indians, the "Electrical Dragon," the "Terror of the Seas," and many others.

Nine-tenths of Lu Senarens' pictured predictions have actually come true. Even the helicopter arrangements in some of his airships—to lift them up vertically—are becoming a realization at the present time. Witness the various helicopter machines that actually have flown in the past few months and which were pictured in this magazine recently.

Nor did Mr. Senarens write vaguely about his wonder machines. Quite the contrary, he described them minutely. While of course no such technical data, as for instance that used by Jules Verne, appears in Senarens' descriptions, still the boy who read the stories had a pretty good understanding of the workings of the mechanism, and his imagination always helped him along so that the machine was pictured by him down to the last nut and to the last binding post. Most interesting is it that Mr. Senarens used electricity as the motive power of most of his devices. At that time, altho there were as yet no electric lights, electricity had just come into use. Electricity in those days could do anything, and the people believed that the marvelous new force was capable of doing the impossible, as indeed it has approximated to since. Not the most interesting part about Senarens' writings were his scientific creations, but the stories themselves were little classics in construction. Mr. Senarens had written since he was a boy fourteen years old, and was a very accomplisht writer with a fervid imagination, not only in things scientific. He was a master of romance, fiction, adventures, plots, and everything.

Mr. Senarens is still at it, and altho he is not writing any more scientific stories just now, he is in the ring and doing work as an editor for a New York publishing house at the present time. Mr. Senarens is fifty-eight years old and is a very active and energetic man who does not show his age. Perhaps at some not too future date Mr. Senarens will give us some more imaginative stories, picturing the world as it will look fifty years hence.

And the other day when we made him dig out a few hundred of his time-worn yellow paper-covered novels, each adorned with an old-fashioned wood cut on its front cover, Mr. Senarens smiled wistfully. He had long since forgotten those "wild impossible dreams" of his younger writing days. Altho he does not like to admit it now, he was actually ashamed to write such "nonsensical wild pipe dreams" those days. In fact, there were many people who thought that his stuff was too fantastic and would actually hurt the young boys. And here nearly everyone of his "pipe dreams" has come true!

"Yes," Mr. Senarens sighed deeply, when we called his attention to it, "truth is indeed stranger than fiction. I believe anything is possible now."

Learn and Work
While You Sleep

Science and Invention, vol. 9, no. 8, December 1921

December 1921 cover of *Science and Invention*.

he normal human life covers a period of some seventy years. During this time, the average adult sleeps eight hours a day, or roughly speaking, one-third of the time. But, the individual when he reaches seventy years of age may be said to have lived only about forty-five years. The other twenty-five years have been taken up in sleep.

While the average length of sleep is about eight hours, it must be taken into consideration, that up to the tenth year, the human body requires more than eight hours of sleep daily, and there are many days during the year, such as holidays, when we sleep considerably more than eight hours. This brings the average up to somewhere near our figures. In other words, over one-third of our lives may be said to be wasted by unproductive sleep. During this time, we are truly dead in the best sense of the word, because sleep is only another form of death.

During our sleeping periods all our usual functions are suspended; of our five senses none remain conscious. We hear no longer; we feel no longer; we see no longer; we smell no longer; we taste no longer. This statement may be qualified at once, by saying that all of the five senses have ceased operating only if conditions are right, or rather normal. To elucidate, a man may be sleeping soundly near a busy railroad, where trains crash along every few minutes, and he will not wake up. *He has become accustomed to the disturbance.* But let a rat or mouse start nibbling in a corner of his room, and he will wake up almost immediately. Why is this so? The reason is that the human body while asleep, is only dead to the accustomed things, and the instant something unusual occurs, the sentinels of the particular sense affected immediately send out its warning.

Thus, while we do not hear the ponderous trains crashing by, the unaccustomed noise of the nibbling mouse awakens us, because it was not foreseen by the sleeper. *Our subconscious self never sleeps.* It is always on the alert. All of the other senses act in the same way, as just exemplified with the sense of hearing.

Thus we sleep along peacefully, but an acrid odor, which is not foreseen, will awaken us. We may turn around and toss about in bed and have every part of our body come in contact with the linen of the bed. It will not awaken us, but a single drop of cold water falling on our hand, will instantly rouse us. Why? Because it was not expected by the sleeper. The same is the case with taste. A man may be snoring along peacefully with his mouth wide open, but if you allow a few drops of sugar-water, or any other unusual solution to fall upon his tongue, he will awaken to nearly every case.

You might think that, when closing your eyes during the night, light would not affect the retina. This is far from true. Try the experiment of switching on an electric light close to a sleeper's face. He will awake almost instantly. Why? Because the lids are translucent, and allow light to pass quite easily. The moment an unaccustomed strong light falls upon the eye lid, we awake.

Why do we not then awaken in the morning when the sun light falls into the room? Very simply because we have grown accustomed by many years of experience to this daily procedure, so it no longer disturbs us. It is the accustomed thing, that fails to disturb the subconscious self, but the unusual affects it strongly.

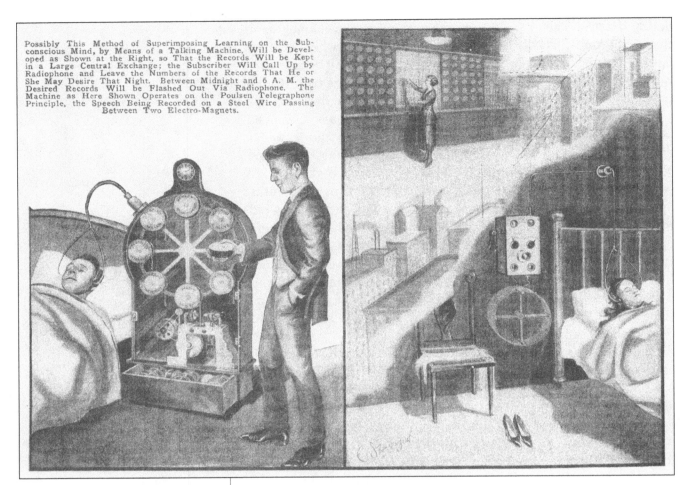

FIGURE 1.

The writer found it necessary to digress from the real purpose of this article, simply to show how the human body acts under various stimuli.

Suppose we could find a way in which we could act upon our sleeping senses in the night time! We would immediately have lifted up the entire human race to a truly unimaginable extent.

Suppose it was possible for you to read, learn, or work while you sleep. Would we not thereby extend the period of our lives by one third?

Suppose it was possible to devise an apparatus whereby you could read a book or study a language while you slept. Would it not be an inestimable boon to humanity? We may be sure that in years to come, we will have arrived at just such a point. There is nothing impossible in science, and our problem is greatly simplified by the fact that when the body is at rest, it is most easily influenced, and impressions are retained best. For instance, when we sit at rest, totally relaxed, and view the latest motion pictures, it makes a lasting impression upon us, *because our mind is not diverted by anything else*. If we dream a vivid dream during the night, or better during the early morning, a lasting impression remains depending greatly upon the intensity of the dream, and it must be remembered that a dream is only the most fleeting thing that can be imagined. There is no outside impression upon the brain. It is only a picture. That is why we do not remember mere thoughts so well. For that reason, most people when they have an idea, find it necessary to write it down, as a thought itself makes practically no lasting impression.

Suppose we were to build a phonographic machine, the sound of which could be conveyed to the ear by two rubber tubes as shown on the front cover of this magazine. Suppose also the music was not grating upon the nerves. What would happen? At the first attempt the sleeper probably would wake up startled because he was not accustomed to it. It probably would take a week before he could accustom himself to wear a head-gear attached to his head by means of a heavy rubber band, and before he could submit to the annoyance of it as well.

After he became accustomed to the head-gear as well as to the unfamiliar noises, it is very probable that his subconscious self would take note of what was going on while he slept. Thus, were it music, or the spoken voice, it seems almost certain that in time, a lasting impression would be made upon the brain through the auditory nerves.

It would be a matter of education. Probably for the first few months, not much impression would be made upon the subconscious self, but by-and-by the hearing sense would be sharpened to such an extent, that the impression would not only reach the brain center, but would be so permanent that the hearer getting up in the morning, would remember something of the nightly procedure. It is remarkable to how many things the human body would respond. It is equally remarkable how many new things the human body is able to absorb. Reading and Writing for instance 300 years ago were considered a science. To-day every school child has absorbed them, and reads and writes unconsciously.

Every time you use a typewriter you write absolutely unconsciously or rather automatically. The expert typist does not have to stop to think of each key when writing. It is done automatically and she pays not the slightest attention to the keyboard while writing. It is a habit formed by the long experience. It may be that the same thing will be true of subconscious learning while you sleep. Once the human system has accustomed itself to the various impressions, there is no doubt that in the morning we will remember everything that we hear during the night. It may take several generations before such a system will be perfected, because human nature would have to accustom itself slowly to the change. In our front illustration, we have tried to show in a practical way the *modus operandi,* and how it can be accomplished.

An ordinary phonograph would, of course, not do. There would be too many grating, jarring noises, which would distract the sleeper's attention. We must have a mechanism that is soft in action, gives no extraneous noises, and will not wake up the sleeper. We have such an apparatus in existence today—the telegraphone.

Its action is based on a steel wire, which is fed across and very near to the poles of an electro-magnet; the magnet is polarized by currents derived from a telephone, and the wire is polarized thereby in an almost infinite number of exactly corresponding poles. On moving the same wire in front of the same or another magnet, the induced telephonic currents are made to act on a transmitter. The telephone receiver has the message repeated in it from the transmitter, without any grating or jarring noise.

By means of a loud talking or amplifying arrangement seen at the upper or left end of the cabinet, the voice, or sound may be amplified to a loud enough degree so that it will give a lasting impression to the sleeper.

The machine which we depict upon our cover contains a number of reels, each of which has enough wire to last for about one hour continuous service. Each reel comes into position automatically as soon as it is "played"; thus eight reels will give the sleeper enough material for a whole night's work.

Whether he wishes to be entertained, or whether he wishes to learn, depends upon himself. It probably would not do to learn history for eight hours at a stretch, for the mind probably would not absorb it all. So we might switch from history to romance, then we might have a concert for an hour to get accustomed to the music of the latest opera, then switch back to mathematics if necessary, and arrange the program as we may see fit, and so as to suit our own individual taste.

Learning, which is remembering, is like any other practice. Some people can remember the contents of a book when only read once. It takes others two or three readings to remember it well. For that reason if the "sleep-reader" thinks that not enough impression has been made by one reading, of any subject, there is no difficulty in repeating it the next night, or as many nights as he desires, until he has absorbed that particular matter.

All of the above is merely theoretical, of course, no machine having been tried but the writer would not be surprised to see one built and in operation in the not distant future.

From *Radio For All*

1922

Preface

In this illustration are shown some of the future wonders of Radio.[1] Several of the ideas are already in use, in an experimental way, and it should not be thought that the entire conception is fantastic.

The illustration shows a business man, let us say, fifty years hence.[2] To the right is a television and automatic radiophone. By means of the plug shown to the right of the machine, the man can plug in any city in the United States he desires; then, by means of this automatic control board he can select any number in that city he wishes, merely by consulting his automatic telephone directory. As soon as he has obtained his number, a connection is made automatically and he not only can talk, but he can see the party whom he calls. At the top of the instrument is a loud-talker which projects the voices of the people, while on a ground-glass in front of him the distant party is made visible. This idea is already in use, experimentally.

Directly in front of the man, we see the "radio business control." By means of another television scheme, right in back of the dial, the man, if he chooses to do so, can load and unload a steamer, all by radio telemechanics, or throw a distant switch, or if a storm comes up, look into the interior of his apartment and then, merely by pressing a

1. *Radio for All* was designed as a primer for the growing lay public curious about the new technology. "The keynote of the book," wrote Gernsback in the preface, is "simplicity in language, and simplicity in radio." With regular broadcast programming beginning to populate the airwaves in 1920, and radio sets going on display in department stores soon thereafter, it was the perfect time for this book, "which rapidly became the bible for amateur radio enthusiasts," in the words of Mike Ashley. Mike Ashley, *The Gernsback Days: A Study of the Evolution of Modern Science Fiction from 1911 to 1936* (Holicong, Penn.: Wildside Press, 2004), 64.

Although most of these early, consumer-friendly radios were vacuum-tube sets, Gernsback chose largely to exclude the tube from his book, noting that it "has been touched upon very lightly and only where it was absolutely necessary. The reason is that the vacuum tube is a highly technical subject, and therefore does not belong in this book. It is a science by itself." Throughout the book, Gernsback uses by then slightly older, outmoded technologies to open up radio's possible futures to a new community of wirelessly connected citizens.

2. After chapters on transmitting, receiving, tuning, aerials, reading radio diagrams, and constructing receiving outfits, Gernsback here paints a picture of radio's future. Toggling between a procedural "how to" mode and brilliant projections of our wireless future, the book's final chapter contains a list of currently registered stations, abbreviations used in radio code, tables of electrical units, and the entire text of the Radio Act of 1912.

FIGURE 1. It is a mistake to think that radio is only good for the distribution of intelligence. As the illustration shows, the great uses of radio have not been touched upon as yet.

key, pull down the windows; all of which can be accomplished by radio telemechanics, a science already well known.

His business correspondence comes in entirely by radio. There is a tele-radio-typewriter. This electro-magnetic typewriter can be actuated by any one who chooses to do so. For instance, if we wish to write a letter to Jones & Company, Chicago, Illinois, we call up by radio, that station, and tell the operator that we wish to write a letter to the Company. Once the connection is established, the letter is written in New York, let us say, on a typewriter, and automatically sent out through space by radio; letter for letter, word for word being written by the other typewriter in Chicago. The letter when finished falls into a basket. Instead of sending our correspondence by mail we shall then do our letter-writing by radio. There is nothing difficult about this scheme, and as a matter of fact, it can be put into use today, if so desired. We have all the instrumentalities ready.

Going further, we find the Radio Power Distributor Station that sends out power over a radius of 100 miles or more. This radio power may be used for lighting, and other purposes.[3]

In front of the bridge we see a number of people who are propelled by Radio Power Roller Skates. On their heads we see curious 3-prong metallic affairs. These collect the radio power from a nearby railing, which, however, is not in view, and which they do not touch. The power is sent through space from the rail to the 3-pronged affair and then is conveyed to the skates, which are operated by small electric motors. They roll at the rate of 15 to 20 miles an hour, and there is no visible connection between the wearer and the Radio Power Distributor.

We next see the crewless ships controlled by radio. This has been made possible today. Indeed, several U.S. battleships have already been manoeuvred over a considerable distance by radio. The time will come when we can direct a ship across the ocean without a human being on board. Future freight will be sent in this manner. The ship, every ten minutes, gives its location by radio, so that the land dispatcher will know at any time where the ship is located. Collisions are avoided by a number of instruments into details of which we need not go here, but which have already been perfected. Collision with icebergs also is avoided by thermo-couples which divert the ship away from the iceberg as soon as it enters water which has been cooled below a certain degree.

The radio-controlled airplane works similarly to the radio-controlled ship, and it will be possible to control such airships very readily in the future. As a matter of fact, John Hays Hammond, Jr., in this country, has done this very thing.[4] Radio-controlled airplanes will play a great role in the next war.[5]

3. For more on Nikola Tesla's World Wireless System, the most fully developed proposal for such an infrastructure at the time, see **The Future of Wireless** in this book. A similar wireless power technology is described in **Our Cover** in this book, referred to as a "Radiofer."

4. Hammond (1888–1965), recipient of the 1963 Institute of Electrical and Electronics Engineers (IEEE) Medal of Honor, was known for piloting a yacht by remote control in 1914 on a 120-mile trip between Gloucester and Boston, Massachusetts. John Dandola, *Living in the Past, Looking to the Future: The Biography of John Hays Hammond, Jr.* (Glen Ridge, N.J.: Quincannon Pub Group, 2004).

5. For more on the idea of full-size, radio-controlled airplanes and their use in war, see **A Radio-Controlled Television Plane** in this book.

It is a mistake to think that radio is only good for the distribution of intelligence. As the illustration shows, the great uses of radio have not been touched upon as yet.

Chapter XI: The Future of Radio

As the author has mentioned often in his various editorials published in *Modern Electrics, the Electrical Experimenter, Science & Invention,* as well as *Radio News,* the radio business may be likened to the amateur photographic business. Within the next few years, we shall see every drug store selling complete radio outfits that can be put on top of the phonograph at home, and which can be worked by your six year old sister. All that is required of you is to manipulate a few knobs, and from a concealed horn, the latest jazz band music will then issue forth. To be sure, this music is broadcasted from a central station which may be a thousand miles away or farther.

Then too, the day of the radio newspaper is quickly coming. Important news of the day will be broadcasted by radio telephone daily at stated intervals, as will be weather reports and other information useful in every community. But of course, the development of radio will not stop at radio telephony alone. Great and wonderful things are coming in radio which are undreamt of today. New uses are constantly being found. New improvements are being made almost over night. We cease to wonder when we hear of some new marvel that is being performed by radio, and simply shrug our shoulders and say "well that was predicted long ago." Thus, recently, a physician several hundred miles inland listened to the heart beats of a man lying unconscious on a ship three miles out on the ocean. Every heart beat was transmitted clearly and faithfully by radio to the physician, who was thus enabled to make a diagnosis.

We now move ships and steer airplanes by radio. Very recently in Germany, radio was used in mines underground to locate ores and coal veins accurately, surely a surprising use for the art! In this invention use is made of a receiving and sending station, both located underground, one signalling to the other. When the signals pass through a coal field, a variation is heard at the receiving end and by triangulation, the exact location of the coal veins can be found.

Recently, in Italy, radio has been used for prospecting metal ores. Here the Italian inventors use very sensitive vacuum tube outfits, and by means of a certain condenser arrangement, it becomes possible to plot accurately the exact location of the future mine. Today, every radio station has its ubiquitous aerial on top of the house. This soon will be a thing of the past. Already two American inventors have demonstrated that far better results may be had by putting the aerial underground. In their experiments, the inventors use the so-called underground loops. Of course, these are necessary for long distance reception, but for your radio outfit on top of the victrola in the parlor, no underground loop, or aerial on the roof is necessary. The aerial will be right inside of the outfit. We are already doing this very thing today, and the outfit need not be larger than a foot square. That gives us sufficient space for the concealed aerial within the box. This is not a dream of the future, as it has already been accomplished for the reception of radio music over distances of several hundred miles.

It is even possible to do away with the loop entirely. There has appeared lately upon the market, an electrical plug which is simply screwed into any lighting fixture in your house. It makes no difference if your lighting current is 110-volt alternating current, or 110-volt direct current, or even 220-volts. This plug consists simply of a condenser arrangement, and the idea of it is as follows: We have seen in former chapters that any aerial wire will pick up radio waves. Now then, every lighting circuit forms a sort of loop aerial itself. This is particularly true in the country where the wires run for great distances out doors. By proper arrangement, as for instance, using such a plug, the radio waves are conveyed right over the lighting circuit without interfering with the electric light bulbs, and other electric appliances in your house. These condenser plugs are already a great success but they do not work under all circumstances. For instance, in apartment houses in which much steel enters into their construction, the results are not so good as in the country where we have a stone or wood house, and where the electric lighting wires run outdoors. These condenser plugs work satisfactorily in nearly all instances, even in apartment houses, in connection with vacuum tube sets, but they do not work well as a rule with crystal receivers. Much experimental work is as yet to be done in this line, but the chances are that ten years from now, the aerial for receiving purposes will be a thing of the past.

Perhaps the greatest development will be the radio power transmission of the future. This is, of course, today, only a dream, but Nikola Tesla has demonstrated that it can be done, and it is interesting to note that this great savant's ideas are coming more and more to the front. Dr. Tesla contends that our radio conceptions are wrong from start to finish. He claims that it is not the radio waves that travel through the ether following the curvature of the earth, but rather currents that travel through the earth, and we seem to be coming to just this. If Tesla is right, power transmission by radio should be a simple thing. It will enable us to tap the earth at any point and receive our energy to light and heat our houses. Of course all this is in the future, but we are surely coming to it.

Another new use for radio is sending pictures, photographs, etc., through the ether, and the author believes that he cannot do better than quote part of his editorial from the November 1921 issue of *Radio News*.[6]

"Recently the signatures of General Foch and General Pershing were sent across the Atlantic by radio on the Belin apparatus. There is no good reason why the amateur cannot do the same thing for smaller distances at any time.

"In the very near future, the amateur in New York will buy the first copy of a New York newspaper, wrap it around his cylinder, and send out a whole sheet by radio. A thousand miles away another amateur will have a receiving machine that will reproduce the printed page, type, pictures, and all in less than a half-hour. This is a thing impossible to do by ordinary wireless telegraphy, if every word must be transmitted. The radio picture transmission solves all this. Thus, in time, a great piece of news 'breaking' in the city, will be sent broadcast by the enterprising amateur, who will send the entire front page of the newspaper, for instance, and the radio facsimile can then be exhibited in a distant town from 10 to 24 hours in advance of the receipt of the actual newspaper.

"All this is not a mere dream, but it already has been accomplished today. It is up to the amateur to make the thing popular."

To go still a step further, the author in a recent article in *Science & Invention* magazine proposed a radio system, which theoretically is sound. It is nothing else than Television by Radio. (Fig. 2). The fundamentals of this proposed scheme are correct, and there is little doubt that we will have radio television within a very few years on a scale that will be tremendous. The idea in short follows:

At the Polo Grounds of New York, let us say, we have a radio transmitting station in a box-like affair of about three or four times the size of a movie camera. We have a box with a lens in front, the back of the camera being composed of a great number of photo-electric cells. These cells have the property of passing more or less electric current, depending upon how much light falls upon the cell. A strong light will pass much current through a cell, while a weak light will pass little current. By means of these cells, we influence a modulator vacuum tube connected to the radio transmitter. This modulator sends out radio waves into space. If a strong light falls upon the electric cell, No. 1, we send out a radio wave of a certain intensity at a certain wave length, let us say $500\frac{1}{2}$ meters. At the receiving end, this wave is

6. Leveling an "off the beaten track" challenge to the radio experimenter to break out into new areas of research, Gernsback describes in this editorial the future of wireless picture transmission. It is quoted directly with no changes in the next three paragraphs. Hugo Gernsback, "The Radio Experimenter," *Radio News* 3, no. 5 (November 1921): 377.

received and is passed through the regulation radio outfit and thence through a condenser, vacuum tube, and audio-frequency transformer. This audio-frequency transformer operates a small magnet which in turn influences a pivoted diaphragm. This diaphragm has mounted upon it a strip mirror about ½-inch long and 1/16-inch wide. Normal and at rest, a light ray from a common source, let us say an electric lamp, directs a single beam of light upon the diaphragm in such a manner that the light ray just misses the mirror. The least vibration of the diaphragm, however, will intercept the light ray and will reflect it upon a ground glass plate. It is evident that the more the diaphragm vibrates, the more the light ray will vibrate back and forward upon the ground glass screen.

If we now imagine at the sending end several hundred of the photo-electric cells and at the receiving end a like number of vibrating mirrors, we can readily see how a picture sent out from the sender can be recomposed and reconstructed at the receiver. It must be understood that our photo-electric cell sends out its own wave-length. Thus, as mentioned before, photo-electric cell No. 1 sends out a wave length of 500½ meters. Photo-electric cell No. 2 will send out a wave of 500¼ meters and so on. All these waves are sent out from the same aerial,

FIGURE 2.

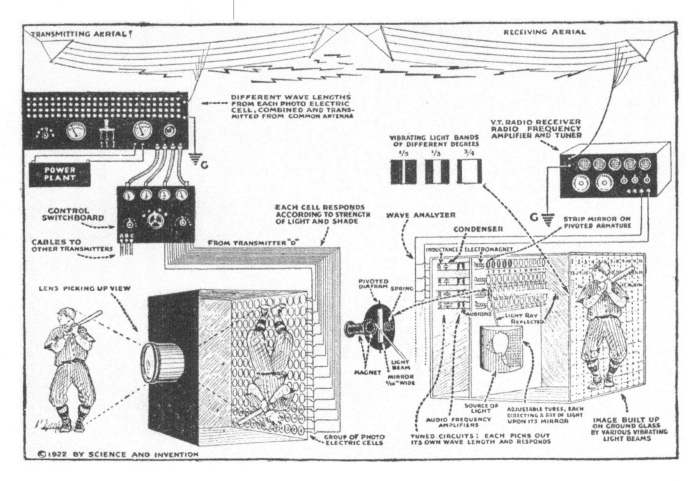

and all the incoming waves are caught upon the same aerial, each wave operating its own electro-magnet and consequently the light beam. We can now see from this how our future audience will be able to witness a baseball game five hundred or five thousand miles distant, as if it were witnessing the game itself. It is, of course, understood that this transmission takes place instantaneously, so we will be enabled in the future to view distant games or other important events at the time they are taking place. This differs from the movies where we are not able to view the events at the time they take place, but always at a later date. In the future there will be the possibility of our seeing the President of the United States make an important speech, and we will be enabled to not only hear every word he utters, but see him in person as well.

Of course, it goes without saying that the scheme here advanced will project the picture in black and white only. In other words, the picture will look just like a movie film with the sole difference that we are witnessing the event at the time it takes place.

Here is an interesting feature of which few people are aware. Some months ago, in one of the writer's editorials he mentioned the fact that radio waves are eternal, as are light waves; they travel according to our present conception out into space at the rate of 186,000 miles a second. We see today the light waves shot off by some far away star, which light may have originated from that star perhaps 10,000 or 100,000 years ago. And those light rays are just coming down to us now. It is the same with radio waves. Any radio message, any broadcasted radio selection that is sent out on radio waves goes out into eternity never to return, never stopping, ever traveling onward. The thought is appalling that while you are listening to a famous operatic star, who is singing from some broadcasting station, her voice may be heard 100,000 years from now on some distant planet belonging to its own little solar system.

For to believe that there is intelligence only on this earth is grotesque and foolish in the extreme. What this superior intelligence, listening to this broadcasted song will think of it 100,000 years hence, is difficult to imagine. But, there seems to be little doubt that this superior intelligence will smile at the idea of our feeble endeavors. This intelligence will probably view our attempts with the same amusement as we look upon children using a string telephone.

At the time this volume is written, radio is just about twenty-five years old. If we have accomplished such wonders in a quarter of a century, who dares say what will be accomplished in twenty-five or fifty years more. Our wildest and most impossible prophesies will seem feeble. When we, therefore, say that one of the coming things is transporting solids through space, that is, sending a carload of coal from Pittsburgh to Paris within a few minutes, all by radio, and all by the invisible self-same waves, we will probably be laughed at by our experts. The thing, however, is perfectly feasible today, as we shoot solid particles through glass walls every time an X-ray picture is taken. X-rays are composed of solid particles which are just as solid as bricks or lumps of coal. When, therefore, we are asked what the future of radio is, we may say in one word, ANYTHING! There seems to be nothing impossible that radio cannot accomplish in the future!

FIGURE 3. The power plant at the radio broadcasting station WJZ, Newark, N.J. We see here five special vacuum tubes, each generating about 50 Watts power. This gives a total power of 250 Watts, or Kilowatt. On top of the case we see the oscillation transformer which has the function of adjusting the wave length at which the broadcasting is accomplished, in this case 360 meters. The waves emitted by this station have been heard several thousand miles away.

10,000 Years Hence

Science and Invention, vol. 9, no. 10, February 1922

If we go back but 100 years, and contemplate how the world looked as, for instance, in the days of Napoleon, we are apt to be amazed at the way in which the world has progressed in a technical sense since then. We believe we need not call attention to the fact that steam, electricity, and up-to-date technic have completely altered not only the face of the globe, but our very lives as well.[1] If such a tremendous change has been possible in a short century, how then will the world appear in a thousand years, or ten thousand years, hence? The imagination fairly staggers at the attempt to picture what our civilization, if it still exists, will look like in the future ages.

The up-to-date scientist has little difficulty in predicting certain things that will happen in ten or fifty years, but ten centuries hence is a large order, even for the most intrepid imagination. That practically nothing of our present civilization will be left after ten thousand years may be safely predicted. We may also prophesy that the human beings, ten centuries hence, will live in entirely altered circumstances from those they now exist in. Captain Lawson, of aerial fame, for instance, not so long ago made the prediction that 10,000 years hence the human race will not live on the surface of the globe at all, but will live far above it. His reasoning is as follows:

He states that at the present time we are living at the bottom of a vast sea—the sea of air—our present atmosphere. We all know that on the surface of the globe this air presses upon every square inch at the rate of 14.7 pounds with a lightly varying pressure. The weight that the human body, for instance, has to sustain is approximately 30,000 pounds, a tremendous figure. We do not come to harm, of course, for the simple reason that the pressure is even in all directions, but our lungs have been accustomed to this pressure and if we suddenly should take this pressure away, our lungs would burst. Even aviators rising only two miles above the surface of the earth have great difficulty in breathing. It is the same with the other great sea, the ocean, which is also but a fluid, just like the atmospheric sea, with the difference that the water is of greater density, otherwise there is little difference, even chemically, between the two seas.

The fish living at great depth (the so-called deep-sea fish) sustain gigantic pressures upon their bodies, and if suddenly brought to

1. For an etymology of the terms "technic" and "technology," see Eric Schatzberg, "Technik Comes to America: Changing Meanings of Technology before 1930," *Technology and Culture* 47, no. 3 (2006): 486–512, doi:10.1353/tech.2006.0201.

the surface, burst like balloons. This is the exact counterpart of the human who wishes to rise to the top of the atmosphere.

Captain Lawson, following his analogy, predicts that centuries hence we will be living at the top of the atmospheric sea instead of at the bottom. In other words, the future human being will not be a deep-sea atmospheric animal, but will reside at the top of the atmosphere, comparatively speaking.

Captain Lawson does not state the advantage of this living miles up, away from the surface of the globe, but we may cite several apparent ones. Most of the human diseases probably are due to bacteria and small micro-organisms floating in our dense air. It may be doubted that such micro-organisms will be found two or three miles above the surface of the earth. Nearly all our diseases, such as tuberculosis, and all other infectious disease arise from micro-organisms, which are carried in the dense air, so by making our future abode two miles above the surface of the earth we would at once remove one of the greatest causes of death that humanity has to contend with.

Another change for the better in the upper atmospheric plane is the obvious one, that we will have continuous sunlight. No rain, no clouds, no thunder storms, no snow are to be contended with, once we rise above the highest clouds; and the latter never rise higher than two miles above the earth. More sunlight, as we all know, is most beneficial to human beings, and having 100 per cent of it all of the time we naturally will be far better off.

It would also seem that the race would be greatly benefited by the rarefied atmosphere as we would be able to move around better, and would not be oppressed by the atmosphere as we are now, particularly on hot days, when the air seems to feel like a thousand tons on our bodies.

These are only a few of the obvious advantages, but there are many more. Thus, for instance, at high altitudes there is no dust to be considered, and dust, as we all know, is highly detrimental to us. We, therefore, may surmise that the human race centuries hence instead of living upon the surface of the globe will live far above it in cities as we have them today, if cities they will then be called.

Our illustration depicts one of the future cities of about the size of New York floating high up in the air, several miles above the earth. The question of sustaining such a large body in a rarefied atmosphere will prove to be of little difficulty to our future electrical engineers. Just as we construct leviathans of the sea today, some of them weighing as much as 50,000 tons, we will construct entire cities weighing billions of tons, which cities will be held in space not by gas balloons, propellers,

FIGURE 1. The City of the Future, 10,000 Years Hence, Will Not Be Located Upon The Surface of the Earth. It Will Be Floating Up Miles High, and Such Things as Snow, Rain, and Storms Will Be Unknown to the City Dwellers of the Future. It Will Have Perpetual Sunlight, and Weather Will Never Bother Our Future Citizens. Just as Our Leviathans of the Sea are Built to Remain on the Top of the Water at All Times, So the Floating City of the Future Will Remain Afloat Continuously, Supported Only by Shafts of Electro-Magnetic Rays, Which, Nullifying Gravity Keep the City Raised Up By Reaction. The City Dweller of the Future Will Not Be Bothered Much With Such Diseases as Tuberculosis, Because All of These are Now Transmitted Due to the High Density of the Air Near the Surface of the Earth. Three or Four Miles Further Up Bacteria are Not So Common as Near the Surface of the Earth.

or the like antiquated machinery, but by means of gravity-annulling devices. Already experiments have been made whereby it has become possible to reduce the weight of substances by electrical forces.

Thus Professor Majorana, in an article printed in this journal three years ago, made it possible to produce negative gravity by reducing the gravitational pull on a lead sphere.[2] Of course, this is but a crude beginning. Centuries hence, when we wish to raise the city of the future high up in the air, we will rely upon an electro-magnetic stream of force which by reaction upon the ether and the earth, lifts the entire city high above the clouds.

Our illustration is but a feeble attempt to show how it may be worked out. Four gigantic generators distributed among equidistant points thru the city shoot earthward electric rays of a nature which as yet we can only imperfectly imagine. These rays, which are not light rays by any means, but are tremendous lines of force, impinge upon the surrounding ether with such stress and speed that the entire city is lifted up to the height desired. These rays may be likened to water streams, which by reaction would hold up the city illustrated. In other words, if we imagine the four rays as shown to be substituted by tremendous jets of water pouring earthward, and provided these jets were continuous, we can easily understand that they would support the entire city by counter-action of the force of inertia of water pressing against the lower part of the city.[3]

By increasing or decreasing the electrical energy of this future floating city it can be lowered or raised as we desire. By directing the rays sideways we will go in the opposite direction. Thus the "captain" of the future city will have a means to steer the city with all its millions of inhabitants to any part of the globe.

Where does this tremendous conglomerate take its energy from? The sun, of course. The city of the future is not dependent at all upon the earth for its power. Solar energy, which is merely another form of electrical energy, will be converted into electricity and stored away covering all the needs of the vast machinery, and that of the populace as well. Also, we should not forget that atmospheric electricity is a power that we only dimly understand today. This power in the future will be turned to the use of mankind, and we will then tap a practically unlimited amount of electrical energy.

As our illustration shows, the city of the future will be entirely roofed over with a substance that is neither glass nor metal. It will be transparent, but as strong as metal and unbreakable. Over this dome-like structure, gigantic towers are placed, which suck in the static electrical energy as well as the solar energy. Within the covered city the atmospheric pressure will perhaps not be more than four or

2. Quirino Majorana was an Italian physicist who published articles from 1918 to 1922 describing how he had successfully used lead and mercury shields to absorb the effects of gravity in a laboratory setting. See Roberto de Andrade Martins, "The Search for Gravitational Absorption in the Early 20th Century," in *The Expanding Worlds of General Relativity*, ed. Jürgen Renn et al., 3–44 (Boston: Birkhäuser, 1998).

3. Mike Ashley writes that Gernsback's floating city is "a concept one might believe inspired two later noted works of science fiction, Edmond Hamilton's 'Cities in the Air' (1929) and James Blish's *Earthman, Come Home* (1955), were it not that neither author knew of the article." Michael Ashley, *The Time Machines: The Story of the Science-Fiction Pulp Magazines from the Beginning to 1950*, The History of the Science-Fiction Magazine Volume I (Liverpool: Liverpool University Press, 2000), 34.

five pounds per square inch, instead of 14.7, as we have it today. That means that humanity will have to accustom itself to such a change, and there is no question that it is possible to do so if the change is made gradually, as it probably will be. It means that the future humanity will have larger chests than they have now.

Naturally it would not do for the future city to descend upon the surface of the earth, as this would be disastrous, not only to the inhabitants, but to the machinery as well. For one thing, the human beings would probably suffocate as they could not stand the high pressure of the present atmosphere. This need not worry us any more than does the present ocean steamer, which is not expected to land on the floor of the ocean, because when it does, all is lost. The steamship has been designed and constructed to stay upon the surface of the ocean, and the future floating city will be designed in the like manner, to stay up; not to go down.

As for flying in 10,000 years, the aeroplane and the flying machine will have disappeared. We will have the individual flyer as depicted on our front cover. The future flying man will be encased in warm fabrics electrically heated and kept hot comfortably. Upon his head he will have a sort of a diver's helmet made of flexible glass, unbreakable. The little atmosphere that he needs—remember he only needs five pounds per square inch—is carried by him in a few small tanks that do not contain air, but chemicals, which are slowly converted into air when required. The power is derived directly from the atmosphere—static electricity converted directly into electro-energy. No wings are used by him, but he propels himself just exactly as the city of the future is propelled, namely, by power rays, as shown. As before mentioned, the energy for these rays is obtained from static electricity reconverted into suitable high potential force, by means of a small converter strapped around the body.

If the flying man wishes to rise, he operates his future "joy"

February 1922 cover of *Science and Invention*.

stick in such a way that the rays point downward. He then rises upward with a speed all dependent on how much energy he furnishes his ray generator. If he wishes to go sideways, or laterally, the ray projector shown on top of his head is put into operation, which moves him sideways at any speed desired. If he wishes to descend, all he needs to do is to reduce the power. The more he reduces it the quicker his descent will be.

Thus the future flying man will be able to travel at prodigious speed without the necessity of using ponderous machinery or heavy apparatus. The future man will also be in touch with his friends or his office, or whatever it may be called then, by means of radio. He will be able to converse with them just as he can do today in a more modest manner, as when speaking to his friends from aboard ship, or in his auto of today. That, however, is not all. He will even be able to eat and drink, not from food which he has brought with him, but from food and drink sent to him through space. It will be possible to de-materialize and re-materialize matter, sending it at any distance desired with the speed of light.

We admit that all of this sounds extremely fantastic, but the truths of tomorrow will surpass our wildest fiction.

Radio Broadcasting

Science and Invention, vol. 9, no. 12, April 1922

great change has come about in the last two months. Overnight, the public seems to have "gone mad" over Radio. Laymen and others who have never paid the slightest attention to Radio before, are storming the Radio supply houses in a frantic search for Radio instruments, only to be disappointed as a rule because there is not enough material to satisfy the tremendous demand. There are now close to eighty broadcasting stations in the United States, new ones being added daily. These stations supply free entertainment to the masses and it is estimated that there are already over 1,000,000 Radio outfits of every sort and description in the United States.[1]

This art is so new that it would be futile to guess what it will be like in ten years to come, but we can make certain prophecies that we are sure will be verified. At present let us say Newark or Pittsburg is sending out some information or entertainment. We sit in our parlor and our friends listen to it by means of a loud speaking receiver. It may be a dry lecture or some other form of entertainment that we do not like. The best we can do at present is to turn off a switch which will silence the loud-talker. In the future we will not be dependent upon just one form of entertainment, but we will be able to choose for ourselves as to whether we should have jazz, grand opera, or a sermon. The present broadcasting stations use a wave length of 360 meters. In the future, each broadcasting station will have a dozen or more laboratories from which different forms of entertainment will be sent simultaneously, each on its own wave length.

For instance, the latest jazz selection may go out on 360.25 meters; a sermon will be broadcasted at 360.87 meters; a grand opera selection will go out at 360.50 meters. In other words, our tuning will be so refined that the different variations, differing by less than a fraction of a meter, will be perfectly distinguished and separated by the recipient, and will be reproduced without interference at the receiving end.

In the next few years it will be possible for us to take our meals with music, if we choose to do so, and after consulting the daily program we will only need to set the knob at the prescribed wave length, in order to get the form of entertainment desired.

If we wish, we can set our alarm clock in the morning, which in turn will set off the Radio outfit, and instead of being disturbed by a harsh

1. Though radio broadcasting became a mass cultural phenomenon in the early 1920s, with "over 2 million broadcast-capable radio sets" sold by the end of 1924, the technical phenomenon of broadcasting (as opposed to point-to-point communication) had been around since at least the early 1910s among small communities of amateurs. Tim Wu: "It was amateurs, some of them teenagers, who pioneered broadcasting. They operated rudimentary radio stations, listening in to radio signals from ships at sea, chatting with fellow amateurs. They began to use the word 'broadcast,' which in contemporary dictionaries was defined as a seeding technique: 'Cast or dispersed in all directions, as seed from the hand in sowing; widely diffused.' The hobbyists imagined that radio, which had existed primarily as a means of two-way communication, could be applied to a more social form of networking, as we might say today. And the amateur needed no special equipment: it was enough simply to buy a standard radio kit." Tim Wu, *The Master Switch: The Rise and Fall of Information Empires* (New York: Knopf Doubleday Publishing Group, 2010), 34–35, quoting the 1913 edition of *Webster's Revised Unabridged Dictionary.*

alarm bell, our awakening will be to any tune that we have selected the night before. One of the important things confronting the broadcasting stations is: How will they be paid for their service?

This applies particularly to broadcasting stations which cannot get indirect results from the sale of Radio apparatus, use of Radio patents, etc. Of course, some such stations, owned by department stores and newspapers, will continue broadcasting from the advertising motive. But let us say that the Metropolitan Opera Company of New York desires to broadcast their entire program. They would wish to be paid for the service. It would be impossible to force everyone who listens to pay, because everyone would listen in anyway, unless means were found to prevent them. Such a means will be found very shortly; as a matter of fact one means exists to-day.

It is very simple. Suppose the Metropolitan Opera Company were to make a charge of say $3.00 per month for listening in to their broadcast. They would not publish in the newspapers or in any other manner the wave length on which they would broadcast. Such information would be sent by mail to subscribers only. After every act the wave length would be changed. Now we are all aware that if we do not know at what wave length a station is sending, it is difficult to quickly tune in. By trial we may succeed, but much of the music would be lost during our struggles with the tuning. Most people would pay a certain amount for the information, which would enable them to listen to the entire entertainment without having to hunt for a certain wave length.[2]

2. For a full elaboration of this idea, see **Grand Opera By Wireless** in this book.

Human Progress

Science and Invention, vol. 10, no. 6, October 1922

When we contemplate the future progress of the human race, as viewed in the light of our present civilization, a beautiful picture is stretched before our eyes. We see the millennium just ahead, man emancipated; in other words, paradise on earth.

If the progress of the human race should go along unabated as it has during the last one hundred years, Science, in five hundred years, would lift the race up to a point where it never stood before. The world then would be a place that even our most fervid imaginations could not conjecture today.

But to the student of history, all does not seem so rosy, and if we really contemplate history carefully we become a good deal more pessimistic in our views as to the future of human progress. We need not look back centuries ago. All we have to do is remember the last world war, which retarded human progress a great deal. Without wishing to be over-pessimistic, we might well tremble for the future of the race, if another such war is let loose among us before a great while.

Our present civilization is but a spider web in strength, and it does not take much to break it. Our economic and our social life is such that a complete cessation of any great industry might cause chaos. Thus, for instance, if some agency should suddenly destroy our transportation means, such as our railroads, automobiles, and ships, for as short a period as one year, civilization would be plunged immediately into a condition akin to that of the Dark Ages. The penalty of our present civilization is that it makes us soft and without resistance. We are not as hardy as our forefathers used to be. The recent war proved this conclusively, where millions of people died, simply because they were not used to the hardships which they were suddenly called upon to face.

October 1922 cover of *Science and Invention.*

1. Henry Adams, confronted with the "supersensual world" of new electrical apparatuses on display at the 1900 Great Exposition in Paris, saw the dynamo as a "symbol of infinity": "As he grew accustomed to the great gallery of machines, he began to feel the forty-foot dynamos as a moral force, much as the early Christians felt the Cross. The planet itself seemed less impressive, in its old-fashioned, deliberate, annual or daily revolution, than this huge wheel, revolving within arm's length at some vertiginous speed, and barely murmuring—scarcely humming an audible warning to stand a hair's-breadth further for respect of power—while it would not wake the baby lying close against its frame. Before the end, one began to pray to it; inherited instinct taught the natural expression of man before silent and infinite force. Among the thousand symbols of ultimate energy the dynamo was not so human as some, but it was the most expressive." Henry Adams, *The Education of Henry Adams* (1918; repr., Mineola, N.Y.: Dover Publications, 2002), 286.

Gernsback literalizes Adams's elevation of the dynamo as the symbol of modernity by proposing the construction of a gigantic, concrete monument to the dynamo. It is illustrated elsewhere in the issue: "In connection with our editorial of this month, we show on this page a monument dedicated to the age in which we are living. Electricity, more than anything else, has made our present civilization what it is, and if this civilization should be wiped out by war or some other cataclysm, nothing would remain to tell what Electricity did for the race during the past century."

Although Gernsback hopes that such a monument would allow future civilizations to "read what has gone on before," he perhaps misses an important point raised by Adams. These new machines amounted to a "break of continuity," an "abysmal fracture for a historian's objects." Adams, himself a historian, foresees that the historian or archaeologist of the future would thus have to use an entirely new set of methods to understand the technological infrastructures of the twentieth century, an endeavor that no concrete replica or monument would help, no matter how massive: "No more relation could he discover between the steam and the electric current than between the Cross and the cathedral. The forces were interchangeable if not reversible, but he could see only an absolute *fiat* in electricity

On the other hand, if we read the past aright, we also know that as a rule history repeats itself. The Egyptians, as well as the Romans, were a highly cultured and civilized people. The Romans build the most wonderful roads in the world, which have lasted for two thousand years and upon which traffic passes every day in Europe at this very minute. They have known how to build and how to do things. The Egyptians were just as highly cultured and, perhaps, if we leave out scientific achievements, they were on a higher plane of civilization than our own. It is a mooted question today how they built their pyramids, and no architect will venture to say how they did it with the tools and facilities at their command in those days. We have never been able to embalm as well as the Egyptians, and we might recall dozens of other examples, but the point we wish to make is that it *did not last*. The Egyptians, as well as the Romans, disappeared, and left the world plunged into gloom, barbarism, and the dark Middle Ages. We might cite many other examples of great peoples who had reached seemingly the pinnacle of civilization, only to be destroyed and plunged into darkness.

In the light of these facts, will any one dare to affirm that our present world may not experience a similar fate in the future? If we take this fate for granted—and it is highly probable—should we not follow the Egyptians' example and build our own monument that would outlast the most severe ravages, just as the pyramids have outlasted not only the fury of the elements, but the destructive powers of man as well?

One of the things that has helped to create our present civilization is the electrical current, and specifically the dynamo. Why should we not build the representation of a 1,000-foot generator in concrete, of such proportions that it would not be easily destroyed, either by man or by the elements? In the interior passages, following the Egyptian example, we might engrave upon the granite walls the principles of modern electricity, so that if our world should be plunged into darkness those that follow would read what has gone on before.[1]

Of course, there are many other suggestions that come to mind to make such a monument not only a lasting one, but a practical one as well. For one thing, if the base of the generator were solid rock there might be great vaulted passages, which would contain a complete electrical museum. A representative piece of apparatus could be placed behind glass in air-proof vaults, illuminated by electricity, either from the outside of the corridor, or from the vault within. The more air-proof we make such walls, the longer the apparatus will last. It is questionable if we should place one of our present-day electric motors in a vault that was not protected against moisture and air currents that it would last more than one hundred years. The insulation would rot very soon, and once rust started its work there would not be left much for future generations to see of this particular motor. It is this way with most of our present-day apparatus and appliances; unless they are placed almost in a vacuum they do not last long.

It is a well-known fact that our present-day books are very short-lived; even if they are kept in an up-to-date library printed books will certainly not last more than five hundred years. The ravages of small micro-organisms, as well as the destructive qualities of modern ink, make our books only of a passing interest. For that reason, anything placed in our electrical monument should be, preferably, engraved upon stone in as few words as possible. After all stone or granite is the only material that will outlast centuries.

as faith. . . . A historian who asked only to learn enough to be as futile as Langley or Kelvin, made rapid progress under his teaching, and mixed himself up in the tangle of ideas until he achieved a sort of Paradise of ignorance vastly consoling to his fatigued senses. He wrapped himself in vibrations and rays which were new, and he would have hugged Marconi and Branly had he met them, as he hugged the dynamo; while he lost his arithmetic in trying to figure out the equation between the discoveries and the economies of force." Ibid., 286–87.

From the October 1922 *Science and Invention.*

Mr. H. Gernsback Has Proposed That We Build a Gigantic Monument to "Electricity." On Some Plateau We Could Erect an Electrical Generator, Molded in Concrete, 1,000 Feet High. Molded of the Finest Concrete, Such a Monument Would Last for Thousands of Years.

It Would Probably Not Be Affected by the Weather and the Climate, and It Is Doubted Whether It Could Be Easily Destroyed by Any Savage Race That Might Come After Us. In the Inside Passages, Along the Walls, Could Be Inscribed, in Diagrams and Otherwise, Electrical Fundamentals, from the First Static Machine Down to the Latest Radio Developments. As New Inventions Come About, These Can Be Inscribed from Year to Year.

© 1922 by Science and Invention

Results of the $500.00 Prize Contest

Who Will Save the Radio Amateur?

Radio News, vol. 4, no. 8, February 1923

n our October issue we published a special $500.00 prize contest, entitled: "Who Will Save the Radio Amateur?" This was in connection with Mr. Armstrong Perry's article, "Is the Radio Amateur Doomed?"[1]

All thinking men, and most intelligent amateurs themselves had long come to the conclusion, ever since the broadcasting popularity started, that the radio amateur was indeed doomed. By "doomed" is meant not that the radiophone popularity was to wipe out the amateur, but, rather, that IT WAS THE AMATEUR WHO DOOMED HIMSELF, and who put himself out of business. The reason is so simple and so obvious that it is difficult to understand why the rank and file of the amateurs have not seen it for themselves long ago.

OF WHAT REAL USE IS THE AMATEUR OF TODAY? What does he really do to make the world a better place to live in? Of what use is he to the community at large? If the amateur will ask himself these questions, and search his heart, he will come to the conclusion that, indeed, his utility is microscopic. It is true that amateurs are sending each other messages, which roughly covers 90 per cent of their utility. A purely selfish pastime! It is true that some amateurs are sending free messages for their friends, to be transmitted and relayed to distant friends, but investigation of the subject shows that this traffic is indeed exceedingly small. It is a fact that out of 100 messages actually filed with amateurs, not 50 per cent reached their destination! The number of such messages actually delivered is exceedingly small. It certainly has never assumed any proportion where the commercial telegraph interests have even felt it necessary to take any notice of such free-message work.[2]

It is true that amateurs have made credible records in sending messages, not only across the continent and further, but have sent messages and are sending them right along, across the oceans. This, certainly, is a very credible scientific undertaking.

Then, too, amateurs, in isolated cases, have helped the police in running down criminals. Such cases, however, do not happen once in six months.

1. Armstrong Perry, who had previously written articles for *Science and Invention* and other periodicals like *Radio Age,* published this piece on amateurism in *Literary Digest* alongside an interview with Sir Oliver Lodge, in which Lodge argues that the theory of the luminiferous ether should be rescued despite the scientific community having decisively moved on from the idea. Armstrong Perry, "Is the Radio Amateur Doomed?" *Literary Digest* (December 1922): 28.

Perry would go on to publish short fiction about radio operators in *Radio News,* albeit of a nonspeculative type. See for instance Armstrong Perry, "The 'Ham,'" *Radio News* 8, no. 8 (January 1927); Armstrong Perry, "Radio Revenge," *Radio News* 9, no. 2 (August 1927).

2. It would have been significant to send a message without a fee, as commercial stations could charge more than a dollar per word for a wireless telegraph transmission.

Also, it is not to be forgotten that during the war amateurs helped in building up a radio force that was of great help in the war. Indeed, the writer himself was instrumental in recruiting the amateurs, and secured over 1,000 enlistments, for which he received a very flattering letter from Ex-Secretary of the Navy, the Hon. Mr. Daniels.[3] No doubt the amateurs would do the same thing again, if called upon, but, as one correspondent in this contest put it *"The amateurs can not rest upon past laurels, particularly when there is not a new war every day."*

Summing up, therefore, the real usefulness of the American amateur in the United States is practically nil. This does not sound very nice, but it is the whole, unvarnished truth. The writer, who has been an amateur, and still is, feels that he knows whereof he speaks. The question simmers down to this: "Is there a real usefulness for the radio amateur?" THERE IS NOT! That is, if the amateur is honest with himself. Sending a few messages to each other, making transatlantic records, and waiting for the next war, to show what we can do, does not enhance our standing with the public. *As far as the public is concerned, the radio amateur does not even exist.* Make the following test, which we made recently in New York, and which some correspondents made in various communities in the United States, and you will get a good idea of what the populace thinks or imagines the radio amateur of today to be.

Stand on any street corner, and ask 100 people who pass by the following question: *What is a radio amateur?* The answers, boiled down, in about 90 per cent of the cases actually tested, will be as follows: *"Oh! A radio amateur is an experimenter who tinkers with radio apparatus."*

This is then what some 95,000,000 or more people in the United States think of us. In other words, we have never sold ourselves to the public—for a very good reason: WE HAD NOTHING TO SELL, for our usefulness in the United States, up to this time, was nil.

Prize Winners

First Prize ($200.00)—Mr. L. W. Gundy (1 BZL), P.O. Box 67, Phillips, Me.

Second Prize ($100.00)—Mr. Jesse Marsten, 909 Beck St., N.Y.C.

Third Prize ($75.00)—Mr. Hugh Wingett, 1205 Stainback Ave., Nashville, Tenn.

Fourth Prize ($50.00)—Mr. E. T. Jones, 3997 Dumaine St., New Orleans, La.

3. Gernsback was a great critic of Daniels, who attempted to maintain Naval control over all civilian airwaves after the war. See **Amateur Radio Restored** in this book.

Fifth Prize ($25.00)—Mr. L. VanSlyck, 123 Hibbard St., Ironwood, Mich.

Sixth Prize ($25.00)—Mr. Stem Anderson, 3257 Q St., Lincoln, Neb.

Seventh Prize ($25.00)—Mr. Frank H. Fanning, 301 Holt St., Ashland, Ky.

Honorable Mention:

Mr. Ernest G. Underwood, Elwood, Calif.
Mr. Allen H. Duncan, 32 Waverly Pl., New York City
Mr. Sumter B. Young, (1 AE) Associate Member I.R.E., formerly
    Chairman Boston Executive Radio Council, Dorchester 24, Mass.
Mr. A. W. Parks, Easton, Pa.
Mr. J. F. Tolley, New Orleans, La.
Mr. Thomas C. Howard, Newport, R. I. (1 AFN)
Mr. L. R. Felden, 979 55th St., Brooklyn, N.Y.
Mr. H. F. Rook, Ridgefield Park, N.J.
Mr. Rex Durant, Cricklewood, London, England.

When radio was young, it was all right for radio amateurs to do just what they were doing, that is, sending each other messages, doing research work, etc., *but the world moves on*—WHILE THE RADIO AMATEUR STANDS STILL. As the writer mentioned before, the radio amateur is in a rut, and deep down in his heart he knows it.

If the great and wonderful art of Radio means nothing more to the radio amateur than sending a few messages, catching a burglar MAYBE, sending a few dots and dashes across the ocean, and waiting for the next war to come along to prove, MAYBE, that he can help—then it certainly would have been far better that Radio had never been invented.

We thought that there must be somewhere, somehow, some way to put the radio amateur on the map so that when the name "radio amateur" was mentioned anywhere, in a crowd, or to any layman, there would be instant attention—not a questioning raise of the shoulders, as is the case now. This brings us to our Contest:

As we had foreseen, and as we mentioned in our columns of the October issue, the two articles, "IS THE RADIO AMATEUR DOOMED?" and "WHO WILL SAVE THE RADIO AMATEUR?" brought a number of letters from the narrow-minded and misguided amateurs, who, in their simplicity, thought that we were "knocking" the game, and trying to put the radio amateur out of business. We were assailed from all sides, with many "brick-bats," and even the mouthpiece of the American Radio Relay League, who certainly should know better, said things about this contest that not even a third-rate, slandering, country newspaper would stoop to utter.[4] That, however was expected. We were even questioned about the $500.00 Prize Contest, and it was darkly hinted that we were chasing the dollar in making this offer, but just how we were to do this was not mentioned. We offered the $500.00 in prizes in good faith, and are paying out these same $500.00 today in the same good faith, cheerfully, because we know that we have accomplished something that will help to make radio amateurism in this country a real force.

The contest, from every point of view, with one exception, was the biggest we ever staged. Over 5,000 replies were received from amateurs all over the world, and we received many wonderful and inspiring letters. A thing that pleased us particularly was that letters from the best amateurs in this country were in the majority, and many hundreds were received from members of the American Radio Relay League, the Radio League of America, and all of the prominent radio clubs in the country. The one fly in the ointment was that nearly one half of the contributors did not take the time to read the conditions, and indeed did not get the spirit of the Contest. On page 795 of our October issue, in the gist of the entire contest, in these words: *"What we wish, therefore, fellow amateurs, is a manuscript of not more than 1,000 words, setting forth your ideas to the best plan to put the radio amateurs on solid footing, where they can perform the greatest good for the community, and for the radio art."* That was our message. Nearly one half of those who answered, evidently did not take the trouble to read this, or, if they did, they thought the contest referred to something else. The fact remains that 50 per cent of the contributors tried to save the amateur by trying to devise new legislation, to protect him! What these contributors did not see at all was that *no force in the whole world can save the amateur except the amateur himself.* Most of these correspondents had an idea that the amateur was doomed on account of the interference which he is making. Nothing can be more erroneous. We never had such an idea in mind, and nowhere did we print a single line about such a thing, or even suggest it.

The truth is that the amount of interference that the public is getting from the amateur is insignificant. There is very much more interference from the commercial stations than from the amateurs, and we believe the public at large knows and appreciates this. Moreover, amateurs are learning not to send during broadcasting hours, and within the next six months *this problem will be solved entirely by the amateurs themselves,* so there is little need of legislation on that score. Even a single circuit crystal set, unless it is right under the shadow of an amateur's transmitting aerial, does not, as a rule, experience much interference from 200 meters. The wave-length of most amateurs does

4. "Brickbats" was a slang term for insults or criticisms that became a favorite among science fiction fans and readers in their letters to the editor later in the 1920s and '30s.

February 1923 cover of *Radio News.*

5. One of the problems with the 1912 Radio Act's restriction of amateur radio operators to wavelengths of 200 meters or less was that early sets "could not always adhere to a wavelength with any degree of accuracy and receivers also experienced drift." Under the leadership of Herbert Hoover, the Department of Commerce sought to further limit not just the range but the content of amateur activities on the airwaves with the following line inserted into amateur radio station licenses in 1922: "This station is not licensed to broadcast weather reports, market reports, music, concerts, speeches, news, or similar information or entertainments." Marvin R. Bensman, *The Beginning of Broadcast Regulation in the Twentieth Century* (Jefferson, N.C.: McFarland, 2000), 41.

not go much above 250 meters, and this, we might say, is exceptional, so why worry on that score?[5]

To resume, our contest has been a great success. The suggestions that are made in the seven prize-winning letters are very substantial, and, if followed, will surely put the amateur on the map in a very short time, but—like your doctor—we can only give the prescription. It is up to the patient, the radio amateur, to take his medicine, which, in this case, is quite pleasant, and, we are certain, effective. The judges of the contest were as follows:

H. Gernsback, Editor of RADIO NEWS

L. G. Pacent, President of Pacent Electric Co.

Robert E. Lacault, Associate Editor of RADIO NEWS

Armstrong Perry, Author, and

L. M. Clement.

The judges were almost unanimous in their decision on the seven prize-winning letters, and there was little divergence of opinion. Mr. L. W. Grundy was awarded first prize, mainly on account of his suggestion to re-transmit broadcast programs over the electric light lines. This, indeed, is one of the best suggestions advanced, and we have more to say about this in another section of the magazine. (See article: "Popularizing Radio.") In addition to the prize winners, there were eight letters which were deemed of sufficient importance to be awarded Honorable Mention. These letters will be published in subsequent issues. Some of the prize-winning letters follow:

Is the Radio Amateur Doomed?

By L. W. Grundy, 1 BZE

First Prize

If he becomes selfish and self-centered he is doomed; but he won't, he will adjust himself to circumstances and be as indispensable as he was yesterday. Time will arrange all things.

Broadcasting is here to stay. The public wants it; they will get it; they ought to have it. If amateur transmission interferes, we must suspend same during reasonable hours. However, the listener to broadcasts must share the air, he cannot expect full control from 10 a.m. until midnight. 8–10 p.m. or 7:30–9:30 is time enough when one considers he has it every evening. The amateur's work is valuable. Most broadcasting is mere amusement. It has its place but it cannot be selfish and "hog the ether."

Let's own up we all like to listen in once in a while. What is to be the amateur's attitude toward broadcasting? We must gracefully accept it and as a public need, help further it. There is chaos in broadcasting now but the inexorable law of evolution, the survival of the fittest will prevail. How can we help? Let's give kindly unbiased advice in regard to selection of sets to cover the required distances. This will prevent the buying of inferior, hence unsatisfactory goods. I butted in once when I heard a misinformed clerk tell a young woman that a crystal set would bring in KDKA six hundred miles away.[6] I simply suggested

6. Broadcasting from Pittsburgh, KDKA was made famous for announcing the results of the 1920 presidential election over the airwaves, an event that many historians of radio see as "the beginning date of United States radio broadcasting." Ray Barfield, *Listening to Radio, 1920–1950* (Westport, Conn.; London: Praeger, 1996), 3.

KDKA was run by Frank Conrad out of his garage and began the first regularly scheduled broadcast programming, "initially called wireless concerts, which consisted primarily of pushing a Victrola up to a microphone and playing records." KDKA became a magnet for operators around the world trying to tune in to distant stations, and inaugurated what Susan Douglas calls the "broadcasting boom" in the early 1920s. Susan J. Douglas, *Listening In: Radio and the American Imagination* (New York: Random House, 1999), 64.

that a regenerative set would do much better. The clerk sold the better set, it worked, everybody happy and satisfied. Let us show novices how to tune their sets, and get the most from them, for few novices can handle a tube set without guidance. Let us tackle receivers, freedom from QRM means freedom to transmit.

Some of the good points brought out in this contest, and which were awarded prizes, were as follows:

—Re-transmitting broadcast programs over the electric light lines, for the benefit of users of cheaper sets, and for those who are out of range of the big broadcasting stations.

—Single Control Receiver, to popularize Radio with the public.

—Signboards in front of amateurs' houses, giving bulletins of all important news to the community.

—Relaying weather, crop, and market reports.

—Doing away with spark transmitters—using C.W. only.

—Nationwide publicity through local newspapers, for amateur activities, performing real service.

—Equipping all transmitting sets with phone, to inform local listeners that amateurs are not interfering with broadcasting.

But we want to transmit. When the new White Radio Bill becomes law let's retransmit some of the excellent programs over the electric light lines or through the air for the benefit of users of cheaper sets.[7] Here's a field experimentation and service. Let us open our homes to the public, let's put our loud speakers in halls, homes, churches, schools, etc., and give them a free concert. The public will appreciate the public-minded amateur. Let us post market and weather reports, news items and other things of public interest. One amateur posted the World's Series' results before they came over the wire. A friend of mine invited in a father and mother to listen to a glee club broadcast four hundred miles away and their son was an accompanist. Did they enjoy it? Just imagine! In some of our small towns no Marine Bands

7. The debate between Congressional Representative Wallace White of Maine and Senator Clarence Dill from Washington over the regulation of the airwaves resulted in the formation of the Federal Radio Commission in 1927, which was renamed Federal Communications Commission in 1934. The White Bill, mentioned here, proposed putting radio broadcasting under the oversight of the Department of Commerce, then led by Herbert Hoover. While "White sought to protect the economic interests of RCA and the rest of the radio industry, . . . freshman Senator Dill aligned himself with the Senate's insurgent Progressives whose goal was to protect the average person from corporate greed. The insurgents wanted the federal government to regulate the industry for common good, but they believed that experts would manage radio better than partisan politicians." Mark Goodman and Mark Gring, "The Ideological Fight over Creation of the Federal Radio Commission in 1927," *Journalism History* 26, no. 3 (2000): 117–23, 118.

nor high class performers ever come. Broadcasting is their blessing.

Let us prepare ourselves for the new era. Let us ban, as quickly as possible, by example and advice, the spark transmitter. Let's push C.W.[8] It's the thing we need. We will experiment with directional transmission and hence avoid interference. Above all we need a wave meter, using it to keep strictly within our wave bands. When we have opportunity for self-policing let us do it efficiently. We should make our relay and DX work of greater value.[9] Let's make the subject matter of the message as important as distance itself. It seems that we may add greatly to the radio art by doing research work.

Another field for service is the realm of the boy. Radio keeps boys at home instead of in mischief-making gangs. It stills a thirst for scientific knowledge and hence stimulates manhood. A few dollars in radio supplies coupled with a friendly interest will save many a lad from going wrong. Boys often lack cash. Let's teach them the possibilities of materials at hand—cardboard tubes,—brass screws, a spool of magnet wire. The writer used, one winter, a regenerative set made of ice cream containers, brass screws for switch points, ink bottle stoppers for knobs. The set cost without battery $14.80 and it covers a range of about 500–800 miles. As I have been writing, a lad has just brought in for my inspection some tubes made of sheathing paper and shellac. But let us teach painstaking care, efficient layout and mechanical thoroughness even with humble materials. We can teach code classes, we can broadcast interesting materials as code practise. We can show that a single amplifying tube can outrange the average spark coil.

If we amateurs will lead the boys of our communities we will make ourselves indispensable and contribute to community and national welfare.

Ours is an educational task. Amateur leaders must make the rank and file see the light. We must work through the various organizations, we might add a home service branch to our Relay League with distinct and definite duties in regard to community and public service.

The writer is a deep-dyed amateur, he has tried to serve. Four hundred people listened in at his home during November and December, 1921. A new cage aerial lately swung into place has elicited commendation and within 24 hours a neighbor "tickled his palm" with a five spot, "in recognition of your willingness to let folks listen in and because you have helped the boys of this town." Let us be patient to serve rather than arbitrarily selfish and we can make ourselves and our cause, amateur radio, positively vital. I do not worry, the amateur is resourceful enough to adjust himself to any condition and he will: and *Long will he live.*

8. An abbreviation for continuous wave, a form of radio transmission that gradually came to replace the noisy and interference-producing spark gap transmission systems favored by amateurs in the 1910s. It is also the name for the second volume of Hugh Aitken's seminal history of radio. For a condensed account of the technical and regulatory paradigm shift from spark transmitters to continuous wave transmitters, see Hugh G. J. Aitken, *The Continuous Wave: Technology and American Radio, 1900–1932* (Princeton, N.J.: Princeton University Press, 1985), 3–27.

9. DXing is the practice of listening for and identifying transmissions from as far away as possible.

A Suggestion for Utilizing the Amateur's Technical Knowledge and Avoiding the Clash with the Lay Radio Public

By Jesse Marsten

Second Prize

The radio industry will develop and flourish in direct proportion as the number of people who take an interest in it increases. Just as the talking machine industry could not thrive if it had to depend for its business on a handful of musicians, so the radio industry cannot depend upon the handful of technical amateurs for its business.[10] It must interest the average man and woman in the street.

To secure this essential interest of the enormous lay public it is essential that the radio instruments, like the talking machine, be models of simplicity as far as operating features are concerned. And simplicity today means a minimum of controls, and preferably a single control. It is the single control receiver which will popularize radio more and more, and it happens that the only type of circuit available to-day permitting of such simplified control is the single circuit receiver.[11]

Naturally the single circuit receiver therefore finds favor with the layman, but its prevalent use has brought in its train difficulties and a big problem. The single circuit tuner is of course less selective than a double or triple digit tuner. As a result, with transmitting amateurs around him particularly in congested districts, the layman has experienced the considerable code interference while receiving the broadcasting. The radio lay public therefore insists that the amateurs keep off the air during broadcasting hours. The amateurs just as staunchly insist what they consider the arrogant interference of an uninformed public which is just entering the radio circle. They generally cite their long standing in the radio community, that they have made radio what it is to-day, and that the single circuit tuner causes most of the layman's interference.

It is true that the amateur has made radio history and is in the vanguard of the radio army. The amateur, however, cannot rest on his laurels or past performances if the layman—who is to furnish the radio industry with practically all of its business—is interfered with. It must be remembered that the average non-technical layman who constitutes the bulk of the nation is being sold on a big, brand-new idea—RADIO, and he has to be catered to and his requirements have to be met. If he is told that his requirements cannot be fully met and he must yield because some amateurs were here first, he will rightly

10. "Talking machine" was the name the Victor Company used for their record players.

11. Christopher Sterling and John Kittross credit John V. L. Hogan, an assistant to Lee de Forest, with the invention of "uni-tuning" receivers in 1927: radios that need only one knob to move between frequencies. Previously, tuning required multiple knobs that controlled several stages of reception, including radio frequency, the detector, and audio frequency amplification. Christopher H. Sterling, *Stay Tuned: A History of American Broadcasting* (Mahwah, N.J.: Lawrence Erlbaum Associates, 2002), 91. As Susan J. Douglas writes, "Tuning was a fine art, requiring endless patience and technical acuity as the listener adjusted four or five knobs to bring in stations. When these were adjusted improperly, he was jolted by earsplitting whistles and squeals. And through the headphones of the crystal set, the human voice sounded like a distant, otherworldly squeak or vibration." Douglas, *Listening In*, 71.

answer in no uncertain terms that he cannot embrace this new idea of radio and that the amateurs can have the field to themselves all of the time. Which means that the radio industry goes back to what it was two or more years ago, a picayune, little industry.

As a matter of single circuit receivers causing the trouble the amateur must remember that Tom Jones, the bookkeeper, the salesman, the clerk, the lawyer and what not, knows as much about technical radio as he does about music. He may appreciate both, but he knows little about either. The reason there is a talking machine in almost every home is that talking machines were not designed for skilled musicians, but were designed for average musically unskilled persons. And there will be a radio set in every home—which is the aim of the radio industry—when radio sets are designed in the same way. This means a minimum of controls on the receiver and to-day the single circuit tuner is about the only tuner that permits such easy control. And the radio industry has the single circuit tuner to thank for whatever popularity radio has achieved to-day.

If the amateur's transmission were of material benefit either to the public or the radio art there might be some ground for their stand on the question. But it serves no advantageous purpose. The mere act of relaying and seeing how far one's transmitter can reach, while the personal interest and pride of the individual amateur, has not any other importance. On the other hand broadcasting is gradually assuming the importance of a public utility, and as such is deserving of the first consideration.

The technical knowledge and ability possessed by the amateur is of tremendous potential value and it would be a great loss to the radio art should anything occur to make the amateur's interest in radio wane. However, from considerations enumerated above his chief radio occupation, *namely relaying, is of no value in itself and must give way to the broadcasting.* The only way out of the dilemma is the opening of *new spheres of activity for the technical amateur.*

From this point of view the writer has a solution which may sound very ambitious, but which has the merits of affording sufficient scope for the amateur's technical interest and affording a minimum of interference with the broadcasting.

Instead of an Amateur Radio Relay League it is suggested than an Amateur Radio Research League be formed, which may be a nation wide organization just as the former is.[12] The various existing amateur stations are potential miniature laboratories which by proper direction can be effectively utilized in securing data and information of great importance to the art. Thus there are a great many problems which are functions of time, locality, and climate; as for example, the

12. Unfortunately, it seemed no such organization was ever formed. In the first half of the twentieth century, Eric Hintz writes, "Independent inventors struggled to form durable professional groups. This rendered them politically impotent and unable to push through certain legislative reforms. Also, without a flagship organization to speak on behalf of the profession, independent inventors were at a disadvantage in their rhetorical battles with industrial researchers, who characterized them as unsophisticated and obsolete." Eric S. Hintz, "The Post-Heroic Generation: American Independent Inventors, 1900–1950," *Enterprise & Society* 12, no. 4 (2011): 732–48.

intensity and directional effect of static, if any; the problem of fading and signals, etc. Range data of telegraph and telephone transmitters are very meagre. By properly co-ordinating the efforts of the various stations throughout the country data of inestimable value may thus be obtained. Besides there are any number of problems which can be taken up and investigated with the facilities the amateurs possess. This would require, of course, very good organization, with central headquarters and districts divided under the supervision of the most capable amateurs. The ability and a considerable part of the facilities are available. What is lacking is the organization and exact plan of procedure. A nation wide organization for systematic relay transmission was efficiently accomplished and with the same talent an organization for this far more important purpose of research can also be accomplished. It is obvious that the purpose of avoiding a clash with the newcomers will be thereby effected and at the same time the ability of the amateur will be expended in a manner far more beneficial to the public and the art than heretofore.

The Radio Amateur

By Hugh H. Wingett

Third Prize

The radio amateur's trouble to-day is that he does not advertise himself in the right way. A few so-called amateurs or "hams" congest the ether by sending more or less meaningless messages, using a kilowatt of power, to a brother "ham" around the corner. The radio public which is trying to listen in on a radio program or musical concert, is of course drowned out and the blame is laid on the shoulders of the amateur, thereby giving the whole body of amateurs over the entire country a black eye, for the misdeed perpetrated by a few "hams" who don't know the meaning of tuning, 200 meters, and broadwave. There is no reason why this condition should exist. The well meaning, hard working amateur should take it upon himself to remonstrate with the owner of such a station or exterminate him, thereby doing away with the nuisance that is menacing their very existence.

This done, the amateur should try to organize the surrounding amateurs, and form a club, having operating rules and regulations. They should, where possible, hold concerts, showing and informing the people of the community or town just what they are doing for the betterment of radio conditions. They should when possible try to influence the amateurs in the neighboring community or town to organize, and thereby perpetuate the once respected order of amateurs.

This organization of amateurs should publish a daily bulletin, consisting of: the weather reports, boxing, baseball and football results, and many other topics of interest that are broadcast. Be of service to those of your community and they will begin to appreciate the amateur. The contents of this bulletin would of course depend largely upon the people of the different communities. This bulletin could be fastened upon a small, neat sign board, placed in front of the house of each amateur, where the neighbors in passing could read it. It would cost next to nothing and would bring in large returns. Not money perhaps, but reinstatement of the amateur in the esteem of the general public, which is a thing to be desired at present. If thought necessary, the amateur could devise some means of defraying the expense of this bulletin, by popular subscription for instance. The people of the community would be glad to aid in defraying the expenses for the service and pleasure rendered. There should be no trouble encountered in this.

Again, the club could issue a weekly or monthly bulletin, costing only a few cents, informing the public, as to the aims and achievements of the amateur in the art of radio. In this they should lay much stress upon the many advancements and inventions brought about by the amateur experimenter.

It would be well also if the clubs would have it known that they would be pleased to have visitors. Also that the radio enthusiast who can tune in a radio concert, but who knows nothing much about the theory and science of radio, may be present at the discussions brought up at the club if he is in earnest and wishes to learn. If he shows interest and intelligence, receive him into your club.

In every club or community there is a person or amateur who is well versed in the theory and rudiments of radio. Let this one act as instructor. The more one knows about the operation of a set, the less likely it is that he will create a disturbance that shall threaten the extermination of the amateur. Do not be afraid to correct mistakes and to offer suggestions to another amateur, just because he happens to be your senior. I do not care how old an amateur is; if he is a real "bug," he won't resent being shown his mistakes, even by one many years his junior.

By now the amateur should recognize his danger and try to organize and cooperate with others in order that he may avoid bringing down upon himself the wrath of the radio public. It will never do to have them turn against us, for they outnumber us and practically control our future, and our very existence. They are in the majority, and they will hardly tolerate anything which they think will deprive them of their amusement and pleasure. We had better squelch these "brass-pounders" who create such a disturbance, and who cause the ire of the pleasure seekers to be directed against us, or the "Ancient Order of Amateurs" is likely to be presented with some such motto as: "Say it with flowers," or "Peace be with you forever more," meaning that we will pass out of existence.

To my mind the remedy lies with the amateurs themselves and it is only through them that the existing conditions can be bettered.

I hope that the above mentioned suggestions will aid in pacifying the misunderstanding at present between we amateurs and the pleasure seeking public.

So fellows, let's band together and fight for our rights as one.

Predicting Future Inventions

Science and Invention, vol. 11, no. 4, August 1923

Every inventor must be a prophet.[1] If he were not, he could not think up inventions that will only exist in the future. For this reason, every inventor must ascend from fact to non-fact. What non-fact will turn out to be, not even the inventor knows beforehand. He prophesies to himself that he can make such and such an invention, all the while thinking about it, and letting his imagination work overtime. He keeps on turning the question or problem over and over in his mind until the subject finally crystallizes itself into a concrete form. All of this takes place in the inventor's mind. He is not working with concrete facts but he imagines and hopes that the particular device upon which he is laboring will turn out to be as he imagines it.

Cover art by Howard V. Brown.

1. This editorial appeared in the special "Scientific Fiction Number" of *Science and Invention,* which featured six short stories in addition to the normal features, departments, and readers' letters. The issue served as a blueprint for the launch of *Amazing Stories* in 1926. The cover image of an astronaut floating through space illustrates a story by the sixteen-year-old G. Peyton Wertenbaker, who in the words of Mike Ashley, "was Gernsback's first important writing discovery. . . . His story is emotionally strong and considers the fate of an explorer who travels through into the macrocosm only to discover he cannot return to Earth because, with time relative to mass, the Earth had grown old and died within minutes of his own subjective time." Michael Ashley, *The Time Machines: The Story of the Science-Fiction Pulp Magazines from the Beginning to 1950,* The History of the Science-Fiction Magazine, vol. 1 (Liverpool: Liverpool University Press, 2000), 47.

The issue also included the nineteenth of the forty-installment Doctor Hackensaw series by Clement Fezandié (1865–1959), who Gernsback later referred to as a "titan of science fiction." Hugo Gernsback, "Guest Editorial," *Amazing Stories* 35, no. 4 (April 1961): 5–7, 93. Like Luis Senarens (see **An American Jules Verne** in this book), Fezandié was another late nineteenth-century author of dime novel scientific tales whose name Gernsback attempted to elevate to the level of Verne, Wells, and Poe within the canon of scientifiction. Fezandié wrote new fiction for Gernsback up until 1926. Each of his Hackensaw stories consisted of a technical description of a new invention by the rogue scientist Doctor Hackensaw and its possibilities, both in terms of practice and profit. Mike Ashley writes on Gernsback's description of Fezandié as a "titan": "That is hard to grasp by today's definition of science fiction, but we have to remember that Gernsback was talking about his own definition. These stories more than any others in *Science and Invention* epitomized Gernsback's model for scientific fiction. They extrapolated from existing known science to suggest future inventions and what they might achieve; and all for the sole purpose of stimulating the ordinary person with a penchant for experimenting with gadgets, into creating the future." Ashley, *The Time Machines,* 34.

According to Sam Moskowitz, the reason for this special issue was "the backlog of science fiction stories piling up at *Science and Invention,* but it quickly persuaded *Argosy* and *Weird Tales* to alter their policies to include stories with a better grounding in science. The beginning of the end for the scientific romance which had been popularized by Edgar Rice Bur-

roughs was brought nearer; the pattern of modern science fiction was in the process of formation. Except for a freakish circumstance, Gernsback would have issued the first science fiction magazine in 1924. That year he sent out 25,000 circulars soliciting subscriptions for a new type of magazine, based on the stories of Verne, Wells, and Poe, to be titled *Scientifiction.* The subscription reaction was so cool that Gernsback did nothing further for another two years, at which time he placed *Amazing Stories,* fully developed, on the stands without a word of notice." Sam Moskowitz, *Explorers of the Infinite: The Shapers of Science Fiction* (New York: The World Publishing Company, 1963), 236.

If the inventor's imaginings were wrong, he is a poor inventor. If they are right, he is a good one.[2]

The art of inventing is to produce something that has not existed or has not been known on earth previously. Of necessity, therefore, it lies in the future. Sometimes an inventor may have a perfectly good idea of a certain machine, which he is convinced will work, if certain conditions were fulfilled. He starts working it out until he finds to his dismay that he cannot produce certain materials or certain articles which he knows are needed, but which have not as yet been developed. For instance; inventors over 150 years back, knew the automobile. Steam automobiles operated on the roads of England in the 18th century capable of running at a fair rate of speed and could carry from ten to fifteen people. Such automobiles failed because the automotive power had not as yet been developed perfectly. The missing link was the gasoline engine, which up to that time was not known. The inventor had had all this in his mind's eye and he was prophetic enough to realize that some day such vehicles would become commonplace, as indeed they are now. Jules Verne in his prophetic books, describes dozens of future inventions, nearly all of which have become realities. Indeed, there are not more than three or four of his imaginations left, and these no doubt will become true very shortly. Consider the submarine which was prophesied in its entirety by Jules Verne long before it made its appearance. He had laid the basis for the present day submarine, and lived to see the day when the first one was actually built and had operated as he had prophesied it would.

There are a certain class of people, and we hear continually from them, who condemn the policy of this magazine because we exploit the future. These good people never realize that there can be no progress without prediction. It is impossible to have in mind an invention without planning it beforehand, and no matter how fantastic and impossible the device may appear, there is no telling when it will attain reality in the future. To illustrate: in the August, 1918 issue of the *Electrical Experimenter,* the writer ran a story entitled: "The Magnetic Storm." This was during the war and was a purely fantastic idea: the suggestion was made to stop the war by burning out all electrical instruments throughout Germany. The idea was to have a tremendously large Tesla coil along the border, which would send a current into all electrical circuits through Germany, burning out armatures, automobile wiring, electric installations of airplanes, telegraph and telephone apparatus, etc. While theoretically possible, the idea was very fantastic. Cable dispatches during the middle part of June of the present year brought the news from Germany that the very thing had actually been accomplished by the powerful Nauen radio station. A number of automobiles were stopped at a distance by the energy sent out from this station.[3]

2. Invention as a form of prophecy became a favorite topic of Gernsback's *Science and Invention* editorials during this period. Distinguishing between the discovery and the invention of a new technical object in "The Mentality of Inventors," for instance, he writes: "An inventor is an individual who rarely makes any discoveries himself. Rather, he takes up something that some discoverer has worked on before, and then makes it practical, which is something which the discoverer never does. For instance, Heinrich Hertz, the inventor of wireless, who was the first to demonstrate what we now term 'radio waves,' was a discoverer. And while he knew, perhaps, more about radio than most of us do today, he did not find or try to find a practical use for it, until Marconi, the inventor, came along, and not only put Hertz's discovery into practical use, but commercialized it as well." Hugo Gernsback, "The Mentality of Inventors," *Science and Invention* 13, no. 7 (November 1925): 603.

Gernsback was also aware that this "mentality" of the inventor, namely the ability to sense future opportunities for innovation, was becoming something of a science among the various new wings of large corporations. "There certainly is nothing new about this, because it has been done for ages, but only in a haphazard manner. Everybody who had a little imagination could take a fling at predicting, and quite often the prediction came true. Your scientific investigator of today, however, tackles the problem in a thorough and often in a mathematical manner, totally different from the guessing predictions that were in vogue heretofore. The largest industrial corporations the world over have people on their staff whose business it is to predict certain inventions or discoveries having to do with their own business. And in nearly all cases, predictions thus made by staff scientists are realized usually within the space of time indicated by these modern prophets. . . . It is all well planned out mathematically, from data available at present." Hugo Gernsback, "Predicting Inventions," *Science and Invention* 13, no. 2 (June 1925): 113.

3. In a piece on two British aviators who disappeared "upon an ordinary reconnaissance over a desert in Mesopotamia" in 1924, the famous writer on occult and paranormal phenomena Charles Fort (1874–1932) attempted to follow up reports that German engineers had been

Then again in this magazine we have for the last ten years exploited television, the faculty of seeing at a distance. We have shown all sorts of television schemes, all of which seemed to belong to the distant future. We have on file a great many letters from critics denouncing us for printing such "foolishness," as they call it, because they said it would ever be impossible to invent a machine, by which a man could see at a distance. During the latter part of June, Mr. Jenkins of Washington, publicly demonstrated before Army and Navy officials a machine, whereby it is possible not only to see at a distance but to project a film on a screen in New York and broadcast it all over the country by radio the same as voice and music is broadcast by radio now.[4]

These are just a few examples among many.

And so it goes. What seems impossible and even ridiculous today becomes an actuality tomorrow. Throughout the ages, the man who looked into the future was usually considered a crank or insane. He is in the same population today. Human nature is such that it opposes changes, particularly if such changes are violent. Anything that tends to pull us out from our daily rut is not welcome, because it means an effort.

When some of our greatest scientific authorities, as late as twenty years ago, proved by mathematics that it was impossible to sustain in the air a machine such as an airplane; when the news of the X-ray was greeted with derision; when the sending of messages by radio was not believed by the populace, when it had already been used for years—it behooves the average man to be extremely cautious in denouncing any idea just because it is new and appears impossible on the face of it.

experimenting with "secret rays" capable of arresting the movement of large objects. "It was said that for some time, at the German wireless station at Nauen, there had been experiments upon directional wireless, with the object of sending out rays, concentrated along a certain path, as the beams of a search-light are directed. The authorities at Nauen denied that they had knowledge of anything that could have affected the French aeroplanes, in ways reported, or supposed. Automobiles can be stopped, by wireless control, if they be provided with special magnetos. Otherwise not. Sir Oliver Lodge was quoted, by the *Daily Mail*, as saying that he knew of no rays that could stop a motor, unless specially equipped. Professor A. M. Low's opinion was that he knew of laboratory experiments in which, over a distance of two feet, rays of sufficient power to melt a small coil of wire had been transmitted. But as to the reported 'accidents' in Ger-

many, Prof. Low said: 'There is a wide difference between transmitting such a power over a distance of a foot or two, and a distance of one or two thousand yards.'" Charles Fort, *The Complete Books of Charles Fort* (New York: Dover Publications, 1974), 955–56.

But the idea Gernsback put forward in **The Magnetic Storm** had definite legs, with stories abounding over the next decade on German experiments with secret rays. For instance, the *New York Times* reported in 1930 that Reichswehr officials denied that what "has been causing automobiles to stall mysteriously on a Saxon road near Czechoslovakia" had anything to do with "a secret ray cast across the unsuspecting countryside during government experiments. . . . And even if they were, it was added, they would be conducted in some isolated region and not on a public highway near the Czechoslovakian frontier." "DENY USING INVISIBLE RAY: Reich

Army Officials Comment on Mysterious Auto Trouble," *New York Times*, October 25, 1930.

Though he was convinced the phenomenon existed, Fort disagreed with Gernsback's hope that any such ray could serve as a deterrent to war: "I am unable to conceive that a power to pick planes out of the sky would be so terrible as to stop war, because up comes the notion that counter-operations would pick the pickers. If we could have new abominations, so unmistakably abominable as to hush the lubricators, who plan murder to stop slaughter—but that is only dreamery, here in our existence of the hyphen, which is the symbol of hypocrisy" (957).

At the very edges of what Gernsback would consider to be true scientifiction, Fort's influence on science fiction writers was "incalculable," according to Damon Knight, the SF author, editor, and historian, who wrote an autobiography of Fort with a preface by R. Buckminster Fuller. "He had twenty-five thousand notes, in pigeonholes that covered a wall; they were not what he wanted, and he destroyed them. He accumulated forty thousand more. Eventually he began to see an unsuspected pattern in them. . . . In four books, *The Book of the Damned, New Lands, Lo!,* and *Wild Talents,* he assembled more than twelve hundred documented reports of happenings which orthodox science could not explain." Damon Knight, *Charles Fort: Prophet of the Unexplained* (London: Littlehampton Book Services Ltd., 1971), 2–3.

Although Gernsback attempted to distance his publications in the 1920s from Fortean obsessions with mysticism and the paranormal, writing several editorials condemning "spiritism" in *Science and Invention,* he did file a "dummy magazine" with the U.S. Patent Office in 1934 with the title *True Supernatural Stories,* containing fiction by Clark Ashton Smith and H. P. Lovecraft. Sam Moskowitz, "The Gernsback Magazines No One Knows," *Riverside Quarterly* 4 (March 1971): 272–74.

4. Charles Francis Jenkins publicly demonstrated his prototype of a mechanical scanning television system with synchronized sound and image on June 13, 1925. The terms he used to describe the image thus transmitted included "radio movies," "radiovision," and "shadowgraph." For more, see Donald Godfrey, *C. Francis Jenkins, Pioneer of Film and Television* (Champaign: University of Illinois Press, 2014).

The "New" Science and Invention

Science and Invention, vol. 11, no. 6, October 1923

This issue of *Science and Invention* witnesses an important and far reaching innovation, not only in the magazine field, but in the technical press as well. For some time we have felt that *Science and Invention* was not fulfilling its mission as it should. When this magazine was first established in 1913, as the *Electrical Experimenter,* it was a highly technical magazine and its circulation reached a figure of not more than 100,000. The reason for this was that there are only a limited number of technical-scientific readers in the country who can support such a magazine. Consequently in 1920 the magazine broadened its scope and changed its name to *Science and Invention.* Sometime previous to the change of the name, the

October 1923 cover
of *Science and Invention.*

magazine had not been strictly electrical nor strictly experimental. It had been found necessary to embrace other scientific lines in order to satisfy the demands of its readers. Then when the magazine became known as *Science and Invention,* the circulation quickly jumped up close to 200,000 copies in 1921. The reason, of course, was that more readers were interested in the sort of matter we published at that time and the material we have been publishing ever since.

But recently we have been aware that we could give our readers more for their money. Along the old and established line of publishing, we printed illustrations and lengthy articles. Often it was necessary to reduce illustrations to such an extent that they did not show up very well. This was necessitated by the fact that we had to run so much text.

A diligent canvas among many readers of *Science and Invention* tended to show that few people actually read the articles. Most of them studied the illustrations and read the captions. We point with pride to our illustrations and particularly to the technical ones. We have been fortunate to secure artists who know how to illustrate difficult subjects in such a way that words really become superfluous.

Moreover, in these days, people do not wish to read if they can help it. A picture or illustration correctly made can tell a story more eloquently than 5,000 words. That is one of the reasons why motion pictures have been so eminently successful; pictures, a few titles, and the story is told far better than any book could do it.

Bearing these things in mind, we have come to the conclusion that the world really needs an entirely new sort of magazine. The answer is simple. *Nothing but pictures, with captions, to explain the story.* The result is the new *Science and Invention* now in your hands. The editorial policy remains exactly the same as before. No changes have been made as you will notice. *Instead, we are offering three to four times as much material as we have ever offered before.*[1] All of our departments, as you will note, have been reduced to pictures and captions, with the exception of our scientific fiction stories which for the time being will run on as usual. The reason is that the average man or woman does not wish to laboriously wander through miles of text of scientific facts, which means concentration and study. With fiction, however, this does not hold true because fiction is an entertainment; it does not require study. On the other hand, *we have tried to reduce all scientific matter to entertainment instead of study.* As to how successful we will be in this issue and subsequent ones, you are the sole judge. At any rate, we believe there is a very large public interest in scientific matters and if such matter can be presented in an easily digestible form, it will, we hope, be welcomed by the multitude.

1. The length of the magazine remained at just over a hundred pages per issue. Gernsback is likely referring here to the quantity of illustrations and the number of topics the redesigned *Science and Invention* covered.

Can We Visit the Planets?

Some of the Problems of Such a Journey.

BY DON HOME.

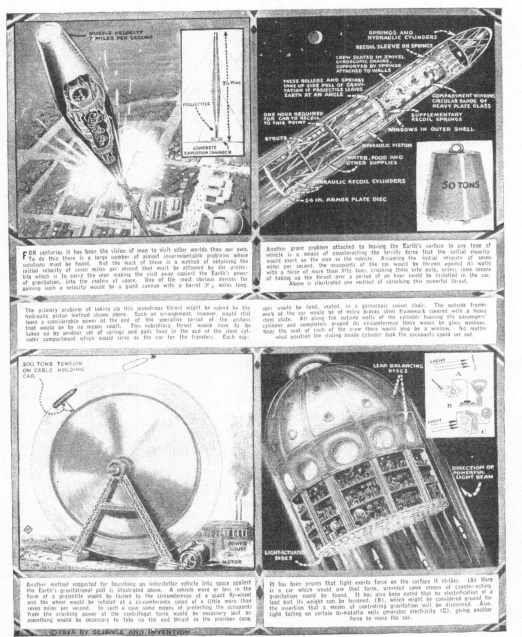

FOR centuries it has been the vision of man to visit other worlds than our own. To do this there is a large number of almost insurmountable problems whose solutions must be found. Not the least of these is a method of obtaining the initial velocity of seven miles per second that must be attained by the projectile which is to carry the man making the visit away against the Earth's power of gravitation, into the realms of space. One of the most obvious devices for gaining such a velocity would be a giant cannon with a barrel 3½ miles long.

Another grave problem attached to leaving the Earth's surface in any type of vehicle is a means of counteracting the terrific force that the initial velocity would exert on the men in the vehicle. Assuming the initial velocity of seven miles per second, the occupants of the car would be thrown against its walls with a force of more than fifty tons, crushing them into pulp, unless some means of taking up the thrust over a period of an hour could be installed in the car.
Above is illustrated one method of absorbing this powerful thrust.

The primary problem of taking up this monstrous thrust might be solved by the hydraulic piston method shown above. Such an arrangement, however, would still leave a considerable power at the end of the operative period of the pistons that would be by no means small. This subsidiary thrust would have to be taken up by another set of springs and pads fixed in the end of the steel cylinder compartment which would serve as the car for the travelers. Each voyager would be fixed, seated, in a gyroscopic swivel chair. The outside framework of the car would be of extra braced steel framework covered with a heavy steel plate. All along the outside walls of the cylinder housing the passengers' cylinder and completely around its circumference there would be glass windows. Near the seat of each of the crew there would also be a window. No matter what position the sliding inside cylinder took the occupants could see out.

Another method suggested for launching an interstellar vehicle into space against the Earth's gravitational pull is illustrated above. A vehicle more or less in the form of a projectile would be lashed to the circumference of a giant fly-wheel and the wheel would be rotated at a circumference speed of a little more than seven miles per second. In such a case some means of protecting the occupants from the crushing power of the centrifugal force would be necessary just as something would be necessary to take up the end thrust in the previous case.

It has been proven that light exerts force on the surface it strikes. (A) Here is a car which would use that force, provided some means of counter-acting gravitation could be found. It has also been noted that by electrification of a lead ball its weight can be lessened, (B), which might be considered ground for the assertion that a means of controlling gravitation will be discovered. Also, light falling on certain bi-metallic cells generates electricity (C), giving another force to move the car.

(Continued on page 1050)

A spread from the redesigned
Science and Invention.

Radio Town Crier

By H. Gernsback.

THE accompanying illustration shows a gigantic radio loud-speaker which probably will be known in the future as the Municipal Announcer. Experiments made for the past few years have shown that it is possible to so amplify the human voice, by means of huge horns and vacuum tubes, that it will be heard over a radius of from three to five miles so that each and every word will be understood clearly. In our Electrical Experimenter of November, 1919 issue, there was illustrated how huge horns placed in a church tower projected sound four to five miles.

The idea of the Municipal Loud Speaker is as follows: Huge sky-scrapers, as shown in the illustration, would be equipped with a number of concrete or non-vibrating metallic horns, pointing downward as shown. In this position, the sounds will be dispersed towards the street and buildings, and will also prevent rain and snow entering the horn itself. In cities like New York and Chicago, such horns would be erected every one or two miles. It goes without saying that present buildings can be equipped with such horns, but in the future probably a new sort of architecture will be evolved to take care of them.

The purpose of these Municipal Announcers will be simple. Any news, either of civic or national importance, can be broadcast so that the whole city can hear every word. By opening your office or home window, it will be possible to listen to every word spoken. Thus a Presidential speech, a talk by the Mayor of the city, or any other important feature can be **instantly** transmitted to the entire populace.

For police work, such a device will be of tremendous importance. Suppose a robbery or a murder is committed; the police headquarters can immediately broadcast this news so that every citizen in the neighborhood will be on the look-out for a certain car and will have a description of the law-breaker. The important part is that the information will be instantaneous, which in the detection of crime, is of utmost importance.

If we make it possible for the man in the street to absorb scientific knowledge without headaches and without trepidation, we believe we shall have accomplished a distinct service to humanity. To the average man, Science means a musty book, sealed with 7 seals. The average man will have none of it. If on the other hand we can show your mother, your wife, or your children that Science is a most interesting subject of which every one in all walks of life should know more, then we believe that we shall have achieved our goal.

Science in all its phases, is the most wonderful thing that the world has ever known and it is just beginning to come into its own. Not so very many years ago, men were burned at stakes and murdered for inventing or thinking up something new. Today the scientist is glorified. Science touches every one of us in one way or another, and is regulating and running our very lives. Science and invention are the direct causes of our huge fortunes, our great industries and the greater part of our national wealth. It will be our purpose to disseminate such scientific knowledge, not only in a palatable manner, but in a way that "he who runs may read."

In conclusion, may we ask a personal favor of you? Please write and tell us frankly how this first issue of the new *Science and Invention* strikes you.

Are We Intelligent?

Science and Invention, vol. 11, no. 7, November 1923

The members of the human race, who think themselves preeminent on this planet, have long since come to the conclusion that the human animal is a superior living being, quite distinct and far above all other living creatures. This conclusion, so general that it seems to be inborn, has been handed down through the ages, mainly for the reason that we imagine all of our acts and thinking are based upon intelligence. For instance, we say that a dog or a cat cannot think and reason. They are, therefore, not endowed with intelligence. Or perhaps to put it in a better way, we think that an animal or insect cannot reason the way we do. Consequently, in our arrogance, we maintain that they are inferior to us.

We have subdued nearly all of the larger animals to our will and have either turned them into domestic servants or else have nearly exterminated them.

But are these proofs that we are intelligent? Intelligence among other things should call for a thorough understanding of every subject. Every human being, however, will admit that we comprehend practically nothing. We cannot fathom the simplest acts of the animals that we class beneath us. Our mentality is such that we cannot even interpret the simplest "thoughts" or instincts of a dog. But when we come to such creatures as insects, which have existed on this planet hundreds and thousands of years longer than we have, we are forced to admit the fact that certain species are far more intelligent than we are.

Take the ants for instance! With them the welfare of the nation is preeminent. Everything is subordinated to this thought; call it instinct if you wish. The ant does not require newspapers to get the news, as all news is transmitted instantaneously by a sort of "radio" system that reaches every ant in a fraction of a second.[1] Similar cases prevail with bees, as well as many other highly civilized insects.

Only because our senses are as poor as they are, do we find it necessary to use such artifices, as the printed page, railroads, the telephone, the telegraph, and nearly every other artifice that you can think of. You can imagine, perhaps, a million years hence, a central "radio" station broadcasting the news of the day, not in the spoken word, but in thought waves, so that everyone on the planet will get the news instantaneously. This does not mean intelligence only, but pictures and everything. In other words, intelligence will be transmitted

1. Jussi Parikka writes on the long tradition of insects serving as metaphorical models for "modes of aesthetic, political, economic, and technological thought." *Insect Media: An Archaeology of Animals and Technology* (Minneapolis: University of Minnesota Press, 2010), xiii. For more on how metaphors influence the design and reception of new media, see Georg Tholen, "Media Metaphorology: Irritations in the Epistemic Field of Media Studies," *South Atlantic Quarterly* 101, no. 3 (Summer 2002): 659–72.

by thought waves which bring into our minds the exact news or information transmitted. Naturally, if we ever reach this stage, the telephone and telegraph will be unnecessary because we will be able to transmit our thoughts direct to our friends in a less cumbersome way than we do today.[2] We need our railroads and transportation systems in the present era, simply because we have not learned to live as does the ant, for instance. We still must roam the planet in order to find the food and clothing we need. We may be sure that 100,000 years hence, such a situation will not prevail. We will be able to convert everything on the spot, for the simple reason that a piece of gold is exactly the same as a brick, and that a drop of water is the same as a piece of granite. Science knows that all kinds of matter are alike and that they appear differently only because their electrons are grouped in a different way. It will even be possible for us to make our own food without first planting the seed, which grows into the plant, which is eaten by the animal, so that we in turn may eat the animal. Synthetic food made from rocks found at our doors 100,000 years hence will be far more palatable, far more nutritious, and less poisonous, than anything we eat today.

The next few centuries will bring about a great era of simplification. Everything we are doing now is too cumbersome. Everything will be freed from complexity. Our lives are entirely too crowded. Where we used to have countless wires and cables, slowly these are making way for radio, where no wires are required. Railroads will be discontinued when aerial navigation comes into general use. Reading and studying of books is already on the wane due to the greater educational and entertaining force of the motion picture. The printed page is being supplanted by the picture everywhere.

In some quarters, it is thought that this is a sign of retrogression of the race. Nothing could be more erroneous. The scientific explanation is very simple. Our lives are crowded to such an extent, that it is impossible to read as much as our grandfathers could. We are constantly being speeded up mentally and if a picture can tell the story in two seconds why should one read a typed story which might occupy 10 or 15 minutes of one's time.[3] This, by the way, is the answer to the unprecedented popularity of the New *Science and Invention* magazine.

2. This vision of thought transmission is not far from the one put forward today by critics and pundits alike who worry that mobile media's blurring of the boundaries between "online" and "real" life will lead to a literal dissolution of interpersonal space. See especially Nicholas Carr, *The Shallows: What the Internet Is Doing to Our Brains* (New York: Norton, 2010). For a refutation of the theory that mobile media entail a cognitive shift or drain, see Betsy Sparrow et al., "Google Effects on Memory: Cognitive Consequences of Having Information at Our Fingertips," *Science* 333, no. 6043 (August 5, 2011): 776–78.

3. The material simplification Gernsback cites in the previous paragraph (from countless wires and cables to complete wirelessness) takes on a cognitive bearing here, where simplification is an evolutionary response to information overload. This passage evokes Alfred North Whitehead's discussion of symbolism and complexity: "Civilization advances by extending the number of important operations which we can perform without thinking of them." These are operations which today we might refer to as "high-level" after those computer languages whose simple, human-readable elements entail a high degree of complexity when compiled by the computer. Alfred North Whitehead, *An Introduction to Mathematics* (London: Williams and Northgate, 1911), 61.

A Radio-Controlled Television Plane

The Experimenter, vol. 1, no. 4, November 1924

*Tomorrow we shall find a new order of things if a war should occur.
Pilot-less Radio-controlled planes fitted with "Television" eyes
will flash back what they see to headquarters.*[1]

On a recent trip to Washington the writer visited the laboratories of C. Francis Jenkins, the well-known experimenter of international reputation. It was Mr. Jenkins who perfected the shutter that made our present-day motion pictures possible. He was paid over $1,000,000 for this invention.[2]

Of late he has been experimenting with television and has already obtained astonishing results. At the time of the writer's visit Mr. Jenkins demonstrated his television machine before a number of Government representatives, including the Chief of the Signal Corps. At that time the writer actually saw his own waving hand, projected by radio over a distance of some thirty feet, the shadow of the waving hand being transmitted to a screen at that distance. Every motion made by the writer's hand was faithfully reproduced on the distant screen. Opaque substances such as a cross, knife, pencil, etc., were also successfully transmitted and projected by the Jenkins Television machine.

It is the writer's opinion that, within two or three years, it will be possible for a man in New York to listen over his radio to a ball game 500 miles away and see the players on a screen before him at the same time. Whether it will be the Jenkins machine or some other machine that will achieve this result is of little consequence. The main thing is that experimenters all over the world are working frantically on television and sooner or later the problem will be solved.[3]

An entirely new age will then be opened up and it is not necessary for the writer to expatiate at length on this phase; as it has been exploited by him in his past writings and by others for some time.

In this article, we shall concern ourselves with the radio-controlled television plane, which will come into being immediately the minute the television problem is put on a practical basis. It should not be construed that the radio television plane is merely a monstrous war machine, but it also has its uses during peace time, as will be

1. This article was republished unchanged in the March-April 1931 issue of *Television News.*

2. For more on Charles Francis Jenkins, see **Predicting Future Inventions** in this book.

3. Two years earlier, an article in *Science and Invention* reported on experiments with telegraphic broadcasts of the 1922 World Series between the New York Giants and the New York Yankees. An observer at the Polo Grounds telegraphed the game's events live to a group operating a number of light projectors that would throw the image of players' movements onto a transparent screen. Earlier incarnations of this setup involved massive baseball diamond boards on display outside of newspaper offices, where crowds would gather to watch light bulbs representing each baserunner, defensive players' positions, and the count. Live radio broadcasts were only a few years away. "Baseball 'Movie' Scoreboard," *Science and Invention* 9, no. 9 (January 1922): 805.

For more on the history of baseball as one of many prime movers in the development of broadcast technologies, see Jules Tygiel, "New Ways of Knowing: Baseball in the 1920s," in *Past Time: Baseball as History* (Oxford: Oxford University Press, 2001).

explained. At the present time it costs great effort, time, and aviators' lives in order to train our perfect flyers.

A radio-controlled airplane has already been demonstrated by the French and American Governments, and it flew for a lengthy period without anyone on board. The entire control was from the ground while the machine was aloft. The plane arose, cut figure eights, volplaned, ascended, descended and went through all the ordinary evolutions; the control being effected entirely and solely by radio. The same kind of a machine is also being experimented with successfully by our own and several other Governments, and it may be said therefore that the radio-controlled airplane has passed the experimental stages and has become practical and feasible for military use.[4]

But the great trouble with radio-controlled airplanes is that the operator must see the plane. If his machine were to make a landing at a great distance he might land the airplane on top of a building or in a river, or it might collide with a mountain.

The "World Series" Baseball Games Were Reproduced, Play by Play at a New York Armory, by Means of This Ingenious Lamp Board and Transparent Screen. The Plays Were Received by Telegraph. The Movements of the Players as Well as the Ball Itself Were All Faithfully Duplicated.

© 1921 by Science and Invention

From the January 1922 *Science and Invention*.

FIGURE 1. The Pilot-less radio television plane, directed by radio; the plane's "eyes" radio back what they see.

4. By the 1940s, experiments had begun in the United States with full-size, remote-controlled aircraft. Joseph P. Kennedy Jr., the oldest brother of President John F. Kennedy, was killed in World War II in one early "drone" program. Operation Aphrodite called for a pilot to fly a B-17 Flying Fortress packed with explosives close to a bombing site and parachute away from the craft before it was remotely piloted into the target using radio control and two television cameras mounted on the dashboard. In the

A Pilot-less Plane Which "Sees"

Imagine now a radio-controlled pilot-less airplane which is also equipped with electrical eyes, which eyes transmit the impulses—or rather what these eyes "see," by radio—to the distant-control operator on the ground. Our illustration on the opposite page, which shows a war machine, depicts this phase. Here we have a radio-controlled airplane equipped with a number of lenses which gather in the light from six different directions, namely, north, south, east, west, up, and down. The impulses are sent to the operator on the ground, who has in front of him six television screens labeled "North," "South." "East," "West," "Up," and "Down." Each screen corresponds to one of the electric eyes attached firmly to the body of the airplane, as shown in the illustration.[5]

Let us now see what happens. The airplane is started from the ground and is sent over the enemy territory. During every second of its fight the control operator, although 50, 100, or possibly 500 miles away, will see exactly what goes on around the plane, just the same as if he himself were seated in the cockpit; with the further advantage that, sitting before a screen, he can scan six directions all at once, which no human aviator can do. If, for instance, an enemy airplane suddenly comes out of a cloud and starts dropping bombs on our machine below, the control operator sees this enemy machine quicker 500 miles away, than if an aviator sat in the cockpit one-quarter of a mile away from or below the enemy bomber. The control operator will send a radio signal that will immediately discharge a smoke screen from his radio television plane, hiding his craft in smoke. He can also make it turn about if such an operation should be necessary, or he can increase its speed if it is desired to escape.

If he outdistances, or otherwise eludes the enemy, the radio-controlled television airplane can then be directed to the spot where it is supposed to drop its bombs. Moreover, the distant-control operator can see exactly when his machine arrives over a given spot. A sighting arrangement can be attached to the plane in such a manner that, when the object to be bombed comes over the cross-wires in the range-finder, the bomb or bombs are dropped at the exact moment. Suppose that the enemy becomes too strong and that a great number of machines attack the radio-controlled plane and that there is no escape from the enemy. In that case the control operator will simply set the radio television plane on fire, bringing it down in flames! Thus it would be useless to the enemy and no lives will have been risked or

case of Kennedy's plane, the explosives detonated prematurely. Ed Grabianowski, "The Secret Drone Mission That Killed Joseph Kennedy Jr.," *io9*, February 21, 2013, http://io9.com/5985733/the-secret-drone-mission-that-killed-joseph-kennedy-jr.

5. Gernsback's use of "electric eyes" isn't just a metaphor. In other contexts, he describes the physiology of the eye as a literal "television apparatus": "The animal eye is the most marvelous television apparatus ever invented. Moreover, it is non-electrical. If we look at an object, the latter is thrown into our eye, which is nothing but a marvelously efficient camera, but instead of a photographic plate, the impulses are thrown up on the Retina which records the object, not only in black and white as does the photographic plate, but the picture is recorded in its natural colors on the retina. From here numerous fine nerve strings interlocked in the retina connect with the optical nerve, which nerve in turn connects with the occipital lobes of the brain, translating the various light impulses, (stimuli) with their component colors into a 'picture,' which is then 'seen' in our mind. We say 'seen' advisedly, because of course the picture is not actually seen in the mind, but the impulses which the retina has picked up are translated into another form, which we experience in turn as the sensation of seeing." Hugo Gernsback, "Television and the Telephot," *Electrical Experimenter* 6, no. 1 (May 1918).

taken—it being cheaper to destroy a machine than the valuable life of a highly trained pilot.

In the future such radio-controlled television planes may be used not only singly but in squadrons as well. They can be used for attacking the enemy if necessary. They can be used in pursuit of the enemy, for taking aerial photographs, and for any other military or peace-time operation, just the same as a present-day plane piloted by an aviator. Suppose the enemy has the same kind of machines, which, of course, he will have. It then becomes a matter of "playing chess," the same as if the machines contained live aviators. The battle, of course, would not be bloody, but practically the same results will be achieved as far as the military maneuver is concerned.

For peace-time purposes it goes without saying that the advantages of such a mechanical and "almost human" airplane are unlimited. It will be possible in the future to send mail planes from one end of the country to the other without a human being on board and such planes will be just as safe letter-carriers, as if they were manned by human beings. Every second of the flight would be watched by a Post Office Department operator and the plane would, of course, be able to defend itself against attack. It could readily be equipped with electrically-operated guns if such should be necessary or desirable. Particularly for transporting mail and the like, the radio-controlled television plane will be invaluable.

There are, of course, hundreds of other applications of the idea which readily suggest themselves to anyone. The writer is certain that such planes will be in existence during the next ten years.

The Dark Age of Science

Science and Invention, vol. 12, no. 9, January 1925

Too many people nowadays are prone to settle themselves back in their chairs and congratulate themselves about the wonderful age in which we are living. With railroads, airships, radio, and other "marvelous" activities abounding around us, it is easy to lull oneself into the belief that surely the millennium has arrived and that the world will never look rosier than it does now.

As a matter of fact, 300 years hence this present age of ours will probably be termed the "dark age of science." This is not merely a catch phrase used by the writer, but he means it in all sincerity, and what is more, he can easily prove it.

Nowadays it makes us laugh when we think of how people in England and other parts of the world some five hundred years ago [lived] on top of coal mines and froze to death. They simply did not know that coal would burn and that it could be made to heat houses and otherwise perform tremendous work. While we laugh, you should soberly consider that we are doing the self same thing ourselves today. All about us there is untold quantities of energy, much cheaper, much better, more sanitary than coal, but we simply have not learned how to use it because we do not know. Every piece of rock, every car full of sand has a potential dormant power locked up within it. We may call this power atomic or by any other name, but the fact remains that we do not use it because we do not know how.

Take a great city like New York, which burns thousands of tons of coal whose products pollute the atmosphere every day, while right at its very feet two mighty rivers, the Hudson and the East Rivers flow by its shores, which rivers every day can actually furnish more power than all the coal burned in a year. Still, this mighty power goes to waste in this dark age of ours.

When you go to the butcher and order a 10-pound roast, you would become highly indignant if he handed you over 9¾ pounds of bones and less than ¼ pound of meat, but when you pay your electric light bill at the end of the month and you send the lighting company your check for $10.00 you do this very thing. The reason is that exactly 98% of the electric power goes up in useless heat, which you do not need and which you do not want but which you must pay for. You actually get 2% of light, and in order to get this 2% you have to pay 98% for something that is a total loss. Of course all this will make people laugh merrily 300 years hence, and they will not be able to understand how we could afford such frightful losses.

Then we go and invent the automobile, another monstrosity of the "dark age of science." In order to propel the same, we generate carbon-monoxide that pollutes the air in our streets, gives us headaches, and otherwise makes life unbearable for us. But this is far from being the worst. Here we go and create the automobile and then build so many of them that they become useless by their very numbers. Instead of transporting us quicker than the old horse-drawn vehicle, we actually find that the latter was much faster, in many instances. If you try, in any of our big city streets, to go about quickly, you will find that there is only one way of doing it and that is you must come right back to prehistoric times and *walk*. In cities like New York and Chicago, you can cover ground much quicker for reasonable distances on foot than by automobile.

Far from living in the millennium, we are living in an age of unspeakable waste. There is hardly anything that we can think of that is not wasted.

We know today the power of waterfalls, the inherent power of the tides, the inherent power of moving rivers, yet—99% of this goes entirely to waste. Wind power, another large source of energy, is hardly touched at all. This power alone is so vast that a small fraction of it, properly applied, would supply the world with sufficient energy to run all the machinery, all trains, and all of our vehicles. Water power and wind power are well understood and can be exploited by us even today, but no real effort in this direction is to be discerned.

We do not wish even to speak of the power derived from the sun's heat, because we have as yet not found the key to unlock this tremendous energy, which is far, far greater than all the others combined.

At the present time we simply use sun power that has been stored up by nature millions of years ago. This is the case with coal, gasoline, and practically every other fuel. All other forms of energy that are lying about us in every direction are not even touched. It is a comforting thought that our great-great-grandchildren will stand on their own legs, for they will know how to unlock a power universe, invisible to us.

The Isolator

Science and Invention, vol. 13, no. 3, July 1925

Perhaps the most difficult thing that a human being is called upon to face is long, concentrated thinking. Whether you are a lawyer, trying to formulate or memorize the pleading of a special case, whether you are an inventor with an intricate problem to be solved, whether you are a playwright trying to hatch out a knotty plot—assiduous concentration on the subject becomes necessary.

Most people who desire thus to concentrate find it necessary to shut themselves up in an almost soundproof room in order to go ahead with their work, but even here there are many things that distract their attention.

Suppose you are sitting in your study or your work room, ready for the task. Even if the window is shut, street noises filter through, and distract your attention. Some one slams a door in the house, and at once your trend of thought is disturbed.

A telephone bell or a door bell rings somewhere, which is sufficient, in nearly all cases, to stop the flow of thought.[1]

But even if supreme quiet reigns, you are your own disturber practically fifty per cent. of the time. You will lean back in your chair and begin to study the pattern of the wallpaper, or you will see a fly crawl along the wall, or a window curtain will be moving back and forth, all of which is often sufficient to turn your mind away from the immediate task to be performed.

The writer repeats that the greatest difficulty that the human mind has to contend with is lack of concentration, mainly due to outside influences.

If, by one stroke, we can do away with these influences, we will not only be benefitted greatly thereby, but our work would be accomplished more quickly and the results would be vastly better.

The writer, who has to perform, almost daily, in connection with his editorial duties, many tasks that involve considerable concentration, has found out that it is almost impossible to keep his mind on a subject for five minutes without disturbance. For that reason, he constructed the helmet shown in the accompanying illustrations, the purpose of which is to do away with all possible interferences that prey on the mind.

The problem was first to do away with the outside noise. The first helmet constructed as per illustration was made of wood, lined with

1. In his magisterial history of autism and its cultural meanings, Steve Silberman connects what Gernsback describes as the particular "radio mind" possessed by amateur radio experimenters to "the curious fascination that many autistic people have for quantifiable data, highly organized systems, and complex machines [that] runs like a half-hidden thread through the fabric of autism research." Moreover, the highly scripted and ritualized conversations conducted via wireless and later ham radio offered "ways of gaining social recognition outside traditional channels. . . . Hams who struggled with spoken language could avoid talking altogether by communicating in code." Steve Silberman, *NeuroTribes: The Legacy of Autism and the Future of Neurodiversity* (New York: Avery, 2015), 240.

According to Silberman, Gernsback himself may have been "an undiagnosed Aspergian": "His peers regarded him as an unsociable figure who remained coolly distant from the communities he created. The people he counted as friends tended to be prominent scientists, influential politicians, and other notable figures with whom he corresponded by mail; historian James Gunn observed in *Alternate Worlds* that he was 'a strange mixture of personal reserve and aggressive salesmanship'" (241).

Silberman refers to the Isolator in particular as Gernsback's "most blatantly autistic creation" (243).

cork inside and out, and finally covered with felt. There were three pieces of glass inserted for the eyes. In front of the mouth there is a baffle, which allows breathing but keeps out the sound. The first construction was fairly successful, and while it did not shut out all the noises, it reached an efficiency of about 75 per cent. The reason was that solid wood was used.

In a subsequent helmet under construction, an air space is included, as per our line illustration, no wood entering into the construction at all. This feature should give almost 90 per cent. to 95 per cent. efficiency, thereby excluding practically all sounds.

It will be noted that the glass windows directly in front of the eyes are black. The construction involved the use of ordinary window glass, the outer glass being painted entirely black. Two small white lines were scratched into the paint, as shown. The idea of this is as follows:

The writer thought that shutting out the noises was not sufficient. The eye would still wander around, thereby distracting attention. By having the two white lines scratched on the glass, the field through which the eye can move is comparatively small. In illustration No. 1, it will be seen that it is almost impossible to see anything except a sheet of paper in front of the wearer. There is, therefore, no optical distraction here.

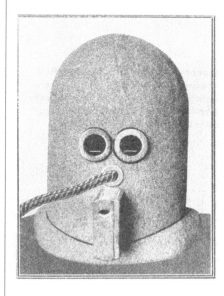

The above photograph shows a close-up view of the Isolator helmet. The oxygen supply enters the helmet via the tube shown.

It was also found that if the helmet was used alone for more than fifteen minutes at a time, the wearer would become more or less drowsy. This is not conductive to hard thinking, and for that reason the writer introduced a small oxygen tank, attached to the helmet. This increases the respiration and livens the subject considerably.

With this arrangement it is found that an important task can be completed in short order and the construction of the Isolator will be found to be a great investment.

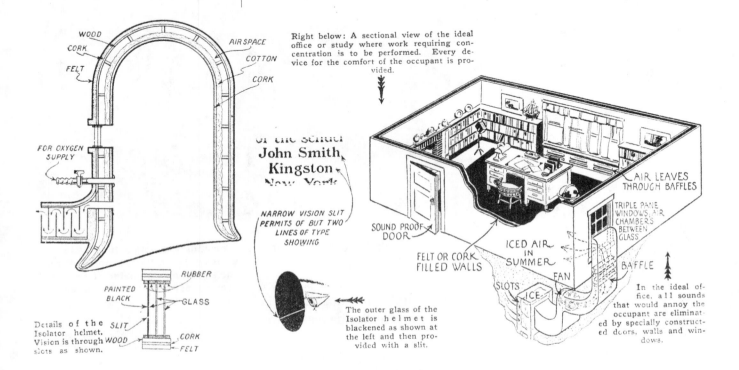

WOOD
CORK
FELT

AIR SPACE
COTTON
CORK

FOR OXYGEN SUPPLY

RUBBER
PAINTED BLACK
GLASS
CORK
FELT

Details of the SLIT Isolator helmet. Vision is through WOOD slots as shown.

Right below: A sectional view of the ideal office or study where work requiring concentration is to be performed. Every device for the comfort of the occupant is provided.

of the sender
John Smith,
Kingston,
New York

NARROW VISION SLIT PERMITS OF BUT TWO LINES OF TYPE SHOWING

The outer glass of the Isolator helmet is blackened as shown at the left and then provided with a slit.

SOUND PROOF DOOR

FELT OR CORK FILLED WALLS

AIR LEAVES THROUGH BAFFLES

TRIPLE PANE WINDOWS, AIR CHAMBERS BETWEEN GLASS

ICED AIR IN SUMMER

SLOTS

ICE

FAN

BAFFLE

In the ideal office, all sounds that would annoy the occupant are eliminated by specially constructed doors, walls and windows.

A sectional view of the ideal office or study where work requiring concentration is to be performed. Every device for the comfort of the occupant is provided.

A New Sort of Magazine

Amazing Stories, vol. 1, no. 1, April 1926

Another Fiction Magazine!

At first thought it does seem impossible that there could be room for another fiction magazine in this country. The reader may well wonder, "Aren't there enough already, with the several hundreds now being published?" True.[1] But this is not "another fiction magazine," *Amazing Stories* is a *new* kind of fiction magazine! It is entirely new—entirely different—something that has never been done before in this country. Therefore, *Amazing Stories* deserves your attention and interest.

There is the usual fiction magazine, the love story and the sex-appeal type of magazine, the adventure type, and so on, but a magazine of "Scientifiction" is a pioneer in its field in America.

By "scientifiction" I mean the Jules Verne, H. G. Wells, and Edgar Allan Poe type of story—a charming romance intermingled with scientific fact and prophetic vision.[2] For many years stories of this nature were published in the sister magazines of *Amazing Stories—Science & Invention* and *Radio News.*

But with the ever increasing demands on us for this sort of story, and more of it, there was only one thing to do—publish a magazine in which the scientific type of story will hold forth exclusively. Toward that end we have laid elaborate plans, sparing neither time nor money.

Edgar Allan Poe may well be called the father of "scientifiction." It was he who really originated the romance, cleverly weaving into and around the story, a scientific thread. Jules Verne, with his amazing romances, also cleverly interwoven with a scientific thread, came next. A little later came H. G. Wells, whose scientifiction stories, like those of his forerunners, have become famous and immortal.

It must be remembered that we live in an entirely new world. Two hundred years ago, stories of this kind were not possible. Science, through its various branches of mechanics, electricity, astronomy, etc., enters so intimately into all our lives today, and we are so much immersed in this science, that we have become rather prone to take new inventions and discoveries for granted. Our entire mode of living has changed with the present progress, and it is little wonder, therefore, that many fantastic situations—impossible 100 years ago—are

1. Frank Munsey is generally credited with publishing the first pulp magazine in 1882, *Argosy,* "printing it on coarse paper and filling it with adventure stories for adults." When Munsey estimated the size of the American magazine reading public, he put the number at 250,000 in 1893 and 750,000 in 1899. By 1947, a national survey of the Magazine Advertising Bureau found 32,300,000 "magazine reading families" in the United States. Theodore Peterson, *Magazines in the Twentieth Century* (Urbana: University of Illinois Press, 1964), 314. This rapid expansion of the magazine market meant that by the mid-1920s, the "several hundreds" of fiction magazines on the market were able to diversify into incredibly narrow niche publications "devoted to any and every genre and topic imaginable, such as *Courtroom Stories* (the first issue featured a cover story on the Oscar Wilde trials), *Football Action, Zeppelin Stories,* and *Gun Molls Magazine*" (Peterson, 75). For an account of how modernist literature worked not in opposition to but within this mass marketplace of popular fiction—"the material product, modernism in the marketplace, as found on the newsstand, in the drugstore, over the counter"—see David M. Earle, *Re-Covering Modernism: Pulps, Paperbacks, and the Prejudice of Form* (Farnham, UK: Ashgate, 2009).

2. Gernsback frequently cites Verne, Wells, and Poe in his fiction magazine editorials as the founding figures of a scientifiction genre. This issue featured work by all three: Verne's *Off on a Comet, or Hector Servadac* (1877), Wells's "The New Accelerator" (1901), and Poe's "The Facts in the Case of M. Valdemar" (1845). But as Gary Westfahl points out—who sees this article as one of the first programmatic definitions of science fiction—Gernsback had a broader understanding of proto-science fictional texts in the nineteenth century outside of these three canonical authors. Gary Westfahl, "'The Jules Verne, H. G. Wells, and Edgar Allan Poe Type of Story': Hugo Gernsback's History of Science Fiction," *Science Fiction Studies* 19, no. 3 (November 1992): 340–53, 342–43.

brought about today. It is in these situations that the new romancers find their great inspiration.

Not only do these amazing tales make tremendously interesting reading—they are also always instructive. They supply knowledge that we might not otherwise obtain—and they supply it in a very palatable form. For the best of these modern writers of scientifiction have the knack of imparting knowledge, and even inspiration, without once making us aware that we are being taught.

And not only that! Poe, Verne, Wells, Bellamy, and many others have proved themselves real prophets.[3] Prophecies made in many of their most amazing stories are being realized—and have been realized. Take the fantastic submarine of Jules Verne's most famous story, "Twenty Thousand Leagues Under the Sea" for instance. He predicted the present day submarine almost down to the last bolt! New inventions pictured for us in the scientifiction of today are not at all impossible of realization tomorrow. Many great science stories destined to be of an historical interest are still to be written, and *Amazing Stories* magazine will be the medium through which such stories will come to you. Posterity will point to them as having blazed a new trail, not only in literature and fiction, but in progress as well.

We who are publishing *Amazing Stories* realize the great responsibility of this undertaking, and will spare no energy in presenting to you, each month, the very best of this sort of literature there is to offer.

Exclusive arrangements have already been made with the copyright holders of the entire voluminous works of ALL of Jules Verne's immortal stories.[4] Many of these stories are not known to the general American public yet. For the first time they will be within easy reach of every reader through *Amazing Stories*. A number of German, French, and English stories of this kind by the best writers in their respective countries, have already been contracted for and we hope very shortly to be able to enlarge the magazine and in that way present always more material to our readers.

How good this magazine will be in the future is up to you. Read *Amazing Stories*—get your friends to read it and then write us what you think of it. We will welcome constructive criticism—for only in this way will we know how to satisfy you.

3. Refers to Edward Bellamy, American author of the best-selling socialist utopia *Looking Backward: 2000–1887* (1888).

4. The cover to this first issue of *Amazing Stories* features an illustration of Verne's *Off on a Comet*. Mike Ashley argues that leading with this novel was probably solely meant to gain name recognition, as it was "arguably one of Verne's least scientifically plausible novels. Gernsback admits so in his introductory blurb: '. . . the author here abandons his usual scrupulously scientific attitude and gives his fancy freer rein.'" Mike Ashley, *The Gernsback Days: A Study of the Evolution of Modern Science Fiction from 1911 to 1936* (Holicong, Penn.: Wildside Press, 2004), 78.

The Lure of Scientifiction

Amazing Stories, vol. 1, no. 3, June 1926

Scientifiction is not a new thing on this planet. While Edgar Allan Poe probably was one of the first to conceive the idea of a scientific story, there are suspicions that there were other scientifiction authors before him. Perhaps they were not such outstanding figures in literature, and perhaps they did not write what we understand today as scientifiction at all. Leonardo da Vinci (1452–1519), a great genius, while he was not really an author of scientifiction, nevertheless had enough prophetic vision to create a number of machines in his own mind that were only to materialize centuries later. He described a number of machines, seemingly fantastic in those days, which would have done credit to a Jules Verne.[1]

There may have been other scientific prophets, if not scientifiction writers, before his time, but the past centuries are so beclouded, and there are so few manuscripts of such literature in existence today, that we cannot really be sure who was the real inventor of scientifiction.

In the eleventh century there also lived a Franciscan monk, the amazing as well as famous Roger Bacon (1214–1294). He had a most astounding and prolific imagination, with which he foresaw many of our present-day wonders. But as an author of scientifiction, he had to be extremely careful, because in those days it was not "healthy" to predict new and startling inventions. It was necessary to disguise the manuscript—to use a cypher—as a matter of fact, so that it has taken many great modern minds to unravel the astonishing scientific prophesies of Roger Bacon.[2]

The scientifiction writer of today is somewhat more fortunate—but not so very much more. It is true that we do not behead him or throw him into a dungeon when he dares to blaze forth with, what seems to us, an impossible tale, but in our inner minds, we are just as intolerant today, as were the contemporaries of Roger Bacon. We have not learned much in the interval. Even such a comparatively tame invention as the submarine, which was predicted by Jules Verne, was greeted with derisive laughter, and he was denounced in many quarters. Still, only forty years after the prediction of the modern submarine by Verne, it has become a reality.

There are few things written by our scientifiction writers, frankly impossible today, that may not become a reality tomorrow. Frequently, the author himself does not realize that his very fantastic

1. Gernsback: "Da Vinci—the Edison of the Middle Ages—is credited with having first imagined the printing press, the breech-loading gun, the mitrailleuse gun, the steam engine, the chain drive, a man-propelled airplane, the parachute, and many others—an amazing array of 'scientifiction'—because he admittedly only imagined these inventions."

2. Gernsback: "In his famous *Opus Majus* he accurately prophesied the telescope. He gave excellent descriptions of the *camera obscura,* and of the burning glass—even the invention of gun powder is accredited to him. He forecast an age of industry and invention, with all prominence given to experiment. As a reward for his immortal work, he was incarcerated for a number of years."

yarn may come true in the future, and often he, himself, does not take his prediction seriously.

But the seriously-minded scientifiction reader absorbs the knowledge contained in such stories with avidity, with the result that such stories prove an incentive in starting some one to work on a device or invention suggested by some author of scientifiction.

One of our great surprises since we started publishing *Amazing Stories* is the tremendous amount of mail we receive from—shall we call them "Scientifiction Fans"?—who seem to be pretty well orientated in this sort of literature. From the suggestions for reprints that are coming in, these "fans" seem to have a hobby all their own of hunting up scientifiction stories, not only in English, but in many other languages.[3] There is not a day, now, that passes, but we get from a dozen to fifty suggestions as to stories of which, frankly, we have no record, although we have a list of some 600 or 700 scientifiction stories.[4] Some of these fans are constantly visiting the book stores with the express purpose of buying new or old scientifiction tales, and they even go to the trouble of advertising for some volumes that have long ago gone out of print.

Scientifiction, in other words, furnishes a tremendous amount of scientific education and fires the reader's imagination more perhaps than anything else of which we know.

3. In his book on fandom as a form of participatory culture, Henry Jenkins provides an etymology: "'Fan' is an abbreviated form of the word 'fanatic,' which has its roots in the Latin word 'fanaticus.' In its most literal sense, 'fanaticus' simply meant 'Of or belonging to the temple, a temple servant, a devotee' but it quickly assumed more negative connotations, 'Of persons inspired by orgiastic rites and enthusiastic frenzy' (*Oxford Latin Dictionary*). As it evolved, the term 'fanatic' moved from a reference to certain excessive forms of religious belief and worship to any 'excessive and mistaken enthusiasm,' often evoked in criticism to opposing political beliefs, and then, more generally, to madness 'such as might result from possession by a deity or demon' (*Oxford English Dictionary*). Its abbreviated form, 'fan,' first appeared in the late 19th century in journalistic accounts describing followers of professional sports teams (especially in baseball) at a time when the sport moved from a predominantly participant activity to a spectator event, but soon was expanded to incorporate any faithful 'devotee' of sports or commercial entertainment. One of its earliest uses was in reference to women theater-goers, 'Matinee Girls,' who male critics claimed had come to admire the actors rather than the plays. If the term 'fan' was originally evoked in a somewhat playful fashion and was often used sympathetically by sports writers, it never fully escaped its earlier connotations of religious and political zealotry, false beliefs, orgiastic excess, possession, and madness, connotations that seem to be at the heart of many of the representations of fans in contemporary discourse." Henry Jenkins, *Textual Poachers: Television Fans & Participatory Culture* (New York: Routledge, 1992), 12–13.

For a classic history of science fiction fans, see Sam Moskowitz, *Immortal Storm: A History of Science Fiction Fandom* (Westport, Conn.: Hyperion, 1974).

4. In its earliest days, the new genre of scientifiction thus involved a concerted effort to dig back into the archives and construct a tradition upon which to build. John Cheng comments on Gernsback's reaction to the fan letters: "Although he suggested surprise, Gernsback's strategy to increase and maintain his readership involved fostering a sense of participation and affiliation among his readers. Another early editorial asked readers directly to help increase *Amazing*'s circulation . . . explaining that an increased circulation would benefit readers because it would allow him to publish a larger magazine with more stories. Notwithstanding the credibility of Gernsback's claims for additional material, his editorial rhetoric included readers by negotiating their responsibility for the magazine. While Gernsback and *Amazing* pledged to 'do our part,' he argued that the success of the magazine depended ultimately on readers doing theirs. If they continued to buy magazines and encouraged their friends to buy them too, they would be the ones to gain with a larger magazine and more material to read." John Cheng, *Astounding Wonder: Imagining Science and Science Fiction in Interwar America* (Philadelphia: University of Pennsylvania Press, 2012), 53.

Fiction versus Facts

Amazing Stories, vol. 1, no. 4, July 1926

A few letters have come to the Editor's desk from some readers who wish to know what prompts us to so frequently preface our stories in our introductory remarks with the statement that this or that scientific plot is not impossible, but quite probable.

These readers seem to have the idea that we try to impress our friends with the fact that whatever is printed in *Amazing Stories* is not necessarily pure fiction, but could or can be fact.

That impression is quite correct. We DO wish to do so, and have tried to do so ever since we started *Amazing Stories.* As a matter of fact, our editorial policy is built upon this structure and will be so continued indefinitely.

The reason is quite simple. The human mind, not only of today, but of ten thousand years ago also is and was so constituted that being merged into the present it can see neither the past nor the future clearly. If only five hundred years ago (or little more than ten generations), which is not a long time as human progress goes, anyone had come along with a story wherein radio telephone, steamships, airplanes, electricity, painless surgery, the phonograph, and a few other modern marvels were described, he would probably have been promptly flung into a dungeon.

All these things sounded preposterous and the height of nonsense even as little as one hundred years ago, and, lo and behold! within two generations we take these marvels and miracles as everyday occurrences, and do not get in the least excited when we read of recent reports that it will be possible, within a year or less, to see as well as hear your sweetheart a thousand miles away, without intervening wires or connections of any sort.

So when we do read one of these to us "impossible" tales, in *Amazing Stories,* we may be almost certain that the "impossibility" will have become a fact perhaps before another generation—if not much sooner. It is most unwise in this age to declare anything impossible, because you may never be sure but that even while you are talking it has already become a reality. Many things in the past which were declared impossible, are of everyday occurrence now.

There are few stories published in this magazine that can be called outright impossible. As a matter of fact, in selecting our stories we always consider their possibility. We reject stories often on the ground that, in our opinion, the plot or action is not in keeping with science as we know it today. For instance, when we see a plot wherein the hero is turned into a tree, later on into a stone, and then again back to himself, we do not consider this science, but, rather, a fairy tale, and such stories have no place in *Amazing Stories*.

Of course once in a great while an author may take some liberties, as happened, for instance, in the conclusion of "A Trip to the Center of the Earth," printed in this issue.

Jules Verne brought back his heroes in a most improbable manner. But this one defect does not detract from the story as a whole, throughout which good science is maintained. It is only when the entire plot becomes frankly impossible, or far too improbable, that we draw the line.

And it should never be forgotten that the educational value of the scientifiction type of story is tremendous.

Mr. G. Peyton Wertenbaker, author of "The Man from the Atom," says this on the same subject:

"*Amazing Stories* should appeal, however, to quite a different public (referring to the sex-type of literature). Scientifiction is a branch of literature which requires more intelligence and even more aesthetic sense than is possessed by the sex-type reading public. It is designed to reach those qualities of the mind which are aroused only by things vast, things cataclysmic, and things unfathomably strange. It is designed to reach that portion of the imagination which grasps with its eager, feeble talons after the unknown. It should be an influence greater than the influence of any literature I know upon the restless ambition of man for further conquests, further understandings. Literature of the past and the present has made the mystery of man and his world more clear to us, and for that reason it has been less beautiful, for beauty lies only in the things that are mysterious. Beauty is a groping of the emotions towards realization of things which may be unknown only to the intellect.

"Scientifiction goes out into the remote vistas of the universe, where there is still mystery and so still beauty. For that reason scientifiction seems to me to be the true literature of the future.

"The danger that may lie before *Amazing Stories* is that of becoming too scientific and not sufficiently literary. It is yet too early to be sure, but not too early for a warning to be issued amicably and frankly.

"It is hard to make an actual measure, of course, for the determination of the correct amount of science, but the aesthetic instinct can judge. I can only point out as a model the works of Mr. H. G. Wells, who hits instinctively recognized, in his stories, the correct proportions of fiction, fact, and science. This has been possible only because Mr. Wells is a literary artist above everything, rather than predominantly a scientist. If he were a scientist, his taste and sense would permit him only to write books of scientific research. Since he is an artist, he has given us the first truly beautiful work in this new field of literature."

These opinions, we believe, state the case clearly. If we may voice our own opinion we should say that the ideal proportion of a scientifiction story should be seventy-five per cent literature interwoven with twenty-five per cent science.[1]

1. For Darko Suvin, whose formulation of "cognitive estrangement" has been one of the most influential critical concepts in science fiction studies, the science that a given work explores shapes the parameters of its "literariness," as Wertenbaker and Gernsback put it here. "Once the elastic criteria of literary structuring have been met, *a cognitive—in most cases strictly scientific—element becomes a measure of aesthetic quality, of the specific pleasure to be sought in SF.* In other words, the cognitive nucleus of the plot [the work's scientifically plausible conceit] codetermines the fictional estrangement [the fictional construction of a world different from our own] itself." Darko Suvin, *Metamorphoses of Science Fiction: On the Poetics and History of a Literary Genre* (New Haven, Conn.: Yale University Press, 1979), 15.

Samuel R. Delany, on the other hand, argues that attempts to locate the proper proportions of science to fiction in the genre are useless, as the variety of forms operating under the banner of SF far outpace any attempts to pin it down. "The presence and interaction of estrangement and cognition in a literary work are simply and blatantly insufficient to produce SF. If they interact in one way, they produce fantasy. If they interact in another, they produce surrealism. If they interact in still another, they produce criticism. And it can be argued that as well as insufficient, they are not really necessary either. There are too many space-operas, as familiar to readers as the last fifty of them read, in which there is no cognitive thrust at all. And if these are excluded—by definition—from the genre, then we have no definition at all. . . . But this notion that SF *is* somehow definable is an idea that haunts the academic discussion of SF as much as it haunts the informal discussion that has filled the fanzines since '39. If SF were definable, then it would be the only genre that was! No one has found the necessary and sufficient conditions for poetry. No one has found the necessary and sufficient conditions for tragedy, for the novel, for fiction. If SF is, as Suvin calls it, 'a full fledged literary genre,' why should it be the single one to *have* necessary and sufficient conditions?" Samuel R. Delany, *Silent Interviews on Language, Race, Sex, Science Fiction, and Some Comics: A Collection of Written Interviews* (Hanover, N.H.: Wesleyan University Press, 1994), 191–92.

1. A similar debate broke out among readers when *Science Wonder Stories* shortened its title to *Wonder Stories* in June 1930. And, once again, the conversation conflated genre and gender. See John Cheng, *Astounding Wonder: Imagining Science and Science Fiction in Interwar America* (Philadelphia: University of Pennsylvania Press, 2012), 116.

2. Mrs. H. O. De Hart of Anderson, Indiana writes: "Well, I've written Mr. Wastebasket a rather lengthy letter this time, but I do not really expect you to clutter up your columns with it. I am only a comparatively uneducated young (is twenty-six young? Thank you!) wife and mother of two babies, so about the only chance I get to travel beyond the four walls of my home is when I pick up your magazine." H. O. De Hart, "A Kind Letter from a Lady Friend and Reader," *Amazing Stories* 3, no. 3 (June 1928): 277.

A reply came a few months later from Mrs. L. Silverberg of Augusta, Georgia: "It is the letter of Mrs. H. O. De Hart in the June issue of your publication that is the cause of my writing my little say. For more than a year I have been a reader of this magazine, and this is the first time I have seen a letter from a woman reader. In fact I was somewhat surprised as I had believed that I was the only feminine reader of your publication. However, it is with pleasure that I note that another of my sex is interested in scientifiction." L. Silverberg, "A Lady Reader's Criticisms," *Amazing Stories* 3, no. 7 (October 1928): 667. Quoted in Justine Larbalestier, *The Battle of the Sexes in Science Fiction* (Middletown, Conn.: Wesleyan University Press, 2002).

The next year, Gloria Rosselli of Hickory Street, Seaford, New York writes: "I am a Senior in a small high school in an adjoining town. Because I am only seventeen (and a girl at that!), maybe I'm not supposed to enjoy the highly educational and scientific stories which you publish . . . but I do." Gloria Rosselli, "A Letter from a High School Senior," *Amazing Stories* 4, no. 3 (June 1929): 286.

And Barbara Baldwin of 566 College Avenue SE, Grand Rapids, Michigan, writes: "This is another letter from a *mere* girl. I am seventeen years old and have been reading *Amazing Stories* for about a year. What we want is more interplanetarian stories and less detective stories. I think that most other girls will agree with me in that respect. . . . It might interest you to know that I have bought a three-inch telescope since I first became

Editorially Speaking

Amazing Stories, vol. 1, no. 6, September 1926

A number of letters have reached the Editor's desk recently from enthusiastic readers who find fault with the name of the publication, namely, *Amazing Stories.*

These readers would greatly prefer us to use the title *"Scientifiction"* instead. The message that these letters seem to convey is that the name really does not do the magazine justice, and that many people get an erroneous impression as to the literary contents from this title.

Several years ago, when I first conceived the idea of publishing a scientifiction magazine, a circular letter was sent to some 25,000 people, informing them that a new magazine by the name "Scientifiction" was shortly to be launched. The response was such that the idea was given up for two years. The plain truth is that the word "Scientifiction" while admittedly a good one, scares off many people who would otherwise read the magazine.

Before the name of *Amazing Stories* was first decided upon, a prize contest was held, but no better name than *Amazing Stories,* out of a list of some 200 names, could be found. The name "Scientifiction" would undoubtedly frighten many readers who would perhaps otherwise be interested in this new type of fiction. After mature thought, the publishers decided that the name which is now used was after all the best one to influence the masses, because anything that smacks of science seems to be too "deep" for the average type of reader.

We knew that once we could make a new reader pick up *Amazing Stories* and read only one story, our cause was won with that reader, and this is indeed what happened. Although the magazine is not as yet six months old, we are already printing 100,000 copies per month, and it also seems that whenever we get a new reader we keep him. A totally unforeseen result of the name, strange to say, was that a great many women are already reading the new magazine.[1] This is most encouraging. We know that they must have picked up *Amazing Stories* out of curiosity more than anything else, and found it to their liking, and we are certain that if the name of the magazine had been "Scientifiction," they would not have been attracted to it at a newsstand.[2]

And after all, we really need not make any excuses for *Amazing Stories,* because the title represents exactly what the stories really are. There is a standing rule in our editorial offices that unless the story is

Readers' Vote of Preference

STORY
REMARKS

(1) ....

(2) ....

(3) ....

I do not like these stories:
Why?

(1) ....

(2) ....

My Name and Address

This is YOUR magazine. Only by knowing what type of stories you like can we continue to please you. Fill out this coupon or copy it on a sheet of paper and mail it to AMAZING STORIES, 53 Park Place, New York City.

amazing, it should not be published in the magazine. To be sure, the amazing quality is only *one* requisite, because the story must contain science in *every* case.

A great many letters are also received, from readers who wish to contribute material to *Amazing Stories*. The formula in all cases is that first the story must be frankly amazing; second, it must contain a scientific background; third, it must possess originality. At the present time we are booked far ahead for long stories of the novel type, and therefore can only accept short new stories. Stories that do not run more than six to eight pages when printed are most welcome providing they fill the above requirements. We believe the era of scientifiction is just commencing. We are receiving a great many fine short stories, and as time goes on we will publish more and more new material besides the classics which we are publishing now, and for which we have many requests from readers.[3]

The Editors also wish it to be understood that this is *your* magazine in all respects; they will always be guided by the wishes of the majority. We will publish from time to time a sort of voting blank in which you may show your preference as to the type of stories published in the various issues. You will find such a blank elsewhere in this issue.

At this point we wish to say that the voting contest we conducted several months ago has now been closed. The vote stood as follows:

Leave the magazine monthly as it is now—498.
Make it a semi-monthly—32,644.

We will probably accede to the wishes of the readers as soon as the circulation of the magazine has become stabilized, which will probably be some time before the end of this year.

interested in astronomy through *Amazing Stories*. Barbara Baldwin, "A Mere Girl Writes Us a Letter, and a Good One," *Amazing Stories* 4, no. 10 (January 1930): 988. Quoted in Cheng, *Astounding Wonder*, 116.

Larbalestier's book shows how letters among women readers allowed for the construction of a community that paved the way for later female science fiction writers who would thematize the terms of these conversations. By April 1931, Leslie F. Stone ("one of the earliest women writers of science fiction") published "The Conquest of Gola" in *Wonder Stories*, the first of an SF tradition Larbalestier calls battle-of-the-sexes texts: "I want to emphasize that the battle-of-the-sexes stories' engagement with debates about the social constructions of women and men and the organization of relations between them is made possible by science fiction's generic rules. Science fiction is not tied to a 'mimetic faithfulness to the world as it is.' The process of imagining a world in which women are the dominant sex immediately exposes many of the processes that normally operate to keep women subordinate; it renders these processes of power *visible*" (8).

3. Among this issue's stories was "The Moon Hoax": reprints of articles published by the *New York Sun* throughout August 1835 that erroneously claimed the astronomer Sir John Herschel had recently discovered vegetation, animals, and an intelligent species that built temples on the moon. The editorial introduction states that it was due to a lack of modern media that the hoax spread so easily: "At that time, when there were no cables and no radio, and communication was slow, it was a simple matter to spring such a hoax, where today it would not last twenty-four hours, because verification or denial would speedily be brought about." Despite the series' obvious falsehood, this introduction calls it "a charming story that will live forever . . . this remains one of the finest pieces of imagination that has ever appeared." Thus the ever-expanding catalogue of scientifiction "classics" included texts that worked upon the reader with a semblance of scientific veracity. But the mark of a truly modern work of scientifiction was the speed with which the truth of the story could be proven or disproven by the community.

Is Radio at a Standstill?

Radio News, vol. 8, no. 3, September 1926

. . . wherein the editor compares the Radio Industry to the Automobile Industry—in which the gradual evolution of radio is sketched— how it is shown that the rate of radio patents is increasing— in which batteries and eliminators are discussed— and why radio sets use five times less energy now than in 1920.

During the course of conversation with any people in all walks of life, the question is frequently asked me, if radio has now settled down, in the same degree as the automobile industry, and whether it has become stabilized?

I have answered, a great many times during the past few years, that we need not look for any revolutionary improvements in radio at the present time. The chances are against any invention that will entirely upset the radio industry. Just as in the automobile industry, we may not look for any revolutionary invention that will upset the entire trend of the automobile—unless it should be a flying attachment, which might be applied to any automobile—and this, while not impossible, nevertheless will not appear in the immediate future.

It is the same with radio. Television, to be sure, is in the offing, but several years will elapse before you will be able to sit before your radio at home and witness a baseball game 100 miles distant. On the broadcasting end, no great and epochal improvements need be expected shortly. While improvements are being made right along, these are now more in the nature of finer touches rather than revolutionary; but we can expect better and better transmission and greater clarity. One of the great troubles in the United States at the present time is the heterodyning between different stations nearly on the same wave-lengths.[1] This is particularly true of the low wave-lengths, where there is serious congestion, and there does not seem to be any immediate remedy for this. Technically, there seems to be no possible way to separate two stations less than 1,000 miles apart and operating on the same wave-length. As Congress has adjourned for some six months, and the Department of Commerce is left with little authority, there seems to be little hope that the heterodyning evil can be done away with in the immediate future.

On the receiving end it does not seem that sets will be altered radically in the next few years. Five- and six-tube sets probably will

1. In Fessenden's case, controlled heterodyning could boost a signal. But with the number of broadcast stations crowding the airwaves in the 1920s, natural heterodyning occurred when two stations operating at closely spaced frequencies produced unwanted interference for each of the signals. For more on the heterodyne principle, see **The Dynamophone** in this book.

prevail for quite a long time to come; although there is always the possibility that a single-tube super-regenerative set, which in output may equal the present 4- and 5-tube set, can be developed. So far the super-regenerative circuit, while admitted to be one of the great possibilities, has been and remains nothing but an experiment. It is, as yet, too tricky and has never left the laboratory stage.

From these remarks no rash conclusions should be reached that radio is stagnant and does not progress. Quite the contrary. During the entire year of 1925 over nine hundred radio patents were issued by the Patent Office; and during the first six months of 1926, almost six hundred radio patents have been issued. As a matter of fact, it will be seen from these figures, our inventors in the various laboratories all over the country are still tremendously busy devising new and better things in radio. It would seem that this activity should keep on increasing rather than decreasing in the immediate future.

Radio in this country goes through various strange cycles. When broadcasting started off with a rush, we were in the crystal-set stage. That prevailed during some six months, until the single-tube epidemic set in, which lasted for a year. With one bound we jumped from the single-tube to the 5-tube set which, even today, is more or less standard. At first the sets were built in a box to put on the table. That continued for about a year, when the industry was affected with the console-set fever, which does not yet seem to have abated.

As to the parts,—components—conditions were much the same. Last year we saw a small epidemic of straight-line-frequency condensers, which have practically displaced the old straight-wave type. Then came the vernier-dial tempest, which is still blowing strong.[2]

This year seems to be an "A" and "B" eliminator year; because more firms are becoming engaged in the manufacture of eliminators

2. Both of these were innovations in tuning knob construction. As more and more broadcasters began to crowd the airwaves in the late 1920s, straight-line frequency condensers made tuning easier by spacing out frequencies at the lower end of the dial so that the distance between stations was uniform from one end of the dial to the other. "At the lower end of the broadcasting band, ten kilocycles do not change the wavelength nearly as much as the same change does at 500 meters, for instance." Straight-line condensers were a solution to this problem. "As the shaft is turned, the plates engage with each other more and more rapidly. The result of this construction is to give slow changes in capacity when the plates are nearly disengaged, and rapid changes when the dial is turned to the higher numbers. . . . This means that as you turn the dial toward zero, the condenser plates move slower and slower in proportion to the amount of motion on the dial." Alfred P. Lane, "New Straight Line Types Separate Stations on Dials," *Popular Science Monthly*, April 1926: 60–62.

Vernier dials allowed for large movements on a knob to result in fine-grained input: "each complete rotation of the control knob causes only a fraction of a revolution of the main shaft, permitting fine and accurate adjustment." "Vernier Dial," in *McGraw-Hill Dictionary of Scientific & Technical Terms*, 6th ed. (New York: McGraw-Hill, 2003). They were named for the vernier scale, which performs a similar function for measurement: a secondary scale on a pair of calipers indicates with more granularity where a measurement lies *between* two marks on the primary scale. It was invented by the French mathematician Pierre Vernier (1580–1637). "Vernier Scale," *Wikipedia, the Free Encyclopedia*, March 2015.

From *Popular Science Monthly*, April 1926.

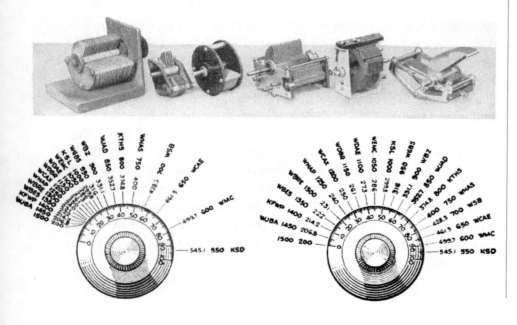

than in possibly any other single radio accessory. There are several million radio sets in use today, and the market for batteries and "A" and "B" eliminators is therefore very large. As in all such phenomena, there is sure to be a race for supremacy between the manufacturers of eliminators and those of batteries. And we may be certain that the battery people are not standing by idly.[3]

When radio first came along, it seemed that the deathknell of the phonograph had definitely been sounded; but the phonograph people merely rolled up their sleeves and went to work producing such phonographs as they had never before believed it possible to build. The immediate result was that the phonograph today is in far greater demand than it was before the advent of radio; and, whereas in 1922 every phonograph manufacturer had 'nerves' every time the word 'radio' was mentioned, he sits back today, complacently, and is not worried at all.

So it will probably come about that the battery manufacturers will be spurred on to meet the invasion and give the eliminator people a stiff battle. Already the storage-battery folk have seen the light, and are putting out radio power plants that connect right to the lighting circuit. These miniature power plants give "A" and "B" battery current with a minimum of attention from the owner. No longer is he required to lug around heavy "A" and "B" storage batteries; now he leaves the unit in the cellar and it is charged automatically.

As to the radio sets, they are getting better and better as time goes on. More attention is being paid now to reduction of losses, shielding, and mechanical perfection, than at any time during the history of radio. It is safe to say that an up-to-date set bought today will be in service for many, many years to come. In the meanwhile, the sets are becoming more sensitive as well, and will have better and better range, by virtue of the improvement in vacuum tubes, which are being made more sensitive every month. Not only are they more sensitive, but they are being made more economical as well. The 5-tube set in 1920 required 5 amperes at 6 volts, which is 30 watts, to light its filaments. It meant, then, recharging your storage battery every few days. The like set today uses only about 1¼ amperes at 5 volts, or 6¼ watts. It is safe to say that the consumption of current by the average radio set, at the end of the next five years, will not be even half what it is today.

During the coming season, the shielding idea seems possibly the greatest advancement in radio receiver building.[4] More and more firms are adopting the shielding system, whereby coils, tubes, and condensers are completely shielded by metallic containers, to do away with stray currents set up not only within the thing itself, but from outside sources. This results in much sharper tuning and very much better reception.

3. "Eliminator" sets, first appearing on the market in 1926, drew their power from household wall outlets rather than batteries. No batteries meant much less wiring and a reduced appliance weight of 40 to 50 pounds. Christopher H. Sterling, *Stay Tuned: A History of American Broadcasting* (Mahwah, N.J.: Lawrence Erlbaum Associates, 2002), 91.

4. Shielding a radio set involved placing electrically conductive or magnetic materials between individual components in order to reduce the possibility of electromagnetic interference.

The Detectorium

Radio News, vol. 8, no. 3, September 1926

In view of the growing popularity of the crystal detector, especially in our large cities, I believe that experimenters all over, particularly the new-comers, will be interested in an instrument invented by me in 1910. The Detectorium, as it was designated by me at that time, was patented June 21, 1910 (U.S. Patent No. 961,855).[1]

During all the years since broadcasting has come into vogue, I do not recall having seen the device described; but it was so good in the old days, and performed so remarkably well, that I feel it my duty to bring it again to light. It was originally described in the world's first radio magazine, which I published; namely, *Modern Electrics,* in the July, 1910, issue.[2]

The Detectorium is interesting chiefly because it does two things at once. Instead of first adjusting the detector and then tuning by means of switches or sliders, in the Detectorium these two operations are performed in one. As the illustration shows, *the detector has become the tuning slider.*

The great utility of a device of this kind will be seen immediately, particularly for sets that are to be transported a good deal.[3]

THE GERNSBACK DETECTORIUM.
Patented June 21, 1910.

This instrument is the outcome of 3 years work to produce an apparatus whereby the two most important instruments are combined into one: namely the Tuner and the Detector. The tuning is done by means of certain detector crystals acting as the slider on the bare convolutions of the Tuner.

Inasmuch as the relationship between Tuner and Detector is very close, it goes without saying that the combination of the two into one instrument, makes it extremely desirable from a scientific standpoint. The combination is vastly more advantageous than using the two instruments separately. The sensitiveness is also greater and the adjustment infinitely better. Instead of adjusting two instruments only one needs to be adjusted, thereby greatly increasing the efficiency.

It is not necessary to adjust the detector when sliding it from one point to another as would be thought, but once adjusted, the detector may be moved back and forward and it will positively not lose its adjustment.

The construction of this instrument is of the highest order, only the best material being used. Hard rubber ends, hard rubber binding posts, and our hard rubber sliders are used. The detector has a double spring (not shown in illustration) and a large adjusting screw, giving extremely fine adjustment. The instrument is made so carefully that the detector may be moved back and forward while a message is coming in and it is quite impossible to lose even a dot. A special feature is, that the cup holding the crystal is detachable by means of the handle below the large knob, and each Detectorium is furnished with one extra cup so that several crystals or minerals may be used and the exchange from one to another can be effected in less than five seconds.

We do not furnish any crystals or minerals with the Detectorium Outfit and leave it to the buyer to use any particular mineral or crystal he fancies. Mostly any of the regular crystals or minerals can be used to advantage except carborundum. Full directions are given with each instrument and we guarantee it fully in every respect.

On account of its great lightness and compactness, this instrument is especially desirable for portable outfits, aeroplanes, etc. Size over all 8"x3½"x3¼". Weight 1 lb., wave length is 600 meters, making it a good all-around instrument. Price $3.50. By mail extra 35 cents.

1. Hugo Gernsback, "Detectorium," June 1910, http://www.google.com/patents /US961855. For more on the multiple sliders and knobs that the Detectorium replaced, see **A Treatise on Wireless Telegraphy** in this book.

2. The Detectorium was sold by Electro Importing, fully assembled, for $3.50 in 1913.

3. In a handbook published by Gernsback's brother Sidney and other Electro Importing employees, one lesson is devoted to the construction of the Detectorium, recommending that a kite be used to lift the aerial to a sufficient height. They write that the apparatus is "very well adapted to the requirements of all portable wireless stations, such as those in mule pack sets, aeroplane airship sets, and in a hundred other places, where light weight and great compactness are prime requisites. . . . This method of using a loading coil is the only way by which so long a wave length may be obtained within such a limited space." Sidney Gernsback, Harry Winfield Secor, and Austin Celestin Lescarboura, *Experimental Electricity Course in Twenty Lessons* (New York: Experimenter Publishing, 1916), 62–63.

Electro Importing Company advertisement in the August 1913 *Modern Electrics.*

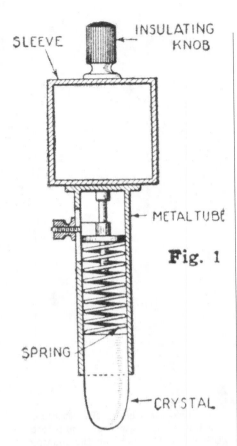

FIGURE 1. A different means for attaching crystal to slider, and spring for regulating tension of crystal.

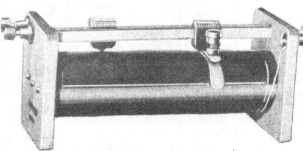

FIGURE 2. View of the complete detectorium, with its two sliders. Note one slider with its crystal cup.

4. B & S refers to the size of the wire. The Brown & Sharpe metal gauge favored by early electrical engineers and contractors to measure the diameter of a wire is still in use today.

In my 1910 experiments I quickly found that the only good minerals were Silicon, Copper Pyrites, Iron Pyrites, Zincite, and Carborundum, in the order named.

A number of circuits showing Detectorium connections are shown on this page.

At this point I wish to say that I believe that our 1910 tuning coils with sliders are still way ahead of anything that is in use now. Very fine tuning can be done with a double-slide tuning coil; much better, in fact, than in most devices used today.

Construction Is Easy

The Detectorium can be readily constructed by an experimenter, and no particular directions need be given here as to sizes. The illustrations show a tuning coil which was common in 1910, but whose range is greater than the present broadcast range. The coil at that time consisted of a two-inch tube about eight inches long, wound with No. 24 enameled or bare wire. The same size tube can be used today, with the exception that the wire should be about No. 18 or No. 20 B & S bare copper, which will cover the broadcast range surprisingly well.[4]

The tube is put in a lathe, or similar winding device, and bare copper wire is used, winding with it, at the same time, a thick thread, to separate the wire convolutions so that they do not touch. The thread can remain, if so desired.

A slider arrangement as shown in Fig. 1, can be used if desired, or otherwise the arrangement shown in Figs. 2 and 3, which was the better arrangement, can also be used. If the slider shown at Fig. 1 is used, it is necessary to obtain a piece of silicon or other crystal of a form somewhat as shown; that is, a bullet-like shape. The ends can be rounded off nicely by grinding on an emery stone and afterwards polishing the crystal perfectly smooth. The end curvature should not be too small, otherwise two turns on the tuning inductance will be short-circuited.

The better way is shown in Figs. 2 and 3. The slider in this case may be a piece of wood with a square-filed hole. Contact is made with the slider rod by means of a small spiral spring, which presses against the metal screw B, which serves to vary the tension of the lower leaf spring, which carries the detector cup A. C represents the wire convolutions, D the second slider rod.

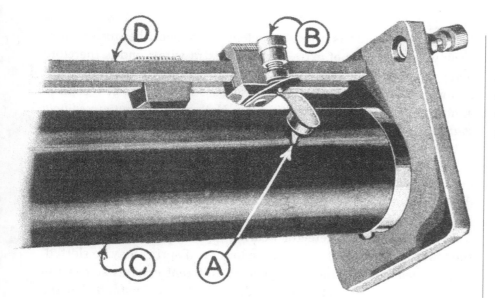

FIGURE 3. Close-up of Detectorium, showing D, slider, B, adjusting screw to bring crystal cup with crystal into contact with wire convolutions, A, crystal, and C, tuning coil.

The detector cup A is a metal cup, in which the detector crystal is held by means of a fusible alloy. The part of the crystal making contact with the wire should not be sharp, but rounded off. If it is sharp it will scratch the wire and stick between the convolutions.

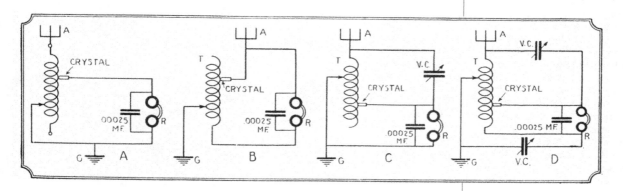

FIGURE 4. Various circuits for Detectorium. Where no sharp tuning is required, circuit A is quite satisfactory. Circuits B and D are used for sharp tuning. Circuit D is particularly good.

Simple, Sharp, Satisfactory

If the Detectorium is constructed with care, a great amount of satisfaction can be had from it; because it places the detector right underneath one's finger, and tuning is done very rapidly. Particularly with silicon and iron pyrites, the tension adjustment is not very critical, and a little more or less pressure does not seem to make much difference, as reception is usually excellent in all cases.

THE DETECTORIUM

The Detectorium is a most efficient instrument, because it does away with a number of extra parts and extra wires; and is therefore really a low-loss detector instrument. If carefully adjusted, it will be found that the Detectorium will surpass in loudness of reception almost any other crystal combination. Not only that, *but exceedingly sharp tuning can be done,* much sharper that you are accustomed to obtain with the usual crystal-detector arrangement.

The circuit diagrams, Figs. B and D, are excellent for sharp tuning; no value is given for the variable condensers, as this depends a great deal upon the construction of the Detectorium. .0005-µf. condensers, however, are satisfactory in nearly all cases.[5]

In Figs. 2 and 3 the detector-bearing cup is shown soldered right to the lower leaf spring. If desired, the lower leaf spring may be slotted, and by means of a screw arrangement, different crystals screwed in or out if different sensitivities are desired. It is understood that the Detectorium uses no batteries of any kind, as the rectified current of the incoming wave is sufficient to operate the telephone receivers.

I shall be glad to hear from those who have constructed the Detectorium.[6]

5. The farad is the standard unit for measuring electrical capacitance. It is named after Michael Faraday (1791–1876), the physicist who pioneered research in electromagnetism. A µf denotes a microfarad, or one millionth of a farad.

6. H. Catchpole from London, England, reported in the February 1927 issue that he had success building the Detectorium using iron pyrite, and that the results were "louder than any crystal set I have made to date." W. M. Cos, EAA., Chatham, England, wrote in a letter, published in the December 1926 issue: "I constructed this instrument in 1918 when I first commenced radio research. Details were then published in your handbook, 'Wireless Telegraph and Telephony.' The instance remains vivid in my memory as the tuner, when incorporated in a portable set was extremely efficient and decidedly unique. The only trouble that I experienced was due to the crystal. Silicon was the most satisfactory although it has a tendency to crumble. This applies to zincite while carborundum cannot be ground and is therefore not smooth in action. The tuning was extremely sharp, and when the secondary was shunted by a .0003-µf. variable condenser results were all that could be desired. I take this opportunity to compliment you on the high standard of your publications, which with all due respect to our English periodicals, are on a much higher scientific basis. Our papers only cater to the broadcast listener, and the experimenter who dabbles in everything electrical is left in the dark."

Imagination and Reality

Amazing Stories, vol. 1, no. 7, October 1926

When reading one of our scientifiction stories in which the author gives free reign to his imagination, providing he is a good story teller, we not infrequently find ourselves deeply thrilled. The reason is that our imagination is fired to the nth degree, and we thus obtain a real satisfaction from the time spent in reading the story, improbable as it often appears at first. I should like to point out here how important this class of literature is to progress and to the race in general.

The human mind is a tremendously complex machine, which often works in a very strange manner. A man sets out to invent a certain house appliance, and while engaged in his experimental work, gets a certain stimulus that takes him in an entirely different direction, so that the first thought of the house appliance may end in the invention of a factory labor-saving device, or perhaps something even more important.

When Alexander Graham Bell was a young man, he occupied himself by devising means of enabling the deaf to hear. This led him into electrical research work, and the apparatus, far from becoming a device by which the deaf can hear, became the present telephone. To be sure, loud-speaking telephones are made today for the use of the deaf, but this is only a by-product and not at all the actual and more important use of the instrument.[1]

Hundreds and thousands of similar instances could be cited. An author, in one of his fantastic scientifiction stories, may start some one thinking along the suggested lines which the author had in mind, whereas the inventor in the end will finish up with something totally different, and perhaps much more important. But the fact remains that *the author provided the stimulus* in the first place, which is a most important function to perform.

On the other hand, many devices predicted by scientifiction authors have literally come true for many generations. There is an old popular saying that what man imagines, man can accomplish. This proverb of course, should be taken with a grain of salt, because not everything that man imagines is possible. For instance, I can imagine that I blow out the sun, or grasp the moon in my hand, or cut off my head without dying. Naturally such things are impossible. On the other hand, many of the so-called wild ideas which we read in our scientifiction

[1]. Alexander Graham Bell was inspired by the work of his father, Alexander Melville Bell, whose phonetic alphabet of "visual speech" served as a pronunciation aid for deaf individuals. On the continued reciprocity between telephone technologies and the naturalization of hearing as a measurable, quantifiable capacity, see Mara Mills, "Deafening: Noise and the Engineering of Communication in the Telephone System," *Grey Room* 43 (April 2011): 118–43. And, for an excellent genealogy of how the assumptions behind another nineteenth-century phonetic alphabet—designed in this case to preserve Native American languages—served as an unexpected origin point for contemporary theories of media, see Brian Hochman, *Savage Preservation: The Ethnographic Origins of Modern Media Technology* (Minneapolis: University of Minnesota Press, 2014).

stories may prove to be not quite so wild if they give an actual stimulus to some inventor or inventor-to-be who reads the story. And as long as there is a stimulus of any sort, we have no reason to complain, because we never realize where progress in any direction may lead us.[2]

There is the well-known story of the inventor who had patented the mouse-trap, and finally sold the patent to a manufacturer, who found that an excellent burglar alarm could be made from the mouse-trap, with but a few changes. Another case of an original stimulus which, perhaps, went wrong, but finally became righted.

We should not, therefore, become too impatient if occasionally we encounter a seemingly impossible prediction or improbable plot. It is beyond our power to foresee what reaction this may produce in some one, and what tremendous consequences it may have in the future. And, strange to relate, the patent offices of most countries follow scientifiction stories pretty closely, because in many of these the germ of an invention is hidden. It is not necessary to actually build a model to be an inventor; often it becomes necessary, for court proceedings and for patent reasons, to find out who really was the original inventor of a certain device; if the inventor is an author who brought out the device, even in a fictional story, this would, in the long run, entitle him to ownership of the patent, always providing that the device is carefully described, as to its functions, its purpose, and so forth.[3]

For instance, in the United States, the inventor would have two years from the publication of the story to apply for a patent. Thus it will be seen that a scientifiction story should not be taken too lightly, and should not be classed just as literature. Far from it. It actually helps in the progress of the world, if ever so little, and the fact remains that it contributes something to progress that probably no other kind of literature does.

2. This separation between the author's stimulus and the inventor's realization is reflected in a collection of essays by science fiction writer Thomas Disch that explores the ways the genre has inspired developments in industrial design, warfare, technology, and fashion from its inception. Thomas M. Disch, *The Dreams Our Stuff Is Made Of: How Science Fiction Conquered the World* (New York: Free Press, 1998).

3. Legal precedent for using works of science fiction as "prior art" in patent law is hard to come by. But scientifiction stories, with their high degree of technical specificity, would theoretically have a better chance given that "enablement"—a requirement at the core of patent law—means that a document must allow a person having ordinary skill in the art (PHOSITA) to make the device as described in order to constitute prior art. Jeffrey H. Ingerman and Drew Schulte, "Should Science Fiction Affect Patentability?" *Law360*, May 2013, http://www .law360.com/articles/440742/should -science-fiction-affect-patentability. In a recent, high-profile example, Samsung cited the use of tablet computers in Stanley Kubrick's *2001: A Space Odyssey* during its patent battle with Apple over the iPad. While Judge Lucy Koh allowed examples such as commercial prototypes designed by Hewlett-Packard and Knight-Ridder to be used in the case, she ruled that Samsung could not use the film as evidence. Jason Mick, "Judge Excludes Samsung's 'Sci-Fi' Tablet Evidence from Trial," *Daily Tech*, August 2012, http:// www.dailytech.com/Judge+Excludes+ Samsungs+SciFi+Tablet+Evidence+ From+Trial/article25329.htm.

The Pianorad, showing the twenty-five tubes and oscillator circuits.

The "Pianorad"

Radio News, vol. 8, no. 5, November 1926

This year marks the second centenary of the creation of the piano. There has been no radical development in the piano since except the advent of the player piano or mechanical piano.

The Pianorad, which is played very much like the piano, by means of a similar keyboard, is a new invention. In this new musical instrument, the principles of the piano as well as the principles of radio are for the first time combined in a single musical instrument.[1]

The Pianorad was first demonstrated on Saturday, June 12, during the celebration of Station WRNY's first anniversary, when it was used to broadcast its music.

1. The Pianorad is an improvement on Gernsback's earlier electronic keyboard, the Staccatone, which debuted in *Practical Electrics*, March 1924, along with instructions on how to build one at home. While the Staccatone consisted of one sine wave oscillator connected to each of sixteen keys—"simple on/off switches [that produced] a sharp staccato note with little control over the attack of the sound," the Pianorad had significantly more control over the pitch and duration of its notes. Thom Holmes, *Electronic and Experimental Music: Technology, Music, and Culture* (New York: Routledge, 2012), 30.

According to musician Simon Crab, the redesigned and rechristened Pianorad "had 25 single LC oscillators, one for every key for its two octave keyboard giving the instrument full polyphony, the oscillators produced virtually pure sine tones. . . . Each one of the twenty five oscillators had its own independent speaker, mounted in a large loudspeaker horn on top of the keyboard and the whole ensemble was housed in a housing resembling a harmonium. A larger 88 non keyboard version was planned but not put into production." Simon Crab, "The 'Pianorad,' Hugo Gernsback, USA, 1926," in *120 Years of Electronic Music: The History of Electronic Music from 1800 to 2015*, 2014, http://120years.net/the-pianoradhugo-gernsbakgermany1926/.

The Pianorad is the invention of Mr. Gernsback, and was built in the radio news laboratories by Mr. Clyde J. Fitch. This is the first time that a musical instrument has been constructed from radio parts; and it should therefore have more than passing attention from all radio interests.

Theory of the Instrument

The Pianorad has a keyboard like an ordinary piano, and there is a radio vacuum tube for each one of the piano keys. Every time a key is depressed, there is energized a radio-oscillator circuit which gives rise to a pure, flute-like note through the loud-speaker connected to the device. It is possible to connect any number of loud-speakers to the Pianorad if it is desired to flood an auditorium with its tones. Also, by arranging suitable outlets for loud-speakers on different floors or different rooms, the sounds of the Pianorad can be heard all over any large building.

The musical notes produced by vacuum tubes in this manner *have practically no overtones.* For this reason the music produced by the Pianorad is of an exquisite *pureness* of tone not realized in any other musical instrument. The quality is better than that of the flute and much purer. The sound, however, does not resemble that of any known musical instrument, the notes are quite sharp and distinct, and the Pianorad can be readily distinguished by its music from any other musical instrument in existence. In the Pianorad one vacuum tube

The Pianorad as operated in WRNY studio.

The loud-speaker horn with its twenty-five units separately connected for eliminating harmonics.

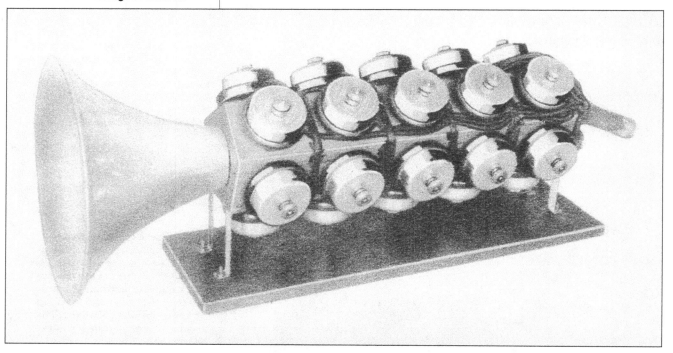

for each key is connected electrically with certain coils (inductances). Any number of notes can be played simultaneously, as on the piano or organ; unlike the piano, however, the notes can be sustained for any length of time. On the ordinary piano you strike the key and the sound quickly dies away, in the Pianorad, the sound remains as long as the keys are depressed.

Electric, Not Sound Waves

The loud-speaker arrangement makes it possible for an artist to play the keyboard while the music emerges, perhaps miles away from the Pianorad. It is thus possible for the pianist to play the instrument in absolute silence while the music is produced at a distance. This requires simply that a wire line must connect the output end of the Pianorad instrument with the loud-speaker at some distance away. It is quite feasible for the Pianorad to be played in New York while the music will be heard at the Chicago end, with any number of loud-speakers connected by amplifiers to a long-distance telephone wire line.

A novel idea is the connection of the Pianorad *direct* to the broadcast-station transmitter. In this case, instead of using a loud-speaker in the studio, the Pianorad is connected electrically to the broadcast transmitter. The artist now plays the Pianorad in the studio in absolute silence. No sound is heard. The radio audience, however, will enjoy the music, although no one in the studio can hear it. In order that the pianist may hear what he is playing, he will wear a set of head receivers attached to an ordinary radio set. The music, therefore, is picked out from the air by the receiver and thus only the artist hears it. In the studio itself, no sound is audible for the Pianorad itself is silent.

Developments Still Continuing

The Pianorad has as yet not entered the commercial stage. The instrument illustrated in this article has 25 keys and therefore, 25 notes. A full 88-note Pianorad has as yet not been constructed, but will be built in a short time. The larger instrument could have been built at once, but it would occupy almost as much space as a piano; and as this amount of room was not then available in the studio of WRNY, for which the first Pianorad was especially constructed, the smaller instrument was built instead.

The Pianorad at WRNY is usually accompanied by piano or violin or both; very pleasing combinations are produced in this manner. At present it uses a single stage of amplification, giving volume enough, in connection with one loud-speaker, to more than fill a fair sized room. By adding several stages of audio-frequency amplification, sufficient volume can be obtained to fill a large church or auditorium.

The Pianorad was first demonstrated publicly Saturday, June 12 at 9 P.M., with a number of brilliant selections played on it by Mr. Ralph Christman; the concert being broadcast over WRNY at The Roosevelt, New York.

The principle embodied in this instrument was first demonstrated in 1915 by Dr. Lee de Forest, inventor of the Audion. At that time Dr. de Forest was able to produce musical tones by means of vacuum tubes, but the radio art at that time had not progressed sufficiently to make possible the Pianorad.

An article by Clyde J. Fitch describing the construction of the Pianorad will appear in the December issue of *Radio News*.

Edison and Radio

Radio News, vol. 8, no. 6, December 1926

... in which the Editor takes issue with Mr. Edison's claim that radio is a failure; yet it is pointed out that the Radio Industry owes Edison a great debt; wherein facts and figures are given to show that Radio is on a steady increase; granting that neither Radio nor the phonograph is yet perfect; how the interest in Radio is steadily increasing, and radio dealers are now making good money.

Thomas A. Edison has recently been quoted in the press as saying that Radio is a dismal failure.[1] The following remarks on the subject are attributed to Mr. Edison: "The radio is a commercial failure, and its popularity with the public is waning. Radio is impractical commercially and esthetically distorted, and is losing its grip rapidly in the market and in the home. There is not ten per cent. of the interest in radio that there was last year. Radio is a highly complicated machine in the hands of people who know nothing about it. No dealers have made any money out of it. It is not a commercial machine because it is too complicated. Reports from 4,000 Edison dealers who have handled radio sets show that they are rapidly abandoning it, and as for its music—it is awful," comments the wizard of Menlo Park. "I don't see how they can listen to it.

"Thousands of people have signed a petition asking that sopranos be kept off the air. Of course most of them don't know that the soprano voice distorts the radio. The phonograph is coming into its own because the people want good music. The fact is that radio has never had a high peak of popularity. In towns where 25 or 30 dealers were handling radio sets, only one or two are now handling them. A farmer five miles from town buys a radio, perhaps on the installment plan. A wire becomes loose. The dealer has to arrange to fix it. This happens time and time again. The business becomes unprofitable for the dealer to engage in. He does not make any money out of it. None of them has. They are giving it up as fast as they can. It is not a commercially successful machine, because it is too complicated."

Turning to the musical side of the question, Mr. Edison chuckled in his characteristic manner, "Static is awful, and the difficulties of tuning out—and now they are stealing each other's wavelengths! It is too bad that the radio has to be so complicated. It was a big and interesting thing and the people responded to it, but they want good

1. "Edison Calls Radio a Failure for Music; Thinks Phonograph Will Regain Its Own," *New York Times,* September 23, 1926: 27.

music and they have found it is not to be had on the radio. That is why the phonograph is reclaiming its own."

Incidentally, this outburst from the dean of modern electricity was in connection with the announcement of Mr. Edison's latest invention, his 40-minute phonograph record—a great achievement, and one that without doubt will be of much benefit to the phonograph industry.

Since the publication of this famous interview with Mr. Edison, the press, and particularly the radio press over the entire country, has been more or less agitated. The following comments of mine, most of which were printed in the *New York Times* of September 26, and the *New York Evening Post* of the same date, were made by me at the time, and are here somewhat amplified:

I have too high a personal opinion of Thomas A. Edison to wish to say anything of a controversial nature, or anything that would even border on discourtesy to the great inventor, but I do believe that Mr. Edison has not been recently in touch with radio sufficiently to appreciate fully the tremendous advances that have been made. Mr. Edison is a busy man, and a tremendously busy inventor. It would be well-nigh impossible for him to be in touch with all of the various commercial phases of radio all over the country; and like other executives he obtains his reports from subordinates, and such reports often as not may be highly colorful and even wrong.[2]

Right here I wish to pay a tribute to Mr. Edison that the radio industry so far has been unwilling to accord him. If it were not for Mr. Edison and the "Edison Effect," radio would not be what it is today. It is the Edison Effect that has made possible our present vacuum tubes, now used universally in radio. Radio, therefore, owes a tremendous debt to Thomas Alva Edison; and I recommend to the radio industry that it acknowledge this debt more frequently in the future.[3]

As to Mr. Edison's remarks, the statements that follow are facts which can be checked up by any one who is unbiased. They are not given with any idea of starting a controversy.

Rather than waning in popularity, it is well known that radio is on the constant increase. Witness, for instance, the recent Third Annual Radio World's Fair, in New York, where the attendance for the week was 228,000, the greatest on record of any radio show, and a tremendous increase over last year's figures.[4] There certainly was no such interest in the phonograph when the latter was but five years old, which is the age of radio, since radio broadcasting started.

The sales of radio apparatus, for the United States alone, will reach $520,000,000 for 1926. The figures for the former years, compiled by the Radio Manufacturers' Association, are given here: 1922, $46,500,000; 1923, $120,000,000; 1924, $350,000,000; 1925,

2. Leroy F. Dyer, managing engineer of the Dyer Radio Manufactory, writes in a letter published in the February 1927 issue of *Radio News*: "I want to commend your lenient attitude in commenting upon the remarks, attributed to Mr. Edison, regarding the comparative merits of radio and the phonograph. In spite of the reverence due our great inventor, most of us could not have resisted a temptation to characterize such statements as propaganda. . . . It is not fair to compare the results of the average phonograph with the average radio receiver in reproducing music. The average phonograph of today is a highly-developed, scientifically-designed, factory-built piece of apparatus, the culmination of many years of experience. The average radio receiver—well, I don't want to hurt anyone's feelings, but to put it mildly, the *average* radio receiver is far from representing the present state of perfection in the radio art."

3. See **Thomas A. Edison Speaks to You** in this book on the Edison effect.

4. With exhibitions from amateurs and corporations alike, the Radio World's Fair was held at the newly-opened, third iteration of Madison Square Garden, beginning September 13, 1926. "For the first time, under a single roof, the public will see all the marvelous new radio equipment that, only a few months ago, was in the experimental stage in a hundred research laboratories. It will not only be a demonstration of the tremendous advance in the radio engineering art, but the sets themselves will be encased in de luxe cabinets that make radio not just a luxury or a necessity in the home, but an actual adornment as well." "New York Radio Show Ushers in Era of New Development and Beginning of New Year," *Brooklyn Daily Eagle,* September 12, 1926: 10C.

$449,000,000. These are not mere estimates, but actual figures. From orders that have been placed, the various radio trade associations know now that the 1926 figure will be exceeded in 1927. The fact is that the popularity of radio is becoming steadily greater rather than less, and no home today is considered complete without its radio set.

Radio's popularity started with the introduction of broadcasting in 1921. In five short years it has accomplished more than the phonograph did in fifteen years. The modern radio set is no more complicated than the automobile when it was five years old; and for best results the radio set should be serviced by radio dealers, just as the modern automobile is serviced by its garage. In the last analysis, radio will probably be handled by radio or electrical stores, whose staff understand the mechanism. The phonograph dealer is not always equipped to do servicing, although quite a good many phonograph stores do so.

As for quality, it is the belief of unprejudiced experts that in many cases the radio, providing it is of a good make, with a good loud-speaker, will deliver quality exceeding that of a phonograph. Neither phonograph nor radio are perfect. The best phonograph is of no avail after a record has been played several dozen times; after which, by no stretch of the imagination, can one call the result music. Furthermore, the scratchy sound produced by every phonograph is highly objectionable and is certainly worse than the few extraneous noises produced in most radio sets today.

The radio and phonograph are two different entities, and should never compete. As a matter of fact, they never do. At the same time, the phonograph has come back *only because the popularity of radio* caused the phonograph makers to turn out a product such as had never existed before.

The radio dealers are making far more money in radio now than ever before. A great number interviewed, in New York and vicinity, claim that their business was never better and is on the increase. There are pretty close to 30,000 radio dealers throughout the country today. It is true that for some time the dealers did not make money, due to the price-cutting evil, but this is rapidly being eradicated.

Some of the best sets of today combine the phonograph and the radio. Each has its particular field. You can not listen to Caruso on the radio, nor can you get the latest presidential address on the phonograph.[5]

The radio industry today is only five years old, and it may safely be predicted that when it becomes as old as the phonograph is today we shall hardly be able to recognize it as the same development. It is admitted that radio is not yet perfect. Neither is the phonograph, nor the automobile, nor motion pictures, nor electric lights; nor, for that matter, a pair of shoes.

5. Enrico Caruso was an Italian opera tenor with a famously "phonogenic" voice, one that drove the sales of phonographs in the early twentieth century according to John Potter, *Tenor: History of a Voice* (New Haven, Conn.: Yale University Press, 2009).

Why the Radio Set Builder?

Radio News, vol. 8, no. 8, February 1927

*. . . in which the Editor recalls the early days of Radio,
before factory-made sets could be obtained—why the manufacturer
of receivers is not unkindly disposed to the set builder—how the latter
is doing some of the most important pioneering in radio design—
why the enthusiastic set builder is also a purchaser of good commercial
receivers—and why this great hobby is appealing year by year to greater
numbers of intelligent people who find in it the highest pleasure. . . .*

When radio was young, in this country, you could not go out in the open market and buy a complete radio set. I refer to the time when radio first came into vogue; that is, after the appearance of the modern vacuum tube, in 1912. At that time such a thing as a radio cabinet was unknown. We used to mount our instruments of various descriptions on our table; and the more room they took up, and the bigger the table was, the better pleased we were.

This state of affairs lasted for a number of years, and possibly culminated about 1923 in the first vacuum-tube sets of the multiple type which then made their appearance. It is true that, beginning with 1918, we had possessed a few self-contained sets of the cabinet variety which, of course, had been used, not for broadcast purposes, but for listening to code.

When broadcasting finally made its appearance, the factory-made set took the country by storm; and, while previously the home-built set had been in vogue, the factory set took the ascendancy immediately. Today, at least in this country, the factory-made radio set for broadcast purposes has far outstripped the home-made set in popular demand. By this I do not mean to imply that the genus of radio constructor who builds his own set has died out. Quite the contrary. There are more sets being built this minute than ever before.

From the best available sources at hand, it seems that there are, at the present time, between 400,000 and 500,000 people who annually build sets, and this figure seems to be on the increase. Large as this figure may seem, it is small compared to the figure of factory-made sets annually turned out in this country (over 2,300,000 at the last census of manufacturers); and it may be said that the manufacturers of ready-made sets today do not worry about the home-built set, but, rather, encourage it. This, at first thought, would seem paradoxical; but it is true, nevertheless, for the following simple reasons:

Radio is an art which changes rapidly, as is well known. While no revolutionary improvements have been made in the past ten years, or are likely to be made soon, changing styles, as well as improvements, keep the trade on the jump. New condensers come out, new dials are devised, new coils are produced. At the present time the shielding idea has attained great favor, almost overnight.[1] Naturally, for this reason, set manufacturers are always anxious to incorporate the latest devices in their receivers.

But once the manufacturer is "tooled up" to turn out the season's supply, it is not always possible or desirable for him to make a change. In the meanwhile the art and progress of radio goes on, and the manufacturer naturally wants to know, in plenty of time, what the tendency will be for next year. By encouraging the set builders he gets a very good idea in what direction the tendency is heading; and he is able, at no cost at all to himself, to get this information, by simply watching the radio press and studying this tendency. When the new season comes along, the manufacturer is, therefore, apt to have a pretty good idea of what will happen, or what may be expected to happen next season. This is not to say that the manufacturer gets all of his ideas from the radio constructors. No such meaning is implied; but he gets valuable information; and for that reason most set manufacturers today openly encourage set building, because, first, they know that it cannot hurt their business and, secondly, because they derive from it valuable information which they would not have if there were no set building going on.

The set builders themselves, in the meanwhile, are having a mighty fine time, building to their hearts' content; in which they are encouraged by the parts manufacturers, who are themselves always ahead of the set manufacturers in bringing out new devices. These new devices are tried out by the set builders, and within six months it becomes known whether a certain device will "take," in the long run, or not.

This has been the case with the straight-line-frequency condensers, as it has also been with the new vernier dials.[2] It is true of shielding the various parts and many other features; none of which would, perhaps, have become incorporated in ready-made sets as soon as they were, if the set builders themselves had not paved the way for such parts.

On the other hand, by encouraging the set builders, the parts manufacturers get, themselves, very valuable experience which they would not obtain otherwise; and, once the majority of set builders have adopted a certain article, the set manufacturers in turn will adopt it as a rule. Such was the case, for instance, with the straight-line-frequency condenser, which was used by set builders for some six to eight months before the set manufacturers adopted this type of condensers.

It may be said, therefore, that the set builders are always ahead

1. Gernsback discusses shielding in **Is Radio at a Standstill?** in this book.

2. For an explanation of straight-line frequency condensers and vernier dials, see **Is Radio at a Standstill?** in this book.

of the game; they are forever pioneering. If you wish to see the latest circuit, or if you wish to see the latest radio wrinkle applied, you will always find it in the best home-made sets. All of this does not mean that the set builder does not use the ready-made set; in most cases he does. There is hardly a radio constructor today worth his salt who does not own two or three sets that are in constant use.

For instance, I myself have two factory-made sets in my home, whereas the set which stands on my study table is one constructed by myself. This particular set probably does not stay there for more than a month at a time, because next month I shall be using a later model; but in the meanwhile the factory-made sets are doing their duty and are being used constantly by the household.

This condition is found all over the country, for it is duplicated in the home of practically every set constructor.

Radio set building may be said to be one of the greatest hobbies that ever came into existence. Unlike most other hobbies, it actually serves to advance a new art, and paves the way for better and bigger things.

To be up to date, under conditions that change as quickly as do those in radio broadcasting, radio receivers must forever be kept up to the minute. Though the changes are gradual, they are constantly taking place, and their effect is cumulative. You would not think of using, in the midst of the heavy traffic on Fifth Avenue or State Street, a 1914-model car that had to be cranked by hand. No more can you expect the set of 1922, built when there were but a few broadcast stations, to give satisfaction, particularly in our congested centers. It is a well-known fact that every time a station changes its transmitter, or increases its power, thousands of nearby sets are immediately found to be inadequate, because they cannot tune sharply enough to cut out the nearby station and get others at will.

Investigation usually shows, on such occasions, that most of these unselective sets are single-circuit or crystal receivers and others of ancient vintage, which are no longer suitable for present-day radio traffic. Furthermore, additional demands are being made right along on the selectivity of radio receivers, because the broadcast stations are continually increasing their power. The set builder, naturally, keeps pace with the evolution of broadcast conditions, and is forever ready to build a new and better set to meet future requirements.

Set building is continuing to increase rapidly, as it has done for five years, in this country; and, now that we stand on the threshold of television, I believe I shall not be contradicted in saying that set building will assume tremendous proportions, undreamt-of today, during the next five years.

New Radio "Things" Wanted

Radio News, vol. 8, no. 11, May 1927

... In which the Editor comments on the present tendency in radio circles to follow the well-known Calf Path—and suggests a few profitable lines for future experiment—what the Super-Regenerator promises—the need for truly portable sets—and the desirability of commercial apparatus to make them practicable—new tubes, new condensers, new coils— and finally adds a practicable "wrinkle" of his own invention—the "Cane Loop," which makes a most compact and portable antenna for any set ...

The progress of radio has been rather slow during the past two years. From the experimenter's standpoint there have been few really new things to occupy him. There have been no revolutionary circuits, nor, as a matter of fact, do we expect them. Radio progress during the past two years may be said to have been refinement and improvement of what we already had. But there are many things still to be worked out, and a great deal of progress, as a matter of fact, the greatest progress, is as yet to come. The trouble with most of us is that we follow the well-beaten path, and as a rule we follow the leader. Very few experimenters and designers have the courage to step out from well-worn radio paths, because, as a rule, they are afraid that the results of their labors will be called freaks or worse.

A condition of this kind, of course, does not worry the progressive man, who, many times, has seen the very thing that was condemned come into favor and acclaimed in the end. When, in 1908, for instance, I published the first book on radiotelephony, entitled "The Wireless Telephone," there was no such thing as a radiophone, because we did not then have the necessary vacuum tube. I described minutely many systems for accomplishing the thing, and I was laughed at by the press and others for my pains, but nevertheless it all came about practically along the lines I had predicted.

When, in 1921, I prophesied the single control,

May 1927 cover of *Radio News.*

multi-tube radio set, it was said that such a thing was impossible of accomplishment, and the experimenters as well as the trade for years refused to work on such sets, but nevertheless they are an actuality today and will be the standard during the next two years.

The random thoughts which I set down here at this time may prove to be of a similar nature and time alone will tell whether the ideas are sound or not. *Radio News*, at my instigation, recently started a set-building contest along the Super-Regenerative lines. The Armstrong Super-Regenerator is one of the most wonderful circuits that we possess.[1] It is believed that in time this circuit will be the one that may yet prevail, because by means of it we can, with one or two tubes, accomplish the same thing that is done today with anywhere from 6 to 10 tubes. Unfortunately the circuit has never been perfected, due to its critical nature, but it is believed that sooner or later a solution will be found which will make this circuit come to the fore.

It certainly deserves this recognition. The Super-Regenerator is the ideal set for portable purposes, and where room is limited, all of this providing it is built so that it can be controlled. Here is a most fertile ground for research and for experimental work, and I suggest to experimenters that they busy themselves with this circuit. Perhaps some new combination will be found that will solve the problem.

Right here I wish to say that it is not always the new and revolutionary thing that is apt to become important. Sometimes an old and forgotten principle can be brought to the fore under new circumstances. For instance, the principle of the Marconi Radio Beam System of today was discovered by Heinrich Hertz in 1888. It was minutely described by him, but nothing much was done for some twenty years, until Marconi picked it up again and is now utilizing it.[2]

The same is the case with many other well-known radio principles, which may be found in text books, in magazines, and in the patent press. These things may have been obsolete ten, fifteen, or twenty years ago, but, due to later and newer developments of other apparatus, are of great importance today, or will be in the future.

At the present time there is need for the following new equipment: Experimenters and manufacturers need a new miniature vacuum tube. Such a radio tube, of the 199 type, should measure about ½-inch diameter by an inch to an inch and a quarter high, over all.[3] This would make it the smallest tube commercially available. It could be equipped with a bayonet socket to take up little more room than the diameter of the tube, and with such a tube it would be possible to make a small portable radio set the size of a box camera.

Radio News has already taken the initiative, and is urging tube manufacturers to bring out such a tube, which we hope they soon will.

1. In 1914, both Edwin Armstrong and Lee de Forest applied "for patents on what became known as the regenerative or *feedback* circuit. One of the longest and most bitter fights in radio history resulted from this conflict over patent priority, leading in 1928 and again in 1934 to the United States Supreme Court." While the courts decided in favor of De Forest, most historians give credit to Armstrong for the invention. Christopher H. Sterling, *Stay Tuned: A History of American Broadcasting* (Mahwah, N.J.: Lawrence Erlbaum Associates, 2002), 37. The super-regenerator was Armstrong's improvement on the original design of his "Audion" tubes.

2. Gernsback is comparing the operative principle behind wireless telegraphy—Hertz's discovery of electromagnetic waves—with its commercial applications. Referencing Marconi here is particularly apt, as "he was a highly competitive businessman whose ultimate goal was to establish a monopoly in wireless telegraphy. This goal was eventually referred to as the Imperial Wireless Scheme; Marconi meant to connect the entire British Empire together by wireless, and he meant to own the only company capable of doing so." Susan J. Douglas, *Inventing American Broadcasting, 1899–1922*, Johns Hopkins Studies in the History of Technology (Baltimore: The Johns Hopkins University Press, 1987), 67.

Marconi's "beam system," or the Imperial Wireless Chain, first opened on April 24, 1922, and was completed June 16, 1928. "Imperial Wireless Chain," *Wikipedia, the Free Encyclopedia*, February 2015.

3. The Type 199 vacuum tube debuted in December 1922 and was much smaller than any others that had come before it. It was "the first tube intended for use in portable receivers as well as in home receivers using dry batteries." "UV199," *Radiomuseum*, 2002, http://www.radiomuseum.org/tubes/tube_uv199.html.

It is felt that miniature radio sets will be in great demand. There is no such thing as a convenient portable set on the market today. Most of the sets made are far too large and too heavy. With these small tubes it should be possible to build a set that does not weigh more than two or three pounds, and that can be slung around the shoulder like a camera, to be taken on long trips, for vacation purposes, and for general traveling.

Furthermore, small sets of this kind can be made for apartment dwellers, and wherever a small set is needed to be carried from one place to another. It may be said that, given such miniature vacuum tubes, we would still need small condensers. It is possible to make such condensers today, to make up a minimum of room, if such condensers are needed. It is known, for instance—a fact which has been forgotten for many years,—that by placing a variable condenser into castor oil, or some other high grade oil, the capacity of the condenser will be quintupled. In other words, by employing the oil immersion, we could make a 13-plate condenser one-fifth as large as we have at the present time for any given capacity. Furthermore, the equivalent of a 17-plate (.00035 mf.) condenser can be made by means of two metallic plates, separated by a sheet of mica. Of course the losses in such a condenser are comparatively high, but it is believed that these losses can be overcome by a greater efficiency elsewhere in the circuit.

It certainly is possible to turn out the equivalent of a 17-plate condenser in a space not larger than a paper book of matches. Inductances can be correspondingly small by means of the spider-web type of coil or even by means of a more efficient cylinder type of coils, wound with small wire such as No. 36 B. & S. gauge, enameled, it may be said that such coils also have losses, but we need not be concerned with this, because it is most likely that the set will have an oscillatory circuit, when it becomes necessary to kill oscillation anyway, and we might just as well have the losses in the coils or condensers as to get the losses by other "doctoring" means.

On the other hand, the future portable radio set, of the 5- or 6-tube variety, will probably not have any variable condensers at all. We may visualize the following system, which, to the best of my knowledge, has never been described so far. Imagine three small stationary spider-web coils. Then imagine three like coils mounted on a shaft, all to be parallel to each other. The three spider-web coils mounted on the rotating shaft swing back and forth approaching the stationary coils, or receding from them. The scheme may be likened mechanically to three variable condensers mounted on one shaft, except that instead of the plates we have six spider-web coils, three stationary, three rotatable. The tuning is then done by means of the rotating shaft. The six coils, of course, will be the radio-frequency transformers functioning as variometers. The stationary coils may be the secondaries, and the rotatable ones the primaries, or vice-versa. Such coils can be made very small, and need not be larger than about 2 inches in diameter. The thickness need not be more than one-eighth of an inch. Shielding may be applied between the various units, if this be necessary.

We have here, then, a condenserless set, which should be excellent for portable purposes, and where there is a minimum of room available. Having disposed of the small tube, the small condenser (which, after all, may not be needed), and the problem of small inductances, you may now rightly ask, "What becomes of the aerial and ground?" Here again there seems to be no difficulty. I have tried, with very good success, a device which I call the "Cane Loop," and which was made up as follows: I wound rather heavy single-conductor lamp cord on a stick 1¼" in diameter by 33" long. This used up approximately 100 feet of wire. The "Cane Loop" can be used like any other loop, and is directive, as is the regular loop, and while it may not be quite as efficient, it still does the work nicely and brings in distant stations very well.

The "Cane Loop" can be used either horizontally or vertically. It may be laid on the floor or stood in a corner, or otherwise tucked out of the way. It has not quite the directive qualities of the square loop, which may be said to be a good thing, because it need not be rotated and turned, as does the usual loop. This, of course, makes it not quite so selective, but for purposes of portability, and where there is a minimum of room, it will prove ideal.

You will still say that such a loop is too big and can not be compressed into the space of a box camera. I already have an answer for this, as well. A collapsible and flexible "Cane Loop" may be constructed as follows: On a smooth broom handle start to wind 100 feet of lamp cord, similar to that described before, but see that between successive convolutions ½" cotton tape is placed in such a way that one turn of wire goes over the cotton tape, the next one under it, and so forth. There should be two such pieces of tape, one on each side of the "Cane Loop." Then, when the loop is finished, the ends of the tape may be sewed to the insulation of the wire, and the broom handle may be slipped out. This gives a flexible sort of "snaky" loop which may be rolled up into a very small compass and placed on the inside of the portable set. When you need it, pull it out and let it hang down, to be used in this position. There are, of course, other ways by which to arrive at the same results, as, for instance, using thin twine instead of tape, etc. Another flexible loop of this kind can be made by winding the wire on a rubber hose, although this is not quite as effective, because it can not be rolled into such a small compass as by the other method described.

After Television—What?

Science and Invention, vol. 15, no. 2, June 1927

Television, which has been in the making for the last twenty-five years, and the perfecting of which has been freely predicted in many technical articles by many writers, as well as by myself, is now a reality. No longer need we look into the future for it. Although not perfected so that it can be attached to every telephone or to every radio set, television is, today, in a state comparable to that of radio, when its principles were first laid down by Heinrich Hertz, in 1888, and to that of Bell's crude telephone, in 1876. It will take a few years to develop the television apparatus out of the laboratory stage, and much work as yet remains to be done. This is always the case when bringing the laboratory product to the final and practical everyday use with any instrument or technical appliance. It may take two years and even five years before every telephone and every radio set is finally equipped with its television attachment, but you may rest assured that this generation will soon personally witness the appearance of this stage of the art. There can be no doubt about it. But, and we may ask this question soberly,—"After television, what next?"

It is now possible to hear and see a person over a wire line, or over the radio. We have, therefore, made it possible to *transport two senses,* so to speak, to a distance, the two senses being sight and hearing.

In these days of wonder and achievement, we should ask ourselves the question, "What other of our senses is it possible to transport a distance, and, from our present-day knowledge of science, is it possible to transport any of them at all?"

The remaining senses are smell, taste, and touch. Now, then, of course nothing can be said to be impossible, although some things are highly improbable. Thus, the next of the senses on the list being smell, is it possible to smell at a distance? I might say that this is not impossible, although highly improbable. From a technical standpoint, it may be quite possible to build an instrument highly sensitive to odors, which instrument would be able to distinguish between the most subtle variations of various smells or odors. The next step would then be to amplify these, which presumably could be done by means of vacuum tube amplifiers. After that, transmission could be effected electrically by many ways not known.

At the receiving side the impulses would be stepped up and some

1. See **A Radio-Controlled Television Plane** in this book.

2. Gernsback's editorials in the previous and subsequent issues of *Science and Invention* argued that modern science could at once exceed and deceive the senses. On the one hand, the question that "philosophers and scientists in general have for centuries been asking themselves, whether it is not possible that, outside of our five senses, there could be still higher senses," had become possible to explore in a rigorous and experimental fashion: "We know so little about our own anatomy, we know so little about certain glands, and certain other component parts located in our bodies, that it is impossible at this time to say that any one of these may not be for a purpose that as yet can not even be comprehended by us. While it seems improbable, it is not impossible that at some future date human beings may be able to communicate with each other by means of some simple apparatus which is neither sound, nor heat, nor radio. That it will be by some vibratory method is not doubted by scientists today, but just how it will be brought about we are entirely too ignorant to know." Hugo Gernsback, "Are New Senses Possible?" *Science and Invention* 15, no. 1 (May 1927): 7.

And yet the very sciences that seemed to promise a new understanding of the human sensorium produced an estrangement from the new phenomena we were sensing when it came to our lived, embodied experience. Science, in effect, deceives our senses when the "magic" of cutting-edge technology renders invisible processes tangible and induces affects that we shouldn't trust: "As time goes on it becomes apparent that our senses are becoming more and more involved directly due to scientific progress. . . . Many new illusions have come about, all illusions, by the way, that strangely tend to aid progress and help in many cases to further elevate the human race. For instance, you pick up a telephone receiver and listen to your friend talk. The engineer knows better and will tell you that you hear no such thing. You do not hear the voice of your friend at all. It is simply an auto-illusion. . . . We should never trust our senses too much in these latter days of scientific progress." Hugo Gernsback, "Modern Illusions," *Science and Invention* 15, no. 3 (July 1927): 201.

means would have to be provided to unscramble the odors. We can imagine, for instance, 5,000 small tanks at the receiving end, each of which would release, upon a contact being made, an amount of odor depending upon how much was wanted, as indicated by the impressed signal. Thus it would be possible to *recreate* at the receiving end, odors or smells similar to those sent out from the transmitter. All perfectly possible, but, and here comes the big question mark, why would any one want to do it? It would cost a million dollars or more to build such an apparatus, and to what good? So I would say, "Not impossible, but highly improbable."

The next sense to be transmitted would be touch. Again I will say, "Not impossible, but somewhat improbable." It should be a simple thing to construct an electrical apparatus operated at a distance, to transport the sense of touch, in some ways. For instance, it is possible, today, to build an apparatus that, by means of television, would enable mechanical fingers to open the combination of a safe. You would watch by television a mechanical hand, of which you would operate a duplicate at the sending end, and you could thus open or close the combination of the safe without much trouble. This is not impossible, nor is it improbable, but, as with the transportation of the sense of smell, there would not be many uses for such a device.

We have with us today the science of *telemechanics*, which means, operating either by wire or by radio an apparatus at a distance. Some years ago, before television was invented, I described the radio-controlled television plane, which will make it possible, in a not-far-distant future, to operate an airplane without a human being on board, and which, being provided with television apparatus, will enable a distant operator to see and guide the plane over enemy territory and drop bombs at any desired instant, although no one be on board the airplane.[1] We may call this "touch at a distance" and, in fact, it is just that. This is not only quite possible, but will be done in the next few years.

But when it comes, for instance, to actually *feeling* the texture of a piece of cloth, at a distance of a thousand miles, this would seem to be highly improbable, at least for practical purposes.

The remaining sense, namely, taste, may be classed with the transportation of the sense of smell.[2] It is not impossible, but highly improbable. A machine can be invented whereby, just like the one explained under odors, certain impressions are made upon certain media, when certain foods or liquids are placed upon it. The tongue, by dissolving certain of the ingredients of the foods or liquids, gives the sensation of taste. The counterpart of an electrical tongue would present no insurmountable difficulties to a clever physicist, and it is possible to transmit such impressions, in the form of electrical impulses, to a distance. Here, at the receiving apparatus, the impulses could release from tanks or some such other apparatus liquids to simulate the transmitted taste impulses. This is not impossible, but the whole thing would be the height of foolishness, because no one would want to do it, as the expense would be entirely too high.

It might be possible for a New York merchant in this way to taste the quality of Chinese tea 6,000 miles from New York, but why would he wish to do it after all? And certainly, if he had to pay the cost of doing it, he probably would think twice before attempting it.

Coming back to television, what application this interesting invention will take in the future can only be dimly guessed at. There was a time when we were talking first about radio telephony, when it was conceded by practically all of us who had a hand in the shaping of its destinies, that the logical thing would be talking by radio to our friends. Thus in the first book ever written on the subject: "The Wireless Telephone," published by me in 1908, before there was a Radiotelephone, I could see only one use for the coming invention and that was a parallel to the wire telephone. I did not dream of broadcasting, nor did any one else.[3]

The same may be said of television. Right now we are glibly talking about television attachments on our telephones, and radio sets. We may be all wrong, and the new art of television may turn into entirely different directions, undreamt of today. Science has the habit of doing the unforeseen, and often throws our best and most logical predictions on the scrap heap.

3. See **From The Wireless Telephone** in this book.

Wired versus Space Radio

Radio News, vol. 9, no. 1, July 1927

*. . . in which the Editor discusses the American system of free
broadcasting and its alternatives—and the possibilities of the proposed
"wired radio"—which is coming soon—and doubts that the latter
can supersede DX radio reception in the favor of true fans—and expresses
the opinion that wired radio will be an immediate challenge to
the well-known ingenuity of myriad radio constructors—why there
is room for both services to the great radio public.*

When broadcasting was established in this country, the universal opinion was that the service would always be free. No one in America has ever seriously considered broadcasting for pay from the listeners. This is in distinct contrast to the European system, whereby every radio set is taxed by the government anywhere from 25 cents a month upwards to pay the broadcasters. This is the custom that prevails in most countries of the world with some few exceptions. Of course, even in the United States, some one foots the bill—that some one being usually the public. But this is indirect taxation, whereas the European system is one of direct taxation on each set.

In America the broadcasters expect to get back, through the returns from good-will programs or indirect advertising, their outlay for broadcasting—in which effort, it may be said, they have been fairly successful. Not every station, however, operates at a profit, nor will probably do so for some time to come. In general, the principle has been recognized in this country that radio should be free for all, so that any one buying a set can listen in to his heart's content, year in and year out. This is the prevailing system of space radio.

There is, however, another system which may shortly go into operation in the eastern part of the United States, and which is known under the name of wired radio. There is nothing new about this, for it is not a new invention by any means.

General G. O. Squier took out patents on wired radio many years ago, but so far the system has not met with much success or encouragement in application to broadcasting; although this can be accomplished by wired radio over any existing lines, be they telephone or telegraph, electric-light or power.[1] It is understood that, for the time

1. Major General George Owen Squier (1865–1934) first became a proponent of wired radio in 1900 after observing that the tactics of the Spanish American War revolved around "coal and cables": "Reliable submarine [i.e. telegraph cable] communications under exclusive control are not only absolutely necessary, but exercise a dominating influence upon the control of the seas, whether in commercial strategy or in military and naval strategy." Eventually, Squier moved from advocating a system of U.S. government cables that networked the newly annexed territories in the Philippines, Cuba, and Puerto Rico, to an agenda closer to home. Forming a company in 1922 called Wired Radio, Squier sold "centralized transmissions within a rationalized system of stimulus codes," or the canned-music crowd control we now know as Muzak, a term coined by Squier. Jonathan Reed Winkler, *Nexus: Strategic Communications and American Security in World War I* (Cambridge, Mass.: Harvard University Press, 2008), 20–21. Susette Min, "Soothe Operator: Muzak and Modern Sound Art," *Cabinet Magazine* 7 (2002): available online at http://www.cabinetmagazine.org/issues/7/min.php.

being at least, the telephone interests will have none of wired radio. On the other hand, one of the largest electric light and power corporations in the country, with networks extending throughout the east, definitely intends to go "on the wire" with wired broadcasting in the near future, probably within six or eight months.

Many technical difficulties had to be overcome to make this possible, but officials of the company sponsoring wired radio now believe that the difficulties have been smoothed out, and that a real service can actually start very soon. Somewhere in the east there will be studios where three different programs will be broadcast simultaneously on different wavelengths over existing wire systems.

By means of a simple switch on a special receiving set, it is promised, the listener renting the instrument from the wired-radio company will be able to select any one of the three programs being fed to the electric-light wires, and this program will issue from a loud speaker. Two models of receivers are planned. One will use a crystal detector, and is intended primarily for headphone reception. The other will include a regular audio amplifier and a loud speaker, all "A," "B," and "C" power being derived from the power line. No aerial and ground will be used, as the receiver picks the programs directly off the power wires.

If one already owns a radio receiver, he can rent the crystal receiver and connect it to his set in such a manner that the audio amplifier in the latter will amplify the signals; the radio loud speaker will then reproduce them as it does space-radio impulses. This is just an outline of the proposal, from the advance information at hand.

Interesting as are the possibilities of wired radio, however, I personally do not believe that it will prove a formidable competitor of space radio.

It may be said, as a matter of fact, that the so-called wired radio really should not be called radio at all, although it uses radio instrumentalities throughout. In any event, wired radio certainly takes the romance and thrill out of radio broadcast reception, unless you are satisfied with one or two local stations. With space radio even a mediocre set has no trouble tuning in any evening at least forty or fifty stations; and if the set is a really good one, as many as a hundred stations can be logged.

This does not mean, of course, that you can enjoy a hundred different programs during that evening, because the time limitation is against this. But the argument remains in favor of space radio; for the simple reason that, if you wish to stay with any one of the programs, you can do so by tuning in the station you wish to listen to and, unless it is an exceptionally bad night, when much static prevails, there is not much difficulty in staying with the station selected. If I do not wish to know what is going on in Chicago, I can listen to Washington or to New York, or to Atlanta. That is, with space radio. With wired radio it would seem that there must be limitation to a very few programs. The fact is, you will have to take what you get. This seems to be a serious disadvantage, and only time will tell whether it can be successfully overcome.

On the other hand, it may be said that, with wired radio, you do not have to contend with static and uncertainties, but you may be assured of a program at all times. How this choice will strike the average listener it is, of course, impossible to predict.

Then comes the most important point under consideration; and that is, *wired radio will not be free.* The apparatus will not he sold, but leased at a certain monthly rental per instrument.

Just how many people will avail themselves of such a service, when general radio entertainment always has been free, remains as yet to be seen. While there can be no doubt that wired radio will in all probability never supplant space radio, it is possible that it will prove an interesting adjunct to space radio. The parallel to this may be found in space radio and the phonograph.

When radio first came into vogue it was freely predicted that the phonograph would speedily be relegated to the scrap heap. I predicted editorially in *Radio News,* early in 1921, when broadcasting first started, that nothing of the kind was apt to happen, and rather that the phonograph would be helped by radio. This indeed proved to be the case, for there are more phonographs and more records being sold today than there were at any time.

I do not believe that it will be at all practicable, as suggested, for the wired-radio companies to establish a method of secret transmission over their lines, so that only the apparatus rented from them will be capable of receiving their programs; because, the moment apparatus is installed and the nature of the device becomes public, every radio constructor will surely try to build a set by means of which he can tune in on the wired radio.

It is in the nature of every radio fan to investigate and the prediction is freely made that, if wired radio comes into universal use, the parts business will take a sudden leap. Every radio fan and every set builder will no doubt try, at one time or another, to build a radio receiving set that will bring in the wired-radio programs. It seems that the wired-radio interests will be powerless to prevent this; because there is no law on the subject, and because the "bootleg" listener would be stealing nothing.[2]

2. Gernsback made a similar point in a *New York Times* editorial earlier that year about the pitfalls of any system that would require listeners to pay for broadcast content, whether directly as subscribers or indirectly through state ownership and taxation. Criticizing the state ownership of radio broadcasting in Europe in an argument that evokes the influential aphorism attributed to Stewart Brand, "information wants to be free," Gernsback writes: "It has always been the contention of leaders in the American radio industry that the European method of paid listeners' programs [taxed in this model, as opposed to the privatized subscriptions of "wired radio"] was economically unsound. . . . The reason is that a sufficient number of listeners will not pay a fee and an entirely too large proportion of European listeners use 'bootleg' radio outfits and listen in secretly." Hugo Gernsback, "Wellsian Opinion of Radio Tinged with Provincialism," *New York Times,* April 17, 1927.

The Electric Duel

Amazing Stories, vol. 2, no. 6, September 1927

A news item from Milan, Italy, reports the strangest duel, probably, that ever was fought between two men. It was supposed to be a contest to the death—the first electric duel in history. The story has it that two young Italians employed in one of Milan's great industrial electric works, became enamored of the Superintendent's daughter and fought many fistic battles over the Titian haired, comely young woman, reported to be one of the belles of Milan. She could not make up her mind whether she would be the future Mrs. Alessandro Fabiano or Mrs. Benedetto Luigi.

Finally the two suitors reached an agreement whereby they were to settle the issue with a fight to the death. This was immediately decided on after a terrific fistic encounter between the two young men.

Both being graduates of the University of Padua in Electrical Engineering, they chose electricity as a new form of duelling [*sic*]. The place of the encounter was chosen some thirty miles from the outskirts of Milan at a spot where a high tension line carrying over twenty

FIGURE 1. Two duelists are provided with a hood or skull cap, analogous to the connection used in the electric chair. These caps are connected by a wire to a high potential electric line. The same line is grounded. With long poles they attempt to push each other off an insulated platform. Whoever touches the ground will be killed.

thousand volts passed through the open country. One of the wires was connected as shown in the illustration so as to be grounded and another wire was attached to one of the feeders which came down to an insulator attached to a pole nearby. A wooden platform which had been used in building a bridge not far away was utilized as an insulating means. The two duelists had brought along from their factory large insulators upon which the platform rested. The wires were then led to the headmasks as shown in our illustration. Three witnesses, as well as a doctor, who had been sworn to secrecy, were also on hand to witness the strange spectacle that was to take place.

The idea was simple in itself. Each of the two was equipped with a pole and a buffer as shown in our illustration. The idea was that the one combatant was to push the other off the platform. The one remaining on the platform would be the winner. The unfortunate one who first touched the ground would naturally be electrocuted the instant his body came in contact with the earth.

The moment arrived when the two combatants at the shot of a pistol started the battle. The two rivals were wary of each other for the first fifteen minutes, and not much headway was made in the dangerous business. First Benedetto, then Alessandro was nearly pushed over the edge of the board only to recover by a supreme effort. At one time when Benedetto was almost on the brink of going over he grabbed hold of the pole of his antagonist and managed to pull himself forward to the other side again. After awhile the men began to fight hard and furious, till finally a most extraordinary thing happened, which neither of them had foreseen. They were rushing at each other, savagely, diagonally across the platform and both caught each other squarely in the stomach at the same instant. The impact was so terrific and so violent that both keeled over the side, one landing on the ground on one side and the other the opposite side, practically at the same instant. There was a bright flash, and the bodies of the poor unfortunates became enveloped in a dense cloud of smoke and were burned by the lightning-like discharge of the tremendous voltage.

The frightfulness of the situation was so great that I myself woke up and promised myself never again to eat a Welsh rarebit before going to bed.[1]

1. A reference to Winsor McCay's comic strip *Dreams of the Rarebit Fiend* (1904–25). In each installment, a character eats Welsh rarebit before bed—a dish of melted, seasoned cheese poured over bread—and has a surreal dream. Edwin S. Porter made a live-action film version of the strip in 1906 titled *Dreams of a Rarebit Fiend*.

Radio Enters into a New Phase

Radio News, vol. 9, no. 4, October 1927

*. . . in which the Editor reviews three eras of economic change,
which the development of radio art has brought about in this industry . . .
and heralds the expected arrival of a fourth, due to the concentration,
in the hands of a few large concerns, of the manufacture of receivers . . .
why this must result in the division of radio set production between
"the great industry" and the building at home, for personal use
or on a custom basis, by constructors . . . how conditions will favor
the professional set constructors in obtaining a good volume of business . . .
and therefore continue to encourage the manufacture of component parts.*

The great revolutions are those that are silent, and almost invisible. Usually, the greatest revolutions are economic in their nature, and therefore not seen immediately, but they are prolonged over a considerable time. The radio industry is now in the midst of one of these major revolutions, which may change its entire aspect during the next few years.

Let me review what happened in radio, even before broadcasting began. Originally, back in 1910, the radio industry, just coming into life, was content to build and manufacture parts which were sold to the radio amateur. The amateur of that day bought the various parts and promptly screwed them down to an old table. This was his radio set, with which he received "wireless" code from other amateurs, from commercial stations and ships.

About 1916, the manufacture of the first complete sets began. They were more or less self-contained, but necessarily rather crude, to our present way of thinking. But still separate parts, in value, formed the greatest proportion of sales by the radio industry.

When broadcasting started, in 1921, the sale of parts to the public immediately jumped to a tremendous figure; and it was some time before the sale of completely-manufactured sets took the lead and exceeded that of parts. Then gradually the ready-made factory-built set came into the ascendancy and greatly eclipsed the sale of separate parts.

October 1927 cover of *Radio News.*

That is the situation which we have at present. Though in 1922 and 1923 every radio dealer stocked an immense number of parts and but few sets, the situation in this country has been reversed since then, and there are now comparatively few dealers who stock parts. Most of them nowadays stock a complete line of sets and accessories.

There is, of course, a great difference between *parts* and *accessories*; parts being usually taken to mean those items that go into the manufacture or construction of a radio set, whereas accessories are the articles that are necessary for the operation of the complete set. Under accessories are listed such articles as tubes, batteries and socket-power units, loud speakers, headphones, etc.

But it is believed that we are now facing a silent revolution and that the order of things is to be changed once more. The reason is purely economic, and found primarily in the patent situation.

The Radio Corporation of America, and its allied interests, who always have been in the lead, as far as radio patents are concerned, have during the course of years acquired practically all of the important radio patents in this country; and they are therefore in a position to impose terms on all who have infringed their patents. Of course the cry of "monopoly!" will, as usual, go up; but the point remains that, after all, every patent is a *monopoly,* and that any one to whom is issued a patent must protect his rights—otherwise he stands to lose them. In upholding the rights to its patents, the Radio Corporation, after all, is within its rights, and will now reap the benefits from the patent situation.[1]

As generally known, practically every one of the large radio manufacturers is now paying a royalty of 7½ per cent to the Radio Corporation, with a clause of $100,000 per annum minimum royalty. This, of course, means but one thing, and that is, the price of radio sets must go up. Furthermore, the small set manufacturers will no doubt go back into the parts business; as the Radio Corporation will probably license only those who are financially responsible (and it may be presumed that most of the smaller ones probably are not in a position to guarantee a minimum royalty of $100,000 per year).

What has been predicted for a number of years has thus come to pass. The radio set business will be in the hands of a few strong corporations, which will control the legitimate set business in the United States on a highly competitive basis. All other reports to the contrary, this certainly is not a general monopoly of the set business; and we believe that, in time to come, it will work out to the advantage of the public.

But what about the parts business? It is believed in many quarters that, because of the conditions just mentioned, the general parts business will come back with a grand rush. Parts manufacturers, of course, sell their merchandise to set manufacturers; and this outlet,

1. The Radio Corporation of America was formed in 1919 as a "private" company that could take over all of the patents that the U.S. government had seized during World War I. Tim Wu: "Structurally the RCA was rather like the BBC, a national champion; but unlike the British company, it was neither established nor sustained with public duties. Rather, in 1919, the RCA was formed mainly in response to the navy's insistence that all vital radio technologies be held by an American firm, in the interests of national security. And so RCA was fashioned out of the existing American Marconi Company to pool and exploit the rights to use more than two thousand patents owned by General Electric, United Fruit, Westinghouse, and AT&T." Tim Wu, *The Master Switch: The Rise and Fall of Information Empires* (New York: Knopf Doubleday Publishing Group, 2010), 79.

For more on the government's patent seizure, see **Silencing America's Wireless** in this book.

frequently, is their largest source of income. But outside of this, they sell their merchandise to radio dealers and to professional set builders. From present indications, there will be a great and immediate demand for parts; because the small set manufacturer, being put out of the way, will leave the road clear for the professional set builder to come into his own.

The man who builds a set now and then, in his attic, is not likely to be worried by any patent situation, nor will the radio interests be much concerned about professional set builders. Quite the contrary, the Radio Corporation has always maintained that it encourages the amateur and constructor. More than ever, no doubt, the big interests now feel that they have little to lose on account of the set builders. Nevertheless, during the next year, a vast quantity of radio sets and power packs will be manufactured and made by these professional set builders, who do a very sizable business in these transactions. It is a peculiar situation, but one very well understood by radio economists, that there is always a healthy demand for the home-made set. It works out somewhat along these lines:

A set builder or an amateur sees a new circuit described in *Radio News*, or elsewhere. He promptly builds it and finds that it works well. A few of his friends or acquaintances see and hear the set and are impressed by it; because, as is usual in such cases, it is the latest thing. They immediately wish to get one like it; and, right away, the set builder is busy making two or three sets for his friends at a nice profit to himself.

The set builder naturally is well able to compete with the manufacturer, for two reasons. First, his time costs him little, and in price, therefore, he can compete easily with the factory-made set. Secondly, he has the jump on the manufactured set for the simple reason that, as like as not, his circuit is the latest out, and, therefore, will have improvements that the manufactured set can not boast for some months to come.

The large set manufacturers, as a rule, change models only once a year, and, therefore, at some time during the year the radio art will have advanced somewhat beyond their facility to follow. Of this situation the professional set builder takes advantage by incorporating the latest advances in his set.

There are always several hundred thousand people in this country who build sets either for themselves or for friends; and this number is likely to increase materially in the immediate future, because of the economic reasons just explained.

Summing up, therefore, these facts, it is reasonable to predict that the radio parts business during the next few years will show a very large and healthy growth.

The Short-Wave Era

Radio News, vol. 10, no. 3, September 1928

nce more a silent, but nevertheless most important revolution is making itself felt in radio. These periodic revolutions in the art of radio are a novelty no longer, but occurrences to which the careful observer has become quite accustomed. Only a short year ago, the radio art began to turn away from battery operation of sets and started to electrify them. Late last year, still another revolution brought the complete A.C.-operated set.[1]

During the past few months, a new era seems to have opened in what will be known hereafter as the "Short-Wave Cycle."

Not that short waves are something new in radio; quite the contrary. This work goes back to 1908, when amateurs first began to converse with each other, by dots and dashes, below 200 meters. The amateurs have kept at this ever since, without making much, or any, impression upon the general public. The reason for this lack of interest, of course, is that, in order to operate either a short-wave transmitting set, you had to be conversant with the "code"; and this is something at which the general public has always balked.

To the average radio listener, to the radio fan, and to the set builder in general, dots and dashes are so much noise and "static," not to be taken seriously at all. Of course, these listeners lose the best part of radio through not being able to understand or work code, and miss many of the thrills that the amateur enjoys in deciphering a message that comes from the Antipodes. Yet, the general radio public, in spite of all its love of novelty, is quite apathetic to these possibilities and the amateurs have failed to gain very much ground. But of late the broadcast listener is becoming very much interested in short waves; not because he wishes to listen to code and its dots and dashes, but because *he can now receive broadcasts on his loud speaker or headphones* from practically any country throughout the entire world.

A reader in New South Wales, Australia, writes us that while he was writing his letter he was listening to WRNY's short-wave transmitter, 2XAL, on a three-tube set; and had to turn down the volume, otherwise he would wake up his family. All this at a distance of some 10,000 miles! Yet 2XAL, it may be said in passing, uses less than 500 watts; a quite negligible amount of power, as power is rated these days.

The radio set manufacturers for the past few years have claimed vociferously that the day of DX fishing and long-distance records is

---

1. Before alternating current (AC)–powered sets were introduced in 1926, which could be plugged into a wall outlet, radios had to be powered by either rechargeable wet-cell batteries, or dry cells that were simply discarded and replaced once they wore out. So-called battery eliminator sets were mediocre at first, until the RCA Radiola 17 went on the market, the first AC set to be mass produced.

past. In their hearts, the manufacturers knew that this is not the case, but that the truth is, the average manufactured set is poor for long distance and is constructed primarily to receive local stations and others not more than 100 to 250 miles away. Manufacturers stress an assertion that set owners no longer wish to receive long-distance signals, on the theory that such reception never is good. We might take issue with the manufacturers on that score and point out that, with the average manufactured set, distant reception is usually not good; but, *if you have a really good set, distant reception certainly is good.* Hence, sooth to say, the set manufacturers to the contrary, the public is still interested in bringing into our everyday, humdrum existence the thrill of covering great distance. If this were not so, how otherwise explain the sudden and tremendous popularity of short-wave broadcast reception?

During the past six months, every time *Radio News* ran a constructional article on a short-wave receiver, set builders and radio fans have responded in the ratio of three to one, compared to those following up the regulation broadcast-set articles. In other words, there still is a thrill in getting distance; because no set builder builds a short-wave receiver unless he is interested in receiving broadcasts, within 100 to 200 miles of a short-wave transmitter, the usual short-wave receiver is hopelessly inefficient, on account of the so-called "skip-distance" effect.[2]

It is quite the thing now for the radio fan to own one or more broadcast sets for the usual upper-wavelength broadcasts on 200 to 600 meters, and also a separate short-wave set which brings in broadcasts from 20 meters up to 200.

And as time goes on, the interest in short waves is becoming greater and greater. It may well be said that we have as yet not scratched the surface. Technicians believe that in due time all broadcasting will be done on short waves; everything seems to point that way. Already many stations are operating two transmitters simultaneously; one in the upper waveband, and the other in the lower waveband. These stations in doing so are simply staking out their claims for what is to come in the future; and the recent scramble for short waves for television purposes points unmistakably in the same direction.

Most radio engineers today are convinced that the final solution—unless an entirely new invention comes along—of television rests in the lower-wave spectrum. Companies and individuals have been awarded licenses to broadcast television on the lower wave channels and, unless a new invention does come along pretty soon, most of the television broadcasting will be done on these lower waves. Such developments, of course, like all revolutions of this kind, are slow and orderly; but they are revolutions nevertheless.

2. Amateurs discovered the skip or skywave phenomenon in the early 1920s: that lower-frequency radio waves can be reflected off the ionosphere, beyond the curve of the earth's horizon. For more on the ionosphere, see **Signaling to Mars** in this book.

It would not surprise me at all if, during the next five years, the broadcasting of both sound and sight will be done completely on short waves; and the upper wave-channels from 200 to 600 meters gradually abandoned, as fast as we can learn more about the short waves.[3] At the present time, the only thing that stands in the way of universal adoption of short waves is the skip-distance effect. Take, for instance, 2XAL, broadcasting on 30.91 meters; within 200 miles of New York, the reception is poor. Beyond this distance it becomes better and better, the further you get away from the transmitter. This is one of the problems that has yet to be solved and, when it has been solved, there is little doubt that all stations will move down into the short-wave part of the spectrum.

In the meanwhile, it is encouraging to note that the radio broadcast listener and the radio set builder have become more and more impressed with the importance of the short-wave situation. The movement is assuming greater proportions every month, and it will not be long before the set builder will desert the upper wavelengths entirely, and construct only short-wave receivers, to the exclusion of all others.

3. Short waves are generally classified as 1.6 to 30 MHz. In the United States, 200 to 600 MHz is now allocated exclusively for federal government use. Digital over-the-air television is broadcast from 470 to 512 MHz. For a browsable visualization of current frequency allocation, see http://reboot.fcc.gov/spectrum dashboard/.

The Killing Flash

Science Wonder Stories, vol. 1, no. 6, November 1929

March 1

Yesterday Lindenfeld signed his own doom, although he did not suspect it.

Years ago it started at college. I stood one sly insult, one injury after another. Always he came out on top. Always he won. I was always the vanquished, the one bested. He beat me in every sport and every competition. So that, at the end of the term, we were bitter enemies.

But, by that queer chance of destiny, it did not stop there, for we both settled down in Beauford; and, by another stroke of Fate—that irresponsible goddess who often governs our lives—we both went into the same line of business, almost at the same time. It took each of us a few weeks to find this out, and then it was too late. My pride forbade me to go into another venture, so I stuck it out. Within the year, Lindenfeld had put me out of business. That, however, could have been endured; but not his last and crowning insult.

Yesterday, he eloped with my fiancée. That signs his death warrant.

The Killing Flash

A SHORT, SHORT STORY

by Hugo Gernsback

I cut him short with: "Take that, you monster!" —and, jumping back, I press the switch. There is a blinding flash.

Of the long list of insults, offenses, and near-crimes, this one must be the last. The sooner the world is rid of this monster, the better.

As soon as he returns, I shall kill him with my own hands, but so subtly, oh, so subtly, that only he will know that I killed him. And no one will know how he met his death.

As a scientist, this should be easy for me. I will use an entirely new method, something new in murder. Simple. Yet subtle. Very. And the police will never find out how it was accomplished, because *I will kill Lindenfeld by long distance.*

This is the plan. I will rent a little loft in the busy east end of the town. Here I will gradually assemble my high-tension apparatus, a 5-kilowatt generator, a 350,000-volt step-up transformer, condensers, and various other electrical paraphernalia. When everything is connected up, I will attach the output of the 350,000-volt transformer to my telephone line. I will then call up Lindenfeld on the telephone and make sure that it is he who answers. I will make it my business to find out that he is home alone. I can readily find that out by one or two extra phone calls to his house. Then, when he answers me, I have but to step back from my phone and press a switch which controls the 350,000-volt high-tension current. The deadly current will leap over the telephone wire to Lindenfeld's house. A long spark will jump between the receiver and the transmitter, and as Lindenfeld's head is between them he will be electrocuted instantly—not, however, without first having heard my voice.

March 15

Everything is in readiness. The machinery is installed. It works beautifully. Of course, I rented the little place under an assumed name, and I wear a disguise. No one can trace me, *even if they knew how the beast was killed.* Most murderers overlook one little detail. I don't. Everything is planned scientifically. Minutely. Painstakingly. I put a phone call in five minutes ago. I asked for Mrs. Lindenfeld, my former sweetheart, in a disguised voice. *He* answered, the cur. She is out.

I call up River 2650. He answers. I say: "Do you know who this is?" There is a momentary pause: "Sure I know," he laughs derisively. "My dear old friend, John Bernard, what gives me . . . ?"

I cut him short with: "Take that, you monster!"—and, jumping back, I press the switch. There is a blinding flash, as the transformer discharges into the telephone line.

I have killed Lindenfeld!

March 22

BEAUFORD TELEPHONE Co. Beauford, N.Y., March 21, 1929.

Mr. John Bernard 16 Locust Ave., City

Dear Sir:

We received your interesting letter, as well as the manuscript entitled, "THE KILLING FLASH."

You ask our advice whether or not it is possible to kill a person over a telephone line as indicated by you in your ingenious manuscript.

For the benefit of the country, we are glad to state that we believe the scheme to be impractical.

While there are isolated instances of people having been killed by high-tension discharges, due to telephone wires having crossed with power lines, such cases are exceedingly rare.

The reason is that the high-tension discharge usually becomes grounded before it has traveled 100 feet over the telephone line.

In your case, the suppositious Lindenfeld would no doubt, have heard a series of loud noises in his phone, but he would not have come to harm.

Sincerely yours,

BEAUFORD TELEPHONE CO.

March 22

(Newspaper account from the afternoon edition of the *Beauford Eagle*.)

John Bernard, 26, of this city, single, manufacturer of patented appliances, was found dead this morning in a loft at No. 627 East Worth Street. His terribly mutilated body was identified by Henry Lindenfeld who, it was divulged, owned the loft.

From an unopened letter, found at his residence, with an enclosed manuscript sent by The Beauford Telephone Co. to him, it is clear that he planned to kill Henry Lindenfeld by connecting a 350,000-volt high-tension transformer to the telephone line. Upon calling up his enemy, he is presumed to have closed the switch which was to electro-cute Lindenfeld.

Bernard, however, never received the letter from the Beauford Telephone Co. telling him his murderous scheme was impractical as, indeed, it proved to be. Lindenfeld at 9:30 P.M. yesterday received a telephone call from Bernard who spoke threateningly over the 'phone, ending up with: "Take that, you . . . ." That was all. Lindenfeld hung up his 'phone and later retired.

But as Bernard pressed the switch, 200 lbs. of high explosives stored on the floor above—no doubt set off by the high-tension discharge traveling along the joint telephone cable to the next floor—blew up and in the wreck Bernard was crushed to death, hoist as it were by his own petard. His funeral will be held from Levitow's Funeral Parlors tomorrow.

April 1

SCIENCE WONDER STORIES
Office of the Editor

Mr. David Friendly
119 W. 46 Ave.
Springfield, Ill.

Dear Sir:

Sorry to find it necessary to return to you your manuscript entitled "THE KILLING FLASH."

It is a nice story and fairly original; but there are a number of weak points which should be eliminated before we can use the manuscript. For instance, the worst is that, in the story, Bernard certainly would not have written to the telephone company telling them about his plot to kill Lindenfeld. That would, of course, have caused his arrest.

Then, too, the story is improbable; because Bernard, if he had any sense, would not have written down the assertion that he killed Lindenfeld. If you can fix these various points, maybe we can use the story.

Sincerely Yours,
SCIENCE WONDER STORIES
By O. Utis, Ass. Editor

THE END.

How to Write "Science" Stories

Writer's Digest, February 1930

In modern detection of crime, the X-ray machine, test-tubes, bunsen-burners, the microphotograph, the spectrograph, the spectrophotometer, and the polarizer are preceding the baton and police whistle in usefulness.[1] As the pioneer in publicizing these advances in criminal-detection, and in educating both police and public, *Scientific Detective Monthly* is performing invaluable duties.[2]

The primary aim of this magazine is to interest and entertain. Apart from the fact that all material must deal with scientific detection of crime, no editorial foibles and policies exist against which the writer so often battles in vain. There is only one editorial dictum—scientific accuracy. That accomplished, the author can give his imagination free reign.

Realizing that *Scientific Detective Monthly,* published at 96 Park Place, New York, is exploring a new field of action, I have prepared for the readers of *Writer's Digest* the following lengthy treatise on the Scientific Detective Story.

Let it be understood, in the first place, that a science fiction story must be an exposition of a scientific theme and it must be also a story. As an exposition of a scientific theme, it must be reasonable and logical and must be based upon known scientific principles. You have a perfect right to use your imagination as you will in developing the principles, but the fundamental scientific theory must be correct.

As a story, it must be interesting. Even though you are making a description of some dry scientific apparatus, invention, or principle, you should never bore your reader by making your description dry or uninteresting. A really good writer arranges descriptions so that they will always be interesting.

The rules that are given here are recommended for your careful consideration.

Scientific detection of crime offers writers the greatest opportunity and most fertile field since the detective first appeared in fiction. Radio, chemistry, physics, bacteriology, medicine, microscopy—every branch of science can be turned to account. The demand for this material is large, the supply is small. But authors who wish to capitalize on this new source of income must be careful to follow certain well-defined principles. These may be explained by setting forth a list of rules: What To Do, and, as the colored character in Octavus Roy Cohen's story says, "What To Don't."[3]

1. This essay was republished with an introduction by Gary Westfahl in *Science Fiction Studies* 63, vol. 21, part 2 (July 1994), and follows his editorial corrections. First appearing in *Writer's Digest,* Westfahl argues that it "qualifies as the first article ever published on how to write science fiction."

2. *Scientific Detective Monthly* was Gernsback's experiment in a new subgenre of science fiction that ran for ten issues only, January–October 1930. The stories it published used "scientific" (which almost always meant technological) deduction and rationality without the need for futuristic predictions or settings. Much of the fiction published in *Modern Electrics* and *Electrical Experimenter*—by writers like Charles S. Wolfe, Jacques Morgan, and Thomas W. Benson—anticipates this style of technoscientific problem solving set in the author's present day. Gernsback argues here that scientific detective stories are simply a more mature, better researched form of detective fiction and anticipates that they will soon overtake the entirety of that genre. With time, all detective work will become "scientific."

3. Octavus Roy Cohen (1891–1959) was a white writer known for his "negro stories" that ran in the *Saturday Evening Post.* Gernsback is referring here to the character of Florian Slappey, a detective whose "comedic dialect" was featured in Cohen's hideous mash-ups of crime fiction and minstrelsy.

Here are some hints that will increase your remuneration very materially, and will insure your manuscripts a thorough reading and prompt report.

(1) A Scientific Detective Story is one in which the method of crime is solved, or the criminal traced, by the aid of scientific apparatus or with the help of scientific knowledge possessed by the detective or his coworkers.

(2) A crime so ingenious, that it requires scientific methods to solve it, usually is committed with scientific aid and in a scientific manner. Therefore the criminal, as well as the detective, should possess some scientific knowledge. You will see that this is not an absolute essential to a good story; a scientific detective can use science in tracing the perpetrator of an ordinary crime, but judicious use of science by both criminal and detective heightens the interest because it puts the two combatants on a more equal plane.

(3) As most of our readers are scientifically minded, the methods used by criminal or detective must be rational, logical, and feasible. Now, this does not limit the author's imagination; he can develop many imaginative uses of science, provided they are reasonable. For example: one author sent us a story of a man who rendered himself invisible by painting his clothes and face with a non-light reflecting paint. By explaining some of the laws of light and color he made this accomplishment sound plausible, as indeed it is. But he forgot to mention the shadow which is naturally cast by any object standing in the light, whether or not it is visible to our eyes. Readers of our magazine pick us up on these little details. To avoid such mistakes in writing, which really arise from lack of thought, consider your story from every angle before you write your final copy.

(4) What description of clouds and sunsets was to the old novelist, description of scientific apparatus and methods is to the modern Scientific Detective writer. Here again the author must remember that his work will be read by competent scientists among our readers; and, without careful reference to the encyclopedia, no descriptions of scientific instruments should be included in your stories. If you are not in touch with a Public Library, it is advisable to buy a few really good reference books. Criminoscientific fiction has come to stay and your investment will pay you dividends.

(5) A scientific crime is, ipso facto, a mysterious one. Do not underestimate the value of mystery and suspense in your stories; but remember that it is not necessary to commit wholesale slaughter in order to obtain these effects. A story is a good story when the reader can imagine himself threatened by the same peril as the characters in the tale. I can imagine myself killed by a diabolical bacteriologist—

I find it harder to visualize wholesale destruction by a mythical organization. The latter is less personal and individual. Your object is to project scientific diablerie into truthful settings.

(6) For your own sake, avoid hackneyed characterization. Keep clear of fair-haired, blue-eyed Irishmen; long, lanky, keen-eyed, dark-complexioned clean-cut Americans, et al. Although good characterization helps a story, better none than poor ones.

(7) With the advancement of science, the criminal-in-fact is turning scientific as well as the criminal-in-fiction. Therefore we prophesy that Scientific Detective fiction will supersede all other types. In fact, the ordinary gangster and detective story will be relegated into the background in a very few years. It is worth your while, then, to study this new development carefully, devoting all your time and efforts towards turning out good stories of this type. Literary history is now in the making, and the pioneers in this field will reap large rewards.

A Few Don'ts must be remembered if you are to turn out a good story. Here are some:

(a) Don't look through your old manuscripts and tack scientific endings to them. A Scientific Detective Story is a particular type, in which the scientific atmosphere is coherent and permeating right through the tale. To write really good fiction, saturate yourself with the required atmosphere. Read scientific books, visit chemical laboratories and electrical engineering shops. When you are charged with scientific enthusiasm, then sit down and write your stuff.

(b) Don't make your professor, if you have one, talk like a military policeman or an Eighth Avenue "cop." Don't put cheap jokes in his mouth. Read semi-technical magazines and reports of speeches to get the flavor of academic phraseology.

(c) Don't drag in television. It is worked to death and there are so many better appliances you can use in your stories.

(d) What you are not sure about—look up at the library. Don't make your criminal or detective sit down at a table and twirl dials and snap switches without an explanation of what these are for, and why they are operated by the character. Your readers want to know about this; and it gives you a good chance to pad your story legitimately from a scientific text book. Scientific Detective Stories are easy to write once you grasp the swing of them.

(e) Don't fall into the misapprehension that, because your story has plenty of science in it, a plot is therefore unnecessary. The science improves the plot—not vice-versa.

(f) Break up your story into action, dialogue, and description. So many lines of one, so many of another. If you have a long descriptive passage to write, interlope some action, as, for example:

"——so the machine works best in an atmosphere of seventy degrees." The Professor crossed the room, closing the copper contact as he passed it. "The higher level of the atmosphere is cold," he continued quickly: "When the machine——" etc.

(g) Don't underestimate the importance of properly-prepared manuscripts. Not only is the easy-to-read manuscript favored by editors; but care in typing and layout will induce careful and orderly thought in your actual writing. Short lines are easier to read than long ones; this is due to a well-known optical law. Therefore, leave a wide margin on the left-hand side of your page. You will find it much more remunerative to write one story well and carefully, than three rapidly and carelessly. Therefore edit and retype before submitting manuscripts. Clean the type bar of your typewriter. Triple spacing is even better than double. Give an accurate word count on the title page. Don't put in your own captions or chapter heads; we do this after the story is in type.

(h) Don't imitate other writers. Many a story is rejected simply because it is too "close" to another one.

(i) Don't name your characters after those in well-known books. Since Van Dine's books appeared, Adas and Sibellas are appearing in every editorial office.[4] We wish to be introduced to some other ladies.

(j) Don't "splurge." Our office is full of stories that are the "greatest, most terrible, fearful, mysterious, world-shaking mysteries of the age." These stories are usually bad; because, in order to make them sensational to the editorial staff, the author has gone beyond the limits of reason. Besides, we cannot fill a book with superlatives. Many (in fact most) scientific murders are little known, are buried deep in public ignorance. Write stories of which the reader will say: "By Gosh! that might have happened right in this town, and no one heard of it." If you have a good idea, in scientific detection of crime, your story will interest us and our readers. That is all we want.

(k) Don't think that Scientific Detective Stories are hard to write. You are working on virgin ground. The whole field of science is your oyster to open with your pen and extract the pearl of steady work and good pay.

Finally, before you mail your manuscript to us, submit it to some local professor or authority on science, or to a physics teacher, to check the scientific principles involved. If you have studied a text book before writing your story, your theme will probably sound logical and sensible.

Remember that short stories should run from 8000 to 20,000 words; serials 50,000 to 60,000 words. The rate of payment is from one-quarter to one-half cent a word, depending on the value of the story. Higher prices are paid for exceptional stories.

4. S. S. Van Dine was the pen name for Willard Huntington Wright (1888–1939), an art critic and literary editor whose career spanned the worlds of high and popular culture in early twentieth-century America. After a stint as editor of the famous literary magazine *The Smart Set* that saw a rise in experimental work (and a decline in readership), a well-received book on the history of modern art (*Modern Painting: Its Tendency and Meaning* [1915]), and the second major book on Nietzsche published in America (*What Nietzsche Taught* [1915]), Wright turned to writing crime fiction. *The Benson Murder Case* (1926) was the first of twelve wildly popular novels to feature the detective Philo Vance. "Wright's transition from 1910s 'intellectual aristocrat' to a provider of 'ephemeral' fiction for the masses speaks to the permeability of literary boundaries, and his early Philo Vance novels—the most popular American detective novels of the 1920s—chart complicated trends and anxieties in this shift between cultural spheres." Brooks E. Hefner, "'I Used to Be a Highbrow but Look at Me Now': Phrenology, Detection, and Cultural Hierarchy in S. S. Van Dine," *Clues: A Journal of Detection* 30, no. 1 (Spring 2012): 30–41.

When you have finished the first draft of your manuscript, hold it for a few days. Then read it over carefully and see if you have left any points unexplained, and threads tangled. Although you must try to avoid "giving away" the secret of the mystery at the start, your finale must clear up everything completely; so that the reader understands just what has happened.

The whole secret of scientific fiction lies in reading about your subject before you start your story. Get an idea of what the murderer is going to do and how he will do it before you even put a word on paper. Then think out what clues the detective will find, and what scientific apparatus or methods he will use to trace the criminal. If you have a mental vision of your story before hand, and the scientific details at your finger tips, the story will almost write itself as you work.

I have gone through this subject at length, because I am very much interested in having our writers become successful. As time goes on, you will see certain writers forging steadily to the front and gaining a reputation and a following. Those are the authors who have spent a good deal of time and effort in the construction of their early stories, making them works of art from every point of view.

Science Fiction
versus Science Faction

Wonder Stories Quarterly, vol. 2, no. 1, Fall 1930

n time to come, there is no question that science fiction will be looked upon with considerable respect by every thinking person. The reason is that science fiction has already contributed quite a good deal to progress and civilization and will do so increasingly as time goes on.

It all started with Jules Verne and his *Nautilus,* which was the forerunner of all modern submarines. The brilliant imagination of Jules Verne no doubt did a tremendous bit to stimulate inventors and constructors of submarines. But then of course, Jules Verne was an exception in that he knew how to use fact and combine it with fiction.

In time to come, also, our authors will make a marked distinction between science fiction and science *faction,* if I may coin such a term.

The distinction should be fairly obvious. In science fiction the author may fairly let his imagination run wild and, as long as he doesn't turn the story into an obvious fairy tale, he will still remain within the bounds of pure science fiction. Science fiction may be prophetic fiction, in that the things imagined by the author may come true in some time; even if this "some time" may mean a hundred thousand years hence. Then, of course, there are a number of degrees to the fantastic in science fiction itself. It may run the entire gamut between the probable, possible, and near-impossible predictions.

In sharp counter-distinction to science fiction, we also have science *faction.* By this term I mean science fiction in which there are so many scientific facts that the story, as far as the scientific part is concerned, is no longer fiction but becomes more or less a recounting of fact.

For instance, if one spoke of rocket-propelled fliers a few years ago, such machines obviously would have come under the heading of science fiction. Today such fliers properly come under the term science *faction* because the rocket is a fact today. And, while rocket-propelled flying machines are as yet in a stage similar to the Wright brothers' first airplane, yet the few experimenters who have worked with rocket-propelled machines have had sufficient encouragement to enable us to predict quite safely that during the next twenty-five years, rocket flying will become the order of the day.

Which is the better story, the one that deals with pure science

Fall 1930 cover of *Wonder Stories Quarterly.*

fiction or the one that deals with science *faction?* That is a difficult thing to say. It depends, of course, entirely upon the story, its treatment and the ingenuity of the author.

Of course, the man of science, the research worker, and even the hard-headed business man will perhaps look with more favor upon science *faction* because here he will get valuable information that may be of immediate use; whereas the information contained in the usual run of science fiction may perhaps be too far in advance of the times and may often be thought to be too fantastic to be of immediate use to humanity. So between science fiction and science *faction* there will always be a great gap—and each will have its thousands and perhaps millions of adherents.

Television Technique

Television News, vol. 1, no. 3, July–August 1931

As the television art advances and tends to become popularized, so the technique of television presentation will advance as well. Radio people will appreciate this fact when they think back how broadcasting first started, and how crude the transmitting technique was.

At the transmitter, somebody spoke into a poor-grade microphone which, like as not, often suddenly went out of commission. It took several years to discover that you should have two independent microphone transmitters, with two independent lines; so that, when one went out of commission, the station was not cut off the air. Then, at the radio transmitter itself, neither the high nor the low notes passed through; and the result at the listening end was consequently poor.

Also, too, there were often long waits, whenever one program went off the air, before another came on. Broadcasters started with a single studio and, when an orchestra went off the air, it took some time to get out of the studio, and there was an interval before the next performer could go on the air. Of course, these things sound ludicrous now; but in those days, when each station had only a single studio, and there was no switching arrangement from one studio to a multiplicity of others, as there is today, the problem did not appear so simple.

When television finally gets running along the lines of broadcasting, we will have all the refinements that we find in audible radio today, plus a good many additional contributions which the television art demands.

Just as we used to have a single microphone in the old days, so we now have a single transmitting scanning mechanism which, as like as not, in the midst of the program, becomes defective; and the image vanishes from the television receiver. In the future, of course, we shall have double and even triple scanning transmitting apparatus, which will overcome possible failure of one of them; and, secondly, this will also give us more intensity, a thing which—in television particularly—is highly desirable.

The other day we were listening in and viewing a television performance from one of the local television stations. After the singing team got through, we were greeted with an announcement that we should stand by for some five minutes, until the carbons were changed on the television transmitter.[1] Of course, such crudities as this will not happen in the future. The "listening" and "seeing" audience will not stand for this in the future; while today, of course, they are glad to see anything.

Then, as we have in present broadcasting, the so-called fade-in and fade-out (that is, where we first hear the music or voice in the background and then hear it come up to full volume) so we will have one screen image merge into another in television. We will have the "fade-out," as we have today in motion pictures. We will have superimposed scenes, and even a multiplicity of scenes at the same time.[2]

We will have all sorts of novelties, once television gets under way; and some enterprising network will soon show us one ship in the Atlantic Ocean and one in the Pacific, both at the same time, side by side. Then, of course, we will have trick television, just as we have trick motion pictures today. It will be possible even to have two well-known actors appear on two different stages in different parts of the country; while, by television, the audience will see the two actors (actually separated by hundreds of miles) together on the television screen, going through their act as if they were actually side by side. Of course, it would not be possible, for instance, for them to shake hands; but they could go through a performance; speak their lines, and act the parts just as if they were together. This naturally would be accomplished by televising the two scenes independently, and bringing them together on one screen.

1. The mechanical television Gernsback discusses here is a flying spot scanner, which produces an image by casting a bright beam of light onto an object and detecting its reflection off the surface of that object. The "carbons" he refers to as frequently requiring replacement are likely the powerful arc lamps projecting that beam of light. In a profile of Charles F. Jenkins's television studio later in this issue, the flying spot scanning system is described as "comprising the beam projector and the photo-electric cell banks. The former is a powerful arc lamp in a large housing, provided with an enclosing scanning disc and with three lenses of different focal lengths; the assembly being mounted on a swivel pedestal resembling the usual barber's chair base. The operator can readily aim the flying-spot beam at the subject and, by using the proper lens, cover the desired area for a close-up, half length, or long shot, without changing the relative positions of either the subject or scanner. The scanner operates on the standard system of 60 lines, 20 pictures per second." D. E. Replogle, "The Jenkins New York Studio," *Television News* 1, no. 3 (August 1931): 170–71.

A working replica of the apparatus was made by the Early Television Museum in Ohio: http://www.earlytelevision .org/fss_camera.html.

2. For more on the comparative development of film and television editing techniques in the late 1920s and early 1930s, see Erik Barnouw, *Tube of Plenty: The Evolution of American Television* (New York: Oxford University Press, 1990), especially chapters 2 and 3.

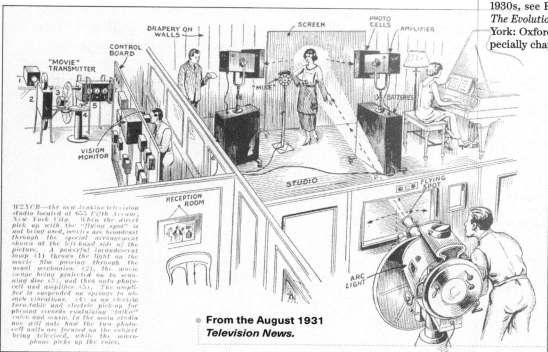

W2XCR—the new Jenkins television studio located at 655 Fifth Avenue, New York City. When the direct pick up with the "flying spot" is not being used, movies are broadcast through the special arrangement shown at the left hand side of the picture. A powerful incandescent lamp (1) throws the light on the movie film passing through the usual mechanism (2), the movie image being projected on to scanning disc (3), and then onto photo-cell and amplifier (5). The amplifier is suspended on springs to absorb vibrations. (4) is an electric turn-table and electric pick-up for playing records containing "talkie" voice and music. In the main studio one will note how the two photo-cell units are focused on the subject being televised, while the microphone picks up the voice.

● **From the August 1931** *Television News.*

Quite a good many tricks which are now being accomplished in motion pictures can be adapted to television as well. We can have the usual close-up; we can have the heroine shedding glycerine tears, while the audience is none the wiser.

We will have our thrillers (like, for instance, Harold Lloyd doing his "impossible" stunts on the face of a skyscraper) because most of these effects can be easily enacted in the studio, without the television audience seeing the technique, any more than the motion-picture audience knows the "inside" of the movie trickery.[3]

We will see pre-view "ticklers" of important motion pictures, giving us a few snatches here and there; similar to what we see in motion picture houses now. The difference, however, is that the enterprising motion-picture directors will have the scenes actually enacted for the benefit of the visionists right at the location, while the sets in Hollywood or elsewhere are still intact; since it will cost very little to have a television transmitter on location and take a few "shots" for the edification of the television "audience" who will wish to see the motion picture in its entirety later on, if the few televised snatches are sufficiently interesting.

Then, of course, we shall have almost immediately the good old sponsored program. We will see Amos 'n' Andy in person; we will see the Cliquot Eskimos doing their stuff; we will see the Armstrong Quakers; the Empire Builders and many others.[4] The television programs enacted by the scenists at the television transmitter are going to be vastly more expensive than the mere aural programs are today. You can fake a train and a galloping horse to perfection by sound alone, at practically no cost, but, when television comes in in earnestness, you can't fake a horse, and you can't fake a collision between trains (except by transmitting motion pictures, which would probably be detected by the audience).

It will be necessary, in other words, for the television broadcasting companies of the future to become practically motion-picture producers on a vast scale, where everything that is thrown on the receiving screen will be backed up by the actual happening at the transmitter.

3. Gernsback is referring to the famous scene from the comedy film *Safety Last!* (1923) in which Harold Lloyd dangles from the arms of a ticking clock atop a skyscraper.

4. The crude minstrel radio program and "most popular show ever broadcast," *Amos 'n' Andy,* did in fact become a television series on CBS in the 1950s, one that was vigorously protested by the NAACP. *The Cliquot Club Eskimos* and the *Armstrong Quakers* were radio variety shows in the 1920s, and *The Empire Builders* (1929–31) was a historical drama about the construction of the Great Northern Railway, sponsored by Great Northern Railway. Mel Watkins, "What Was It About 'Amos 'N' Andy'?," *New York Times,* July 7, 1991.

Wonders of the Machine Age

Wonder Stories, vol. 3, no. 2, February 1931

STATEMENT OF POLICY

*In the present editorial, Mr. Hugo Gernsback not only sets forth
an interesting discourse on the Machine Age but, at the same time,
dispels a number of erroneous views held by various so-called authorities.*

*Incidentally, the subject is intimately linked with Science Fiction,
and we know that it will be of more than passing interest to our readers.*

*We would very much like to learn from our readers how they feel
on this important subject.*

THE PUBLISHERS.

It is a curious failing of the human race, that it has never been able to look to the future and to apply the lessons of past history to its own benefit.

Yet, we all know that the future is made by the past; and the trite saying, "History repeats itself," is a proved fact.

At the present time, when humanity finds itself in the throes of a world-wide depression, everybody is looking for light on the subject; and people ask themselves what is the cause of the depression, and particularly of unemployment.

Of late, a certain school of thought has cried persistently that all our present troubles, particularly unemployment, are directly traceable to our "Machine Civilization." This attitude has been taken up by many economists and indeed, many so-called industrialists; and many books have been written on the subject. By reading this abundant literature, the reader very often will come to the conclusion that there must be something in their beliefs and that, indeed, the "machine monster" is beginning to swallow humanity and, pretty soon, we will be all at the mercy of the machine.[1]

We need not get very excited about the statements of these "authorities"; because, in the past it has been found that many authorities were usually wrong in their outlook on the future. Only about 25 years ago one of our greatest economists made the statement that by this time (that is, around 1930) there would not be enough wheat grown to keep the world's population from starving. Everybody can see the

1. Marx's chapter on "Machinery and Large-Scale Industry" in volume 1 of *Capital* begins with an analysis of the machine as a new mode of production that replaced handicraft or specialized manufacture during the Industrial Revolution. He writes that the machine constitutes an intensification of the organic limitations of the human: "The number of tools that a machine can bring into play simultaneously is from the outset independent of the organic limitations that confine the tools of the handicraftsman" (495). And when the skills of an individual laborer are turned into the specialized "organs" of a machine, the conventional division of labor between differently trained workers "no longer exists in production by machinery. Here the total process is examined objectively, viewed in and for itself, and analysed into its constitutive phases" (501).

But for Marx, the debate over whether the machine steals jobs or creates new ones overlooks the fact that the nature of that laborer's work is fundamentally changed under the conditions of mechanical production. Differences in mode of production are important to study because they create differences in the social relations of production. In this way, he sees Luddite revolts in which workers destroy factory machines (and, one would imagine, arguments that the "machine monster" caused the Depression) as a "crude form" of revolt: "It took both time and experience before the workers learnt to distinguish between machinery and its employment by capital, and therefore to transfer their attacks from the material instruments of production to the form of society which utilizes those instruments" (555).

See especially the section on "The Struggle between Worker and Machine" and "The Compensation Theory," in Karl Marx, *Capital, Volume 1*, trans. Ben Fowkes, 553–75 (New York: Vintage Books, 1977).

foolishness of that statement, because we have today more wheat than we have ever had.

Twenty years ago, an internationally-known scientist predicted that by 1925 the available petroleum would be exhausted. Yet, in 1931, we find the price of petroleum falling because we have too much of it.

When the automobile first came along, experts all over the world told us that it was the death-knell of the horse. Yet, in the United States in 1922 more farm horses were living and in service than there were before the advent of the automobile in 1900. (1900—18,267,000. In 1922—18,564,000.) Since then horses have declined somewhat, but the automobile will never make the horse extinct.

Then, it was confidently predicted also that the automobile was putting the railways out of business; and there are, even today, many authorities who still believe these fairy tales when, as a matter of fact, the railways are continually gaining ground and are now using the auto to truck themselves as a valuable adjunct to their business!

In the early '30's, a well-known Patent Examiner in Washington threw up his job because he had become convinced from his studies that everything worthwhile had been invented, and that there was no future for him in the Patent Office. That was before the day of the telephone, the X-Ray, the automobile, the airplane, radio, and thousands of other inventions made in the last sixty years. Yet, here was an expert in his line who could not see further than his nose, but he was certain that he was right; otherwise he would, of course, not have made such a colossal fool of himself.[2]

I have given these few examples only to show why people should keep a level head in these days of stress and not become unduly excited about the future. The Machine Age and Applied Science, far from destroying humanity, will now, and will forever be, humanity's servant.[3]

I will go on record and state that, with very few exceptions, practically all *useful* inventions and *useful* machines, so far invented, have not only helped the human race socially, but HAVE BEEN THE DIRECT CAUSE OF KEEPING MILLIONS OF PEOPLE EMPLOYED. It is the common talk of the misinformed, as well of the so-called informed classes, that the present unemployment situation is due to the machine. The argument runs something as follows:

A factory gives employment to a thousand men. A new invention is made and new machinery is installed in this plant, which then does twice as much work with half the men. Half of the men are thrown out of employment; consequently, the new machine has been destructive, in that it put 500 men out of work.

This argument is pure foolishness; for, if its proponent stopped to

2. For more on the myth of the Patent Office examiner who thought everything that could be invented, had, see **Imagination versus Facts** in this book.

3. Here, Gernsback begins to develop ideas inspired by Technocracy, an American political movement that blossomed during the Depression. Technocracy argued that engineers and skilled technologists should replace politicians, who were not able to rationally and effectively address the material needs of the American public. Gernsback's short-lived *Technocracy Review* attempted to capitalize on the movement and serve as a forum for the exchange of Technocratic ideas. It lasted for only two issues, and Gernsback himself attempted to remain neutral in its pages save for the argument repeated here, that new inventions and machines produce employment. The inaugural issue of the *Technocracy Review* (February 1933) contained articles by Howard Scott ("Technocracy Speaks"), Paul Blanshard ("A Socialist Looks at Technocracy"), William Z. Foster ("Technocracy and Communism"), and Gernsback's assistant editor David Lasser ("Technocracy—Hero or Villain?").

According to Sam Moskowitz, Lasser was the prime mover behind the *Technocracy Review*. Lasser had become a member of the Socialist Party in 1933, inspired by its efforts to fight unemployment. But eventually, Gernsback was "made uncomfortable by the company he was keeping." In an interview with Moskowitz, Lasser recalls that Gernsback called him into his office and said, "If you like working with the unemployed so much, I suggest you go and join them." Sam Moskowitz, *Seekers of Tomorrow: Masters of Modern Science Fiction* (Cleveland, Ohio: The World Publishing Company, 1966), 357. This also meant Lasser lost his post as managing editor of *Wonder Stories*, a role in which he greatly increased the quality of fiction found in the magazine. Mike Ashley: "during the period of his involvement with science fiction, 1929 to 1933, he was without a doubt the best of the magazine editors. It is almost entirely due to Lasser that science fiction continued to develop and not stagnate during the years of the Depression. Science fiction's Golden Age really began with him." Mike Ashley, *The Gernsback Days: A Study of the Evolution of Modern Science Fiction from 1911 to 1936* (Holicong, Penn.: Wildside Press, 2004), 141.

De Witt Douglas Kilgore writes on David Lasser's participation in amateur rocket societies and interest in "the rocket

think about it, he could reason that the argument of necessity was wrong. There is not a single useful machine or invention which has not given actual employment in the course of time, and actually created employment where none existed before.

In *Collier's, the National Weekly,* lately, was a most interesting article which I heartily recommend to all. It is entitled "The Job Making Machine" by John T. Flynn. The author takes a single machine, the automobile, and cites the following facts, which are not fancy or hearsay, but can be easily proved from reports of the Department of Commerce, and other industrial bodies, for all who care to look into the facts:

There are over a million men employed in the production of automobiles and their components: 427,000 automobile factory workers; 250,000 automobile parts workers; 135,000 tire workers; 72,000 in blast furnaces and steel mills; 18,000 in the production of copper and other metals; 18,000 lumbermen and wood workers; 76,000 in the production of textiles, glass, and other materials; 7,000 producing coal and power. There are 170,000 making and marketing automobile accessories.

Then, too, there are 370,000 dealers and salesmen; 420,000 garage and service men; 650,000 chauffeurs and cabmen; 1,500,000 truck and bus drivers.

The total of those who derive their employment from the automotive industry and automotive operation reaches *four and half millions;* an increase of 1,275,000 in the past five years! The automobiles use more steel than any other industry—more than the United States produced thirty years ago; their tires use more cotton than was used for all purposes thirty years ago. There are 60,000 men employed in the production of gasoline; 125,000 in road building and maintenance. On the railroads, 90,000 are employed through the transportation of motor cars. Vast amounts of building construction have been brought about, by the extended radius of travel which the automobile has given to the public.

"I have chosen the automobile industry," says Mr. Flynn, "to illustrate what is going on in industry as a result of the machine; because its effects are visible on a large scale. But what is true of the automobile is true in a smaller way of many other industries."

Now then, the automobile is only one machine; but the same case could be built up for any other useful machine, be it the steam engine, the printing press, the radio or thousands of others.

It is perfectly true that a new and revolutionary invention or labor-saving machine may throw out of employment, TEMPORARILY, some people. No one denies this; you cannot have revolutions without

as an instrument of political reform" after his editorial career with Gernsback: *Astrofuturism: Science, Race, and Visions of Utopia in Space* (Philadelphia: University of Pennsylvania Press, 2003), 31–48.

a temporary loss of some kind. But the point is, that the 500 men thrown out of employment, whom I mentioned above, are not going to be out of employment forever; they will find other jobs, *probably created, directly or indirectly, by the very machine that threw them out of work originally.* Our present civilization is so interdependent, as a little thought on the subject must convince the most skeptical, that in the final analysis the Machine and Applied Science will make employment for them.

What most people are apt to forget is that there were unemployment crises long before there was a Machine Age; and that there have been unemployment cycles from the earliest recorded civilization down to the present day. More than 150 years ago, there was certainly no Machine Age; yet there were world-wide depressions and unemployment cycles then, just as we have them today.

Famines, pestilence, and other scourges were the usual thing long before the Machine Age, when there was no machine to put the blame on. Today, thanks to the Machine Age, we no longer have country-wide famines of the severity of the past; and thanks to science, we no longer have the scourges and pestilences that our ancestors had to contend with. Quick communication by rail, water, and air, tends to do away with both acute famine and widespread diseases.[4]

It is admitted, as I said before, that new machines, new inventions, may *temporarily* throw people out of employment; but, within a few years, this situation rectifies itself to the benefit of all concerned. It is even conceivable that the Machine Age in the end will do away with business cycles. However, since business fluctuations are caused principally by human nature, it may take centuries, and indeed, thousands of years, before the millennium is reached—if ever.[5]

All nature runs in cycles. Just as the sun has its sunspots in a regular cycle; just as the earth has its cycles of earthquakes and cycles of drought; just so economics will have its cycles—its ups and downs. There are many causes to which the present unemployment situation is attributable; for they are world-wide and are just as acute in the non-mechanized countries as they are in the more industrial ones. We need not go into these causes, because I really believe that no one knows all of them.

Certainly, the Machine Age is not an important cause and, if we did not have our machine civilization, it is quite certain that the depression would be far more severe than it is today. It is certain that the cycle would, as it has done in past ages, run for a longer period than it does today.

The reason is, that the causes of all of our troubles lie not in the machine, but are really found in human nature. When people start

4. The Technocrats took this argument even further. An early pamphlet reads, "In Technocracy we see science banishing waste, unemployment, hunger, and insecurity of income forever. . . . we see science replacing an economy of scarcity with an era of abundance. . . . [And] we see functional competence displacing grotesque and wasteful incompetence, facts displacing guesswork, order displacing disorder, industrial planning displacing industrial chaos." Ralph Chaplin, "Foreword," in *Science versus Chaos* (New York: Technocracy Inc., 1933). Quoted in Howard P. Segal, *Technological Utopianism in American Culture* (Syracuse, N.Y.: Syracuse University Press, 2005), 122.

5. This is a core tenet of Technocracy: that cycles of economic boom and bust are *human* in origin (whether due to greed, mismanagement, or the alienation of workers) and that if we allow the economy to be run more logically, objectively, and mechanically, the Depression could be ended overnight.

to hoard their money, when they are afraid of their own shadow, and when they tremble at the future for no reason at all, the machine certainly cannot be blamed.

For my part, I have always felt that the present depression is purely psychological rather than physical, and in the last analysis, it probably will be found so.

What has all of this got to do with SCIENCE FICTION? Just this:

Science fiction is based upon the progress of science; THAT IS ITS VERY FOUNDATION. Without it, there could be no science fiction.

On the other hand, science fiction is supposed to portray and mirror the future as reasonably as it is possible to do from our present perspective.

If you admit that Machines and Science are all wrong, and that they are destroying humanity, then there should be no such thing as science fiction; and it would be useless to preach the gospel of science.

I feel most strongly on the subject because during recent months, we have received a number of science fiction stories, probably fostered by the unemployment atmosphere, which I have rejected because they distorted the facts and, in many cases, were pure out-and-out propaganda against the Machine Age.[6]

Some of the authors, who should know better, maintained in their stories that, little by little, the machines and science are becoming a Frankenstein monster, and finally humanity will rise in revolt and destroy all the machines, and go back to the Middle Ages. The usual underlying plot is that, because of capitalistic concentration of wealth, the machines will ultimately be controlled by a few powerful men, who will enslave the entire world to the detriment of humanity.[7]

I have gone to great length, in my opening paragraphs, to show that this situation has never arisen as yet; and, from past experience, we know that it cannot arise. And it is for this reason that *Wonder Stories* will not, in the future, publish propaganda of this sort which tends to inflame an unreasoning public against scientific progress, against useful machines, and against inventions in general.

It is conceivable, and indeed is proved by history, that nations are born, grow to maturity, and die. This has happened through our entire recorded history. It probably will repeat itself indefinitely in the future. Which nations will survive, we have no means of foretelling. Of course, when it happens, the Machine Age will be blamed again, and the authors of the assertions will be blissfully ignorant of the fact that other nations, living alongside of them, prosper and grow while living in the self-same Machine Age.

It is, indeed, quite within the bounds of possibility that humanity will at some time find itself back in the Middle Ages—or worse. This

6. One author who was able to incorporate a degree of skepticism into otherwise favorably technocratic fiction was Nathan Schachner. "The Robot Technocrat" was the cover feature for the March 1933 issue of *Wonder Stories,* which also included an editorial by Gernsback on "The Wonders of Technocracy." In a parable of collective decision making after disaster, American society lays in ruins after a Luddite revolt against the machinery of production in the year 1954. Onto this scene steps Hugh Corbin, leader of a movement known as the Reconstructionists who seek to rebuild the infrastructure and machinery that, in their eyes, once constituted the promise of the American experiment.

Meanwhile, a Russian scientist named Anton Kalmikoff creates a massive computer that can predict the future given a range of political, economic, and social variables, and runs simulations on all possible forms of government. It is a Depression-era story in which scientists battle politicians for control of the country, but one that nevertheless weighs the comparative benefits of different political systems.

7. Years later, after the devastation of a second World War, Luddism seemed to offer a much more viable politics for works of science fiction, argues Thomas Pynchon in an essay that refers to the genre not just as "a nearly ideal synthesis of the Two Cultures" but also "one of the principal refuges, in our time, for those of Luddite persuasion." "Modern Luddite imaginations have yet to come up with any countercritter Bad and Big enough, even in the most irresponsible of fictions, to begin to compare with what would happen in a nuclear war. So, in the science fiction of the Atomic Age and the cold war, we see the Luddite impulse to deny the machine taking a different direction. The hardware angle got de-emphasized in favor of more humanistic concerns— exotic cultural evolutions and social scenarios, paradoxes and games with space/ time, wild philosophical questions—most of it sharing, as the critical literature has amply discussed, a definition of 'human' as particularly distinguished from 'machine.' Like their earlier counterparts, 20th-century Luddites looked back yearningly to another age—curiously, the same Age of Reason which had forced the first Luddites into nostalgia for the Age of Miracles." Thomas Pynchon, "Is It O.K. to Be a Luddite?" *The New York Times,* October 28, 1984.

IN THIS ISSUE — THE ROBOT TECHNOCRAT

NOW 15¢

WONDER Stories

HUGO GERNSBACK
Editor

A GERNSBACK PUBLICATION

March

Other Science Stories
In This Issue

"DWELLER IN MARTIAN DEPTHS"
by Clark Ashton Smith

"WANDERERS OF TIME"
by John Beynon Harris

"THE MAN WHO AWOKE"
by Laurence Manning

also has happened in the past, and may conceivably happen again. If it does happen, the causes will probably be due to great cataclysms, such as floods and earthquakes or wars, but the Machine Age will have very little or nothing to do with it.

Humanity will have its ups and downs in the future as it had in the past.

I am, however, of the firm opinion that, as the past 150 years have shown, because of Science and the Machine Age, the ups and downs of humanity will be less severe than they were before the Machine Age; and that is the reason why I have no patience with those who tend to preach the evils of the Machine Age, which, in the long run, are non-existent.

Illustration for Nathan Schachner's story "The Robot Technocrat" (1933).

1. This editorial note introduces John W. Campbell's short story, "Space Rays." The ad for the story in the previous month's issue announces: "'Space Rays' is a new and rather intriguing story by that popular author. Mr. Campbell probably had an unusual idea in writing this story. He not only has written a gripping adventure of space, but has at the same time put his finger on one of the weaknesses of science fiction. You will be amazed at the man who can throw a screw driver with the speed of a pistol bullet, and who can whip a dozen men at once. These feats, strange as they seem, are explained by our author scientifically, in a battle of one man against desperate odds."

Campbell would, of course, soon become famous as the editor of *Astounding Science Fiction* from 1937–1971, playing a decisive role in the development of the genre throughout the twentieth century. Here, Gernsback takes the opportunity to remind his readers (and potential contributors) of the empirical plausibility and educational aims of scientifiction as he first formulated it. Although Campbell himself would later demand the same sort of empirical rigor from his authors, this is evidently not what he was after in this story of space piracy and a hero from Jupiter with the strength of ten men. Commenting on Gernsback's tone here, Gary Westfahl writes, "Clearly, Gernsback was imposing his own didactic message on a story with no satiric intent, but no doubt thought this was the gentlest way to inform Campbell and his colleagues that their colorful adventure stories were not the sort of science fiction he admired." Gary Westfahl, "The Mightiest Machine: The Development of American Science Fiction from the 1920s to the 1960s," in *The Cambridge Companion to American Science Fiction*, ed. Gerry Canavan and Eric Carl Link, 17–30 (New York: Cambridge University Press, 2015).

Reasonableness in Science Fiction

Wonder Stories, vol. 4, no. 7, December 1932

When science fiction first came into being, it was taken most seriously by all authors. In practically all instances, authors laid the basis of their stories upon a solid scientific foundation. If an author made a statement as to certain future instrumentalities, he usually found it advisable to adhere closely to the possibilities of science as it was then known.

Many modern science fiction authors have no such scruples. They do not hesitate to throw scientific plausibility overboard, and embark upon a policy of what I might call scientific magic, in other words, science that is neither plausible, nor possible. Indeed, it overlaps the fairy tale, and often goes the fairy tale one better.

This is a deplorable state of affairs, and one that I certainly believe should be avoided by all science fiction authors, if science fiction is to survive.

In the present offering, Mr. John W. Campbell, Jr., has no doubt realized this state of affairs and has proceeded in an earnest way to burlesque some of our rash authors to whom plausibility and possible science mean nothing.[1] He pulls, magician-like, all sorts of impossible rays from his silk hat, much as a magician extracts rabbits. There is no situation that cannot easily be overcome by some sort of a preposterous—(as he terms it)—gimmick. Hurling fifteen million horsepower from one space flyer to another means nothing. If he has left out any colored rays, or any magical rays that could not immediately perform certain miraculous wonders, we are not aware of this shortcoming in his story.

In other words, the author proceeds to burlesque science fiction—not only science fiction, but he burlesques his own abilities, to show us what really can be accomplished when the bounds of reasonableness are overstepped.

Yet, all in all, he has spun a delightful tale which will, nevertheless, entertain you, and keep your interest throughout. We are tempted to rename the story "Ray! Ray!" but thought better of it.

I have gone to this length to preach the sermon in the hope that misguided authors will see the light, and hereafter stick to science as it is known, or as it may reasonably develop in the future.

Inlaid editorial commentary for John W. Campbell's story "Space Rays" in the December 1932 *Wonder Stories*.

Index

Note: Entries in bold refer to articles by Hugo Gernsback reprinted in this book.
Page numbers with an n indicate footnotes.

experiments, 78–79; use in warfare, 136, 172, 332n

telegraphone, 150–51, 236

telepathy, 44–45, 105–7, 155n, 276–77

telephone (wired), 13, 83–84, 90, 164, 200, 303n1; use in Hugo Gernsback's experiments, 39, 95, 115n3, 236

Telephot, 84–89, 101, 103–4

television, 90n3; amateur experimentation with, 37; definitions of, 11, 83–84; predictions for, 237–38, 278–81, 319–21; prototypes of, 20–22, 41, 83n1, 241–43, 278, 344–46

Television and the Telephot (1909), 83–89

Television News [magazine], 37

Television Technique (1931), 344–46

Telimco, 1–2, 11, 15, 39, 42–43, 63–64

10,000 Years Hence (1922), 245–50

Tesla, Nikola, 15, 129n2, 214n; fictional representations of, 174–76, 183–84; speculative inventions by, 160, 201, 240–41

Thibault, Ghislain, 160n3

Thomas A. Edison Speaks to You (1919), 202–13

tinkering, 10, 39, 44–49, 165–67, 312–14, 327–29. *See also* amateur experimenter communities; craft; Gernsback, Hugo: theories of media and technology; science fiction: as a form of invention

Treatise on Wireless Telegraphy, A (1913), 115–28

tuning coil, 115, 117n5, 121, 300–301

Twain, Mark, 12

upward mobility, 1, 9, 45–46. *See also* class; Gernsback, Hugo: political views of; labor movements

vacuum tube, 15, 35, 70n4, 212, 244; and amateur experimenters, 48, 312; Hugo Gernsback's opinions on, 132n6, 237n1, 298, 306, 315–17

Vail, Theodore Newton, 159

variable condenser. *See* condenser.

Veblen, Thorstein, 31

Venus, 57, 214, 216n3

Verne, Jules, 12, 54n136, 229n5, 287–88, 292; as a prophet, 139–40, 270

Verrill, A. Hyatt, 50

War and the Radio Amateur (1917), 168–70

Wells, H. G., 12, 200n1, 210n3, 287n2, 292–93

Wertenbaker, G. Peyton, 7, 52, 269n, 292, 293n

Westfahl, Gary, 9, 51, 55, 101n, 337n1, 354n

What to Invent (1916), 163–64

Wheeler, Edward L., 56, 153n2

Whitehead, Alfred North, 277n3

Who Will Save the Radio Amateur? (1923), 256–68

Why "Radio Amateur News" Is Here (1919), 194–5

Why the Radio Set Builder? (1927), 312–14

Wicks, Mark, 51

Wilson, Woodrow, 171–72, 192n

Winner, Langdon, 32

Wired versus Space Radio (1927), 322–24

Wireless Age, The, 6

Wireless and the Amateur: A Retrospect (1913), 110–12

Wireless Association of America, The, 27, 69–74, 92, 110, 111n8

Wireless Association of America, The (1909), 69–72

Wireless Joker, The (1908), 67–68

wireless power transmission, predictions of, 113–14, 159–60, 201

wireless telegraphy: communication protocols, 27, 67–68, 81–82, 116, 123–24, 127; legislation on, 26–27, 34–35, 122–24; nautical applications of, 16, 69, 73, 111n8, 136–37; public perception of, 1–2, 35, 38–39, 69, 73, 100n3, 115n1

Wireless Telephone, The (1911), 93–98

wireless telephony [early radio, or "radiotelephony"], 117, 159–60, 190–92, 237–44; early experiments in, 93–98, 100; public perception of, 65–66, 237n1–2. *See also* radio (broadcast era)

Wonder Stories [magazine], 24, 50, 58, 348n3

Wonders of the Machine Age (1931), 347–53

World War I, 135–37, 153n1, 168, 172n, 174–87; fictional depiction of, 174–89. *See also* wireless telegraphy: legislation concerning

Wright, Willard Huntington, 340n4

WRNY, 41, 305–8, 330

Wu, Tim, 41n, 194n, 251n, 328n

X-ray technology, 61n5, 86

Yaszek, Lisa, 227n2

Zielinski, Siegfried, 46

HUGO GERNSBACK (1884–1967) was a Luxembourgish-American inventor, writer, editor, and publisher who founded the first science fiction magazine, *Amazing Stories,* in 1926. The annual Hugo Awards for the best works of science fiction and fantasy are named in his honor.

GRANT WYTHOFF is a postdoctoral fellow in the Society of Fellows in the Humanities and a lecturer in English and comparative literature at Columbia University.